THE MACAQUES: STUDIES IN ECOLOGY, BEHAVIOR AND EVOLUTION

About the editor . . .

Donald G. Lindburg is Associate Professor of Anthropology at the University of California, Los Angeles. A research anthropologist and former investigator at the California Primate Research Center for over eight years, Dr. Lindburg has served as a member of the Committee on Primate Conversation (National Academy of Science) and as a consultant to the Zoological Survey of India. His field research includes work on the socioecology of *Macaca fascicularis* in Indonesia and the social behavior of rhesus monkeys in northern India. The author/co-author of a wide range of articles on primatology, Dr. Lindburg is a Fellow of the American Anthropological Association and an Organizing Member of the American Society of Primatologists.

Van Nostrand Reinhold Primate Behavior and Development Series

AGING IN NONHUMAN PRIMATES, edited by Douglas M. Bowden

CAPTIVITY AND BEHAVIOR: Primates in Breeding Colonies, Laboratories, and Zoos, edited by J. Erwin, T. Maple and G. Mitchell

ORANG-UTAN BEHAVIOR, by Terry L. Maple

GORILLA BEHAVIOR, by Terry L. Maple

CHIMPANZEE BEHAVIOR, by Terry L. Maple

BEHAVIORAL SEX DIFFERENCES IN NONHUMAN PRIMATES, by G. Mitchell

THE MACAQUES: STUDIES IN ECOLOGY, BEHAVIOR AND EVOLUTION

Edited by

Donald G. Lindburg, Ph.D.
Department of Anthropology
University of California
Los Angeles, California

Van Nostrand Reinhold Primate Behavior and Development Series

VNR **VAN NOSTRAND REINHOLD COMPANY**
NEW YORK   CINCINNATI   ATLANTA   DALLAS   SAN FRANCISCO
LONDON   TORONTO   MELBOURNE

Van Nostrand Reinhold Company Regional Offices:
New York   Cincinnati   Atlanta   Dallas   San Francisco

Van Nostrand Reinhold Company International Offices:
London   Toronto   Melbourne

Library of Congress Catalog Card Number: 79-19668
ISBN: 0-442-24817-2

Manufactured in the United States of America

Published by Van Nostrand Reinhold Company
135 West 50th Street, New York, N.Y.   10020

Published simultaneously in Canada by Van Nostrand Reinhold Ltd.

15   14   13   12   11   10   9   8   7   6   5   4   3   2   1

Library of Congress Cataloging in Publication Data

Main entry under title:

The Macaques: studies in ecology, behavior, and
   evolution.

   Includes bibliographies and index.
   1. Macaques.   I. Lindburg, Donald G., 1932-
QL737.P93M28   599'.82   79-19668
ISBN 0-442-24817-2

List of Contributors

Dr. Irwin Bernstein, Department of Psychology, University of Georgia, Athens, and Yerkes Regional Primate Research Center of Emory University, Atlanta, Georgia.

Dr. Rebecca Cann, Departments of Anthropology and Biochemistry, University of California, Berkeley, California.

Dr. Carolyn Crockett, Regional Primate Research Center, University of Washington, Seattle, Washington.

Dr. John E. Cronin, Department of Anthropology, Peabody Museum, Harvard University, Cambridge, Massachusetts.

Dr. Eric Delson, Department of Anthropology, Lehman College of City University of New York, and Department of Vertebrate Paleontology, American Museum of Natural History, New York, New York.

Dr. Wolfgang P. J. Dittus, National Zoological Park, Smithsonian Institution, Washington, D.C.

Dr. Ardith Eudey, Department of Anthropology, University of Nevada, Reno, Nevada.

Dr. N. A. Fittinghoff, Jr., Department of Anthropology, California State University, Fullerton, California.

Dr. Jack Fooden, Field Museum of Natural History and Chicago State University, Chicago, Illinois.

Dr. Barbara Glick, Department of Anthropology, University of California, Los Angeles, California.

Dr. Thomas Gordon, Department of Psychology, University of Georgia, Athens, and Yerkes Regional Primate Research Center of Emory University, Atlanta, Georgia.

Dr. Colin Groves, Department of Prehistory and Anthropology, Australian National University, Canberra, Australia.

Dr. Donald Lindburg, Department of Anthropology, University of California, Los Angeles, California.

Dr. Thomas Richie, School of Medicine, University of Pennsylvania, Philadelphia, Pennsylvania.

Dr. Vincent Sarich, Departments of Anthropology and Biochemistry, University of California, Berkeley, California.

Dr. Charles Southwick, Department of Pathobiology, The Johns Hopkins University, Baltimore, Maryland.

Dr. David Taub, Department of Comparative Medicine, Bowman-Gray School of Medicine, Winston-Salem, North Carolina.

Dr. Henry Taylor, School of Medicine, Harvard University, Cambridge, Massachusetts.

Dr. Jane Teas, School of Public Health, Harvard University, Cambridge, Massachusetts.

Dr. Bruce Wheatley, Department of Anthropology, University of Oregon, Eugene, Oregon.

Dr. Wendell Wilson, Regional Primate Research Center, University of Washington, Seattle, Washington.

elucidate macaque evolution. An updated version of Fooden's schema, presented in the first chapter, serves as a framework for presenting results of recent studies in paleontology, molecular genetics, and paleoclimatic effects on macaque dispersal. These findings, respectively the subjects of Chapters 2–4 (Delson; Cronin *et al.*; Eudey), may in fact be regarded as independent tests of Fooden's concept, and while the answers given here are of necessity qualified, it is apparent that areas of substantial agreement exist. Chapters 5 and 6 deal with more restricted problems in macaque evolution. In the former, Groves revises the taxonomy of the macaques of Sulawesi as a result of studies of distribution over their range of certain morphological traits. In Chapter 6, Bernstein and Gordon report results of hybridization studies using eight different species, as well as conclusions regarding behavioral isolating mechanisms based on mixed taxa social groupings.

Chapters 7–12 present findings on five different species from recent field studies. In the course of surveying the pig-tailed and crab-eating macaques of Sumatra, Crockett and Wilson documented demographic parameters and habitat preferences, and used these data to characterize interspecific competition and resource partitioning. The basis of a refuging pattern of range utilization by crab-eating macaques in Bornean riverine habitats is the subject of investigations by Fittinghoff and Lindburg in Chapter 8. The same population is discussed by Wheatley in Chapter 9 from the perspective of ranging activity in relation to food resource availability and distribution in space and time. Teas *et al.* extend studies of the rhesus monkey to habitats in Nepal, including in their analysis a profile of behavioral activities for groups occupying temple habitats (Chapter 10). Building on findings from his long-term studies of toque macaques in Sri Lanka, Dittus in Chapter 11 explores social regulators of population parameters in these and other primates for which suitable ecological and demographic data are available. In Chapter 12, Taub advances a theory on sex roles in mate selection through a careful study of mating activities in a wild population of Barbary macaques. In the final chapter, Glick examines related reproductive phenomena in bonnet macaque males by focusing on growth and physiological development, social maturation, and seasonal effects. Examination of these parameters in a context where restraint and manipulation were essential demonstrate approaches which can facilitate the understanding of reproductive behavior in more natural settings.

The inspiration for this volume came from years of interacting with a variety of macaques in both field and captive settings, an activity which raised intriguing questions of interrelationships and evolutionary history. It is much too early to describe that history with certainty. But as the survival of some species becomes increasingly uncertain, and as world politics affects the access of field workers to others, it is important that the process of assessment not be further postponed. It is hoped that this volume will aid in identifying the priorities for further study and will inspire additional investigative effort as remaining opportunities allow.

I am grateful to the contributors to this volume for their enthusiasm for the concept, their cooperation and patience in its production, and for their high standards of scholarship and reportage. Many colleagues, students and friends provided editorial encouragement and advice at various points, and I am much in their debt. I especially thank those who contributed directly to this effort: the editors (Terry Maple, Ashak Rawji and Alberta Gordon), Susan Shideler for advice and help with preparation of the index, and Vicky Unpingco for assistance in typing and editorial details.

D.G. Lindburg
San Diego, California

Contents

xi

THE MACAQUES: STUDIES IN ECOLOGY, BEHAVIOR AND EVOLUTION

Chapter 1
Classification and Distribution of Living Macaques (Macaca Lacépède, 1799)

Jack Fooden

Available evidence concerning the classification and distribution of living macaques (Genus *Macaca*) provides a basis for broad interpretation of the main lines of morphological evolution and geographic dispersal in this genus. As discussed subsequently, this evidence also reveals a consistent and somewhat enigmatic pattern of taxonomic differentiation within each of the major groups of macaques.

Living species of macaques are readily divisible into four clearly defined subgroups: (1) *silenus-sylvanus* group, (2) *sinica* group, (3) *fascicularis* group, (4) *arctoides* group (Table 1-1; Fooden, 1976, p. 225). Each of these species groups is characterized by a distinctive form of the glans penis (Fig. 1-1). In the *silenus-sylvanus* group (Fig. 1-1a) the glans is bluntly bilobed and broad; in the *sinica* group (Fig. 1-1b) it is apically acute and broad (sagittate in dorsal view); in the *fascicularis* group (Fig. 1-1c) it is bluntly bilobed and narrow; and in the *arctoides* group (Fig 1-1d) it is apically acute and elongate (lanceolate in dorsal view). Judging from penial morphology in other cercopithecids (Pocock, 1926, p. 1557), the bluntly rounded form of glans probably is primitive. The apically acute form appears to have been independently derived in the *sinica* group, and in the *arctoides* group, since the urethral orifice opens dorsal to the apex of the glans (and protruding distal end of the baculum) in the former, and ventral to the apex of the glans (and baculum) in the latter.

Female genital tract morphology is similar in the *silenus-sylvanus* and

1

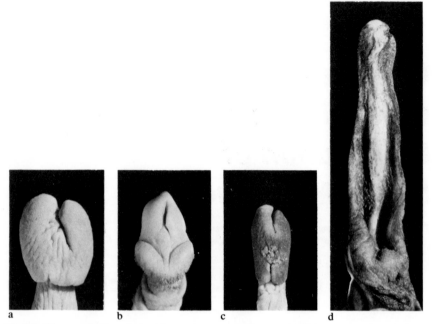

a   b   c   d

Fig.1-1. Glans penis (dorsal view) in species groups of macaques (*Macaca*): *a, silenus-sylvanus* group (*M. nemestrina leonina*): *b, sinica* group (*M. sinica*); *c, fascicularis* group (*M. fascicularis*); *d, arctoides* group (*M. arctoides*). For specimen numbers, see Fooden, 1976, Fig. 1. X 1.5 (Photos by John Bayalis, Division of Photography, Field Museum of Natural History).

fascicularis groups, but it is distinctive in the *sinica* group and *arctoides* group (Fig. 1-2). In the *silenus-sylvanus* group (Fooden, 1975a, Fig. 12) and *fascicularis* group the uterine cervix and cervical colliculi are moderately large (Fig. 1-2a). In the *sinica* group these structures are enormously hypertrophied (Fig. 1-2b). In the *arctoides* group cervical colliculi are absent and a unique vestibular collicle is present (Fig. 1-2c).

Copulatory behavior patterns also may distinguish the *sinica* group and *arctoides* group from each other and from the combined *silenus-sylvanus* and *fascicularis* groups (Chevalier-Skolnikoff, 1975, p. 203; Fooden, 1975a, p. 36). Species in the *silenus-sylvanus* and *fascicularis* groups (with the possible exception of *M. fascicularis*, where available evidence is ambiguous) apparently are multiple-mount ejaculators in which a series of about 10 mounts precedes ejaculation (but see Taub, this volume, for copulatory pattern in *M. sylvanus*). Species in the *sinica* group evidently are normally single-mount ejaculators with about 15–20 intromissive thrusts per mount. The single species in the *arctoides* group also is a single-mount ejaculator, but it averages 60–65 thrusts per mount. Differentiation of male and female reproductive structures,

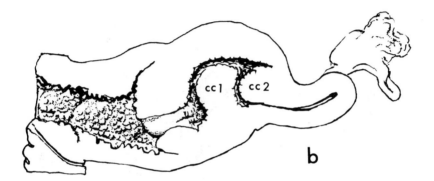

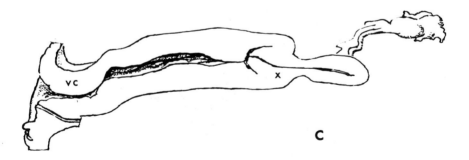

Fig. 1-2. Female reproductive tract morphology (sagittal section) in species groups of macaques (*Macaca*): *a, silenus-sylvanus* group and *fascicularis* group (*M. mulatta*), showing moderately large uterine cervix and cervical colliculi (CC1, CC2); *b, sinica* group (*M. assamensis*), showing hypertrophied uterine cervix and cervical colliculi (CC1, CC2); *c, arctoides* group (*M. arctoides*), showing absence of cervical colluculi (X) and presence of vestibular collicle (VC). Modified from Fooden, 1967, Fig. 4, and 1971b, Fig. 4. X 1.0 (Illustration by Kevin Royt).

and differentiation of copulatory behavior patterns presumably are both related, either as cause or effect, to evolutionary development of reproductive isolation between ancestral stocks of the four macaque species groups.

The composite geographic range of living macaque species groups includes southern and eastern Asia and extreme northwestern Africa (Fig. 1-3). Ranges of all four species groups overlap in the Indochinese Peninsula; other parts of southern and eastern Asia are inhabited by one, two or three of these groups

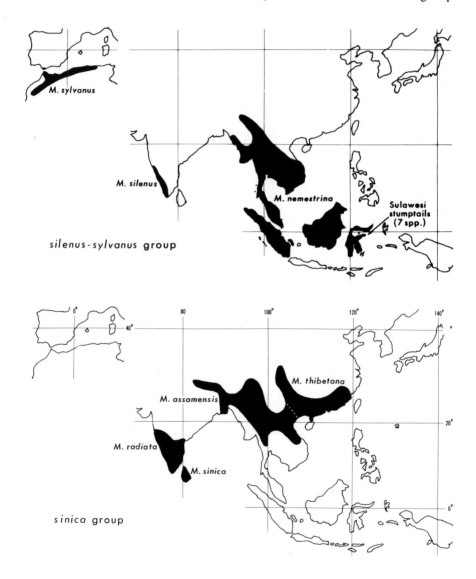

in various combinations, and northwestern Africa is inhabited by only one group. The range of the *silenus-sylvanus* group is broadly disjunct, suggesting early dispersal and subsequent intermediate disappearance; early dispersal of

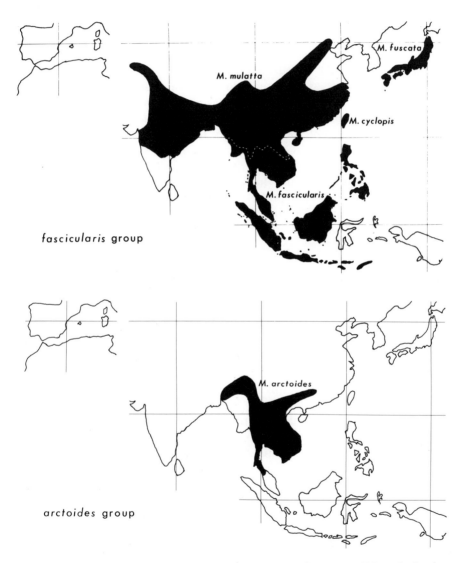

Fig. 1-3. Approximate limits of distribution in living species of macaques (*Macaca*), showing successively less disjunct distributions of *silenus-sylvanus* group, *sinica* group and *fascicularis* group and relatively small compact distribution of *arctoides* group. For references, see Fooden, 1976, Fig. 2, and 1979b, Fig. 1.

this group is likewise implied by the fact that this is the only macaque group native to Sulawesi (Celebes) and the Mentawai Islands, island groups which seem to have been relatively inaccessible to recently dispersed mammals (Fooden, 1975a, p. 69). Dispersal of the *silenus-sylvanus* group to the Greater Sunda Islands, enroute to Sulawesi and the Mentawai Islands, evidently had occurred by mid-Pleistocene, as indicated by macaque fossils of that age in Java. This probably is true regardless of whether these somewhat problematic fossils are assigned to the *silenus-sylvanus* group or the *fascicularis* group (Fooden, 1975a, p. 64), because the former species group presumably dispersed before the latter, as discussed below. The range of the *sinica* group is less broadly disjunct than that of the *silenus-sylvanus* group, suggesting somewhat more recent dispersal. The extensive range of the *fascicularis* group is essentially continuous, except for present-day (mostly postglacial) water gaps, which suggests still more recent dispersal and also suggests persistent occupation of territory. Complementarity of the *fascicularis* group and *sinica* group in northern peninsular India suggests a competitive relationship between these two groups and probably implies that the inferred recent spread of the *fascicularis* group in this area has been directly responsible for the retreat of the *sinica* group and disjunction of its range. The relatively small compact range of the *arctoides* group, taken together with the highly specialized structure of its genitalia and the advanced degree of tail reduction, may indicate that this species group evolved and dispersed more recently than any other macaque species group. An opposite and alternative hypothesis is that this group may have a relict distribution.

With differentiation of species within macaque species groups, tail length evidently has undergone independent parallel reduction in the *silenus-sylvanus* group, the *sinica* group and the *fascicularis* group (Table 1-1). In general, tail length is inversely correlated with body size (measured as combined length of head and body); however, this relationship is not maintained in stumptail species with relative tail length of 0.10 or less. In the *sinica* group and *fascicularis* group relative tail length tends to decrease (Allen's rule) and body size tends to increase (Bergmann's rule) with increasing northern latitude of ranges of species (Fig. 1-3). It is highly significant that in both of these species groups the northern limit of species in which relative tail length is 1.00 or greater is approximately the same, about 20° N lat., which is close to the tropic of Cancer. In the *fascicularis* group it has been shown experimentally that *M. mulatta* and *M. fuscata*, species with relatively large body size and low relative tail length, are more tolerant to cold stress than *M. fascicularis* (Tokura et al., 1975, p. 173). Assuming that a long tail is primitive in macaques, the orderly relationship between tail length and latitude in species of the *sinica* group and *fascicularis* group suggests that both of these groups originated in the south and dispersed northward. In the *silenus-sylvanus* group, tail length decreases from west (*M. silenus*) to east (Sulawesi stumptails; Fooden 1969, p. 15; 1975a, p. 68)

Table 1-1. Classification, Nomenclature, Head and Body Length (HB), and Relative Tail Length (T/HB) in Living Species of Macaques.

SPECIES GROUP/SPECIES[1]	COMMON NAME	EXTERNAL MEASUREMENTS: MEANS[2] FOR ADULT MALES		
		HB (MM)	T/HB	N
1. *silenus-sylvanus* group				
M. *silenus*	Liontail macaque	550	0.65	4
M. *nemestrina*[3]	Pigtail m.	574	0.36	38
M. *tonkeana*[4]	Tonkean m.	636	0.09	15
M. *maura*[4]	Moor m.	657	0.08	3
M. *ochreata*[4]	Booted m.	590	0.06	1
M. *brunnescens*[4]	Muna-Butung m.	485	0.10	2
M. *hecki*[4]	Heck's m.	604	0.04	16
M. *nigrescens*[4]	Gorontalo m.	---	---	0
M. *nigra*[4]	Celebes ape	552	0.04	4
M. *sylvanus*	Barbary ape	641	0.03	1[5]
2. *sinica* group				
M. *sinica*	Toque m.	468	1.22	21
M. *radiata*	Bonnet m.	539	1.11	5
M. *assamensis*	Assam m.	594	0.46	5
M. *thibetana*	Père David's stumptail m.	609	0.10	2
3. *fascicularis* group				
M. *fascicularis*[6]	Crab-eating m.	502	1.09	15
M. *cyclopis*	Taiwan m.	560	0.74	3
M. *mulatta*	Rhesus m.	535	0.46	16
M. *fuscata*	Japanese m.	571	0.18	22
4. *arctoides* group				
M. *arctoides*[7]	Bear m.	566	0.09	7

1. For author and date citations and key to external characters, see Fooden, 1976, p. 226.
2. Sample means provide a convenient basis for rough interspecific comparisons. However, it should be noted that these means are not statistically robust; in many or most macaque species the distributions of head and body length and tail length are known to be clinal, bimodal and/or polymodal, *not* normal. References: McCann, 1933, p. 801; Fooden, 1969, pp. 15, 74, 84, 96, 110, 118, 126; 1971a, p. 38; 1971b, p. 70; 1972, p. 310; 1975a, pp. 5, 80, 86, 168; 1975b, p. 102; 1979a, p. 111.
3. Including subspecies *M. n. nemestrina*, *M. n. leonina* and Mentawai *M. n. pagensis* (Fooden, 1975a, p. 85). Tenaza (1975, p. 67) and Wilson and Wilson (1977, p. 216) have recently proposed that the Mentawai macaque should be regarded as a full species, closely related to *M. fascicularis*; however these authors present no concrete evidence for their decision, and known data concerning cranial morphology, pelage color and pattern and tail morphology in the Mentawai macaque all tend to relate this monkey to *M. nemestrina*, not *M. fascicularis*.
4. Sulawesi stumptail species.
5. Specimen No. 15296, Museum of Comparative Zoology, Harvard.
6. Synonym: *M. irus*.
7. Synonym: *M. speciosa*.

and also is reduced in northwestern Africa (*M. sylvanus*); the somewhat complex geographic pattern of tail length distribution in this group may be related to its inferred early origin and dispersal, as discussed above.

Although geographic ranges of all four macaque species groups are partly overlapping (sympatric) in the Indochinese Peninsula (Fig. 1-3), ranges of component species within each of the three multispecies groups are strictly nonoverlapping (allopatric). Evolutionary differentiation within all three multispecies groups evidently has reached a level at which interbreeding at species boundaries is minimal or absent, judging from the rarity of morphologically intermediate specimens. But this differentiation has not yet advanced to a higher level where neighboring species can coexist in the same area. Because of this marginal or intermediate level of evolutionary differentiation within species groups, it is difficult in macaque taxonomy to distinguish between the status of recognized well-defined subspecies such as *M. n. nemestrina* and *M. n. leonina* (Fooden, 1975a, p. 85) and that of recognized allopatric species such as *M. fascicularis* and *M. mulatta* (Fooden, 1971a, p. 24). Hence, taxonomic decisions to allocate such forms either to subspecific rank or specific rank are somewhat arbitrary. However, such taxonomic allocations are less important than recognition that borderline subspecific/specific differentiation is the rule, not the exception, within macaque species groups. The biological and evolutionary significance of the regularity of this interesting and unusual level of differentiation within macaque species groups merits future consideration and research.

ACKNOWLEDGMENTS

This research was partially supported by National Institutes of Health MBS Grant No. RR-08043 to Chicago State University. I thank Henry Dybas, Division of Insects, Field Museum of Natural History, for constructive criticism of the manuscript.

REFERENCES

Chevalier-Skolnikoff, S. Heterosexual copulatory patterns in stumptail macaques (*Macaca arctoides*) and in other macaque species. *Arch. Sex. Behav.*, **4**: 199–220 (1975).

Fooden, J. Complementary specialization of male and female reproductive structures in the bear macaque, (*Macaca arctoides*). *Nature*, **214**: 939–941 (1967).

_____. Taxonomy and evolution of the monkeys of Celebes (Primates: Cercopithecidae). *Bibl. Primat.*, **10**: 1–148 (1969).

_____. Report on primates collected in western Thailand January–April, 1967. *Fieldiana: Zool.*, **59**: 1–62 (1971a).

_____. Female genitalia and taxonomic relationships of *Macaca assamensis*. *Primates*, **12**: 63–73 (1971b).

_____. Male external genitalia and systematic relationships of the Japanese macaque (*Macaca fuscata* Blyth, 1875). *Primates*, **12**: 305–311 (1972).

_____. Taxonomy and evolution of liontail and pigtail macaques (Primates: Cercopithecidae). *Fieldiana: Zool.*, **67**: 1–169 (1975a).

_____. Primates obtained in peninsular Thailand June–July, 1973, with notes on the distribution of continental Southeast Asian leaf-monkeys (*Presbytis*). *Primates*, **17**: 95–118 (1975b).

_____. Provisional classification and key to living species of macaques (Primates: *Macaca*). *Folia Primat.*, **25**: 225–236 (1976).

_____. Taxonomy and evolution of the *sinica* group of macaques: 1. Species and subspecies accounts of *Macaca sinica*. *Primates*, **20 (1)**: 109–140 (1979a).

_____. Taxonomy and evolution of the *sinica* group of macaques: 2. Species and subspecies accounts of the Indian bonnet macaque, *Macaca radiata. Fieldiana: Zool.*, in press (1979b).

McCann, C. Notes on some Indian macaques. *J. Bombay Nat. Hist. Soc.*, **36**: 796–810 (1933).

Pocock, R. I. The external characters of the catarrhine monkeys and apes. *Proc. Zool. Soc. London 1925*: 1479–1579 (1926).

Tenaza, R. R. Territory and monogamy among Kloss' gibbons (*Hylobates klossii*) in Siberut Island, Indonesia. *Folia Primat.*, **24**: 60–80 (1975).

Tokura, H., Hara, F., Okada, M., Mekata, F. and Ohsawa, W. Thermoregulatory responses at various ambient temperatures in some primates. *Contemporary Primatology, 5th Int. Congr. Primat., Nagoya 1974*, S. Kondo, M. Kawai and A. Ehara (eds.), pp. 171–176. Basel, S. Karger (1975).

Wilson, C. C., and Wilson, W. W. Behavioral and morphological variation among primate populations in Sumatra. *Yearb. Phys. Anthrop.*, **20**: 207–233 (1977).

Chapter 2
Fossil Macaques, Phyletic Relationships and a Scenario of Deployment

Eric Delson

INTRODUCTION

The macaques, genus *Macaca,* comprise all of the living cercopithecine Old World monkeys of Eurasia and northern Africa. Today, there are between 13 and 19 species which may be arranged in four or five larger units (species groups or perhaps subgenera), as discussed especially by Fooden (1976 and this volume) and below. The modern zoogeographic pattern is by no means typical for all of the Pleistocene, as macaques have inhabited Europe for most of the past five million years, and one or two large-bodied highly terrestrial relatives of *Macaca* (*Procynocephalus* in Asia and *Paradolichopithecus* in Europe) are also known from the earlier Pleistocene and Pliocene. All three genera may be classified as the subtribe Macacina, in the tribe Papionini of the subfamily Cercopithecinae[1]. After a brief review of some aspects of skeletal morphology, this chapter will concentrate on the distribution and relationships of past and present species of macaques.

MORPHOLOGY

In many ways, *Macaca* is a conservative genus, retaining numerous features which characterized the early cercopithecines, as well as they can be recon-

---

[1] In addition to specifically cited publications, Szalay and Delson (1979) provide detailed information on skeletal morphology, fossils, distribution, adaptation and phylogeny of macaques and other primates mentioned here.

structed. Delson (1975a) considered the basic adaptations of the Cercopithecidae and its subtaxa, following Napier's (1970) suggestion that the Old World monkey "bilophodont" dentition originated as a response to a habitat in which the ability to eat leaves when fruit was less plentiful may have been selectively advantageous. It is clear that the cheek teeth of colobines (with their derived deep lingual notches and short trigonids on lowers, and overall greater crown relief) are even better adapted to the processing of leaves than those of cercopithecines (see also Walker and Murray, 1975; Kay, 1978), which lends credence to Delson's view that the latter represent the ancestral condition for the family.

The cheek teeth of macaques (Figs. 2-1 and 2-2) were used by Delson as a model for those of early monkeys, as they present neither the increased lateral "flare" of some mangabeys and baboons, nor the reduced flare and loss of M_3 hypoconulid of the tribe Cercopithecini. On the other hand, macaques share with other papionins relatively enlarged incisors and reduced lingual enamel on lowers. In this area of the dentition, the colobine pattern of small incisors with "conical" I^2 and fully enameled lowers was probably ancestral for cercopithecids, although macaques are less "specialized" than some baboons and mangabeys.

The cranium of *Macaca* (Figs. 2-3 and 2-4) is also conservative among cercopithecines, lacking facial fossae or great elongation, but presenting such typical features of the subfamily as a vomerine contribution to the medial orbit wall, lacrimal fossa completely within the lacrimal bone, narrow interorbital area and high, narrow choanae. Modern macaques are mostly arboreal to semiterrestrial animals, and their skeleton reflects this behavior pattern both in proportions and morphology. The smaller macaques probably are rather similar in size, proportions and joint morphology to the earliest cercopithecines, which may have separated from the ancestral colobines about 15 million years ago (MYA).

PALEONTOLOGY

Miocene Papionin Diversification

Little is known of the earliest stages of cercopithecine radiation, but Delson (1975a, b) has suggested that by about 10 MYA the ancestral, semiterrestrial papionins had diverged from early cercopithecines, which entered the higher canopy to compete with the arboreal colobines. Other, more terrestrial colobines had by this time entered Eurasia. Two sets of probably Late Miocene (11–5 MYA) fossil monkeys are known which may relate to the origin of the macaques, although both consist of dental remains only, and as has been said, the teeth of *Macaca* are conservative ("primitive") within the Papionini. A single M_3 was reported by Hooijer (1963) from Ongoliba, Zaire, in a horizon

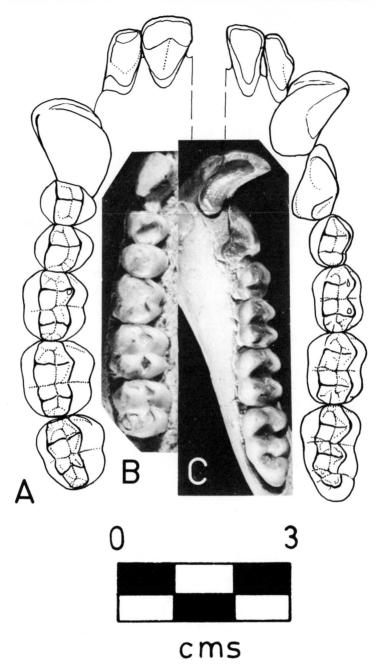

Fig. 2-1. Occlusal views of macaque dentition. A: *M. thibetana*, modern, drawing of male right, upper (to left) and lower (to right) I1–M3; B: *M. anderssoni*, Honan, China, right C¹–M³ of male face (see fig. 2-3B, E); C: *M. sylvanus florentina*, Upper Val d'Arno, Italy, right C₁–M₃, male.

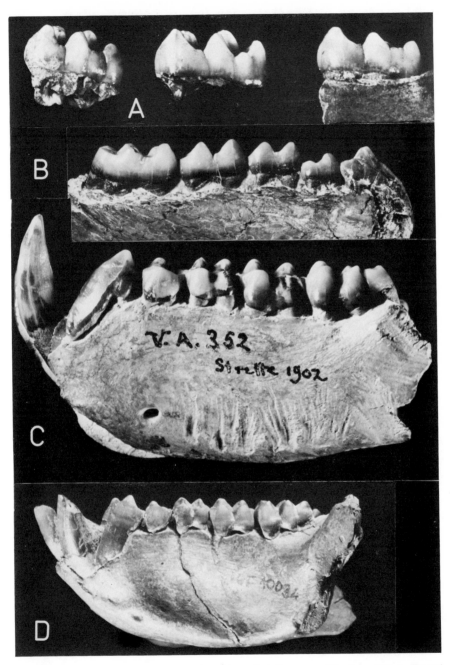

Fig. 2-2. Side views of macaque lower teeth and jaws. A: ?*Macaca* sp., Marceau, Algeria, lingual view of three right M₃ (indicating morphology and size range; B, C, D: *M. sylvanus florentina*, Upper Val d'Arno, Italy, lingual view of male left P₃–M₃, left lateral views of two male mandibles with C₁–M₃. Note lack of mandibular corpus fossa in C, presence in D (holotype). Not to same scales.

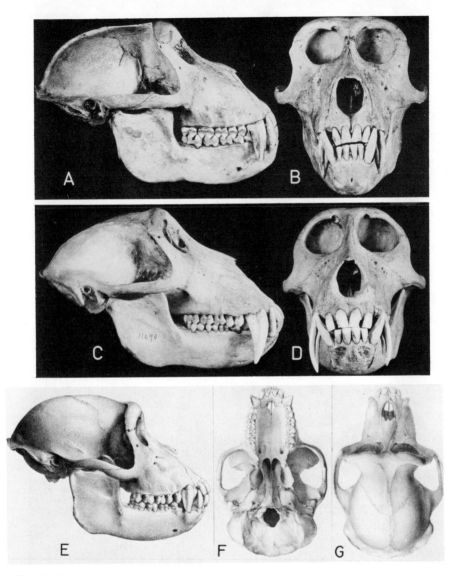

Fig. 2-3. Crania of modern macaque species. A, B: *M. thibetana,* right lateral and frontal views (from Elliot, 1913); C, D: *M. nemestrina,* right lateral and frontal views (from Elliot, 1913); E, F, G: *M. arctoides,* right lateral, basal and dorsal views (from de Blainville, 1839).

which is probably Late Miocene in age. Originally identified as "Cf. *Mesopithecus* cq. *Macaca*", it is clearly of cercopithecine rather than colobine morphology, but is not readily allocated beyond the tribe Papionini. From the

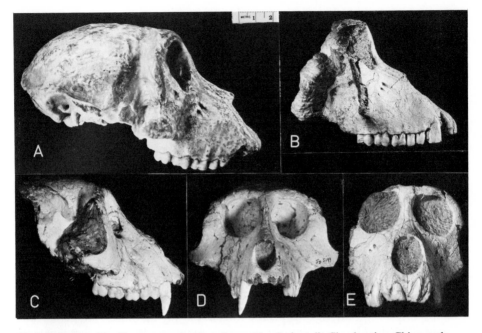

Fig. 2-4. Crania of fossil macaques. A: *M. anderssoni* (= *"robusta"*), Choukoutien, China, male cranium (cast) in right lateral view; B, E: *M. anderssoni,* Honan China, male face in right lateral and frontal views; C, D: *M. ?sylvanus majori,* Capo Figari, Sardinia, male face in right lateral and frontal views.

probably later Turolian (ca. 7 MYA) locality of Marceau, Algeria, Arambourg (1959) reported a number of isolated cercopithecid teeth which he named *Macaca flandrini*. The type and several other specimens are colobine, from a taxon which Delson (1975a) has provisionally called ?*Colobus flandrini*, using the modern term as a "form-genus" for African colobines. In the same way, the majority of the Marceau teeth (about 40) can be termed ?*Macaca* sp. It is possible that more than one species is represented, as there is a wide range of size present (e.g., in M_3, see Fig. 2-2A), but my detailed study of the material is not yet completed. The Ongoliba specimen is similar in form to those from Marceau, and the two populations might represent what Delson (1975b) has postulated to have been a wide-ranging, possibly polytypic species of early papionin present in much of Africa in the Late Miocene.

At the end of the Miocene, the Mediterranean almost completely dried up as a result of tectonic movements, and this desiccation might have aggravated the general trend toward aridity seen in southern Europe earlier in the Late Miocene, eventually leading to the formation of a semidesertic barrier across the Sahara (see extended discussion and references in Delson, 1975b). Such a barrier might have led to the differentiation of the Papionini into the three known groups (subtribes): *Theropithecus* in wet lowlands; *Papio, Cercocebus* and extinct allies in more forested regions, eventually invading the plains; and *Macaca* to the north, in a variety of habitats. The earliest known papionins (other than *Theropithecus*) south of the Sahara are nearly impossible to allocate generically without cranial material, and the distinctive crania of the conservative *Parapapio* are in many ways similar to those of *Macaca*, lacking the steep anteorbital drop of *Papio* and the infraorbital and mandibular-corpus fossae of modern African papionins.

North African and European Fossil Macaques

The fossil macaques of the circum-Mediterranean region have been given about a dozen nominally distinct species designations, but in fact only a few, if any, are significantly distinguishable morphologically from the living *M. sylvanus*, let alone from each other. It seems probable at present that almost all of this Mediterranean record reflects the sampling of a single evolving lineage, with no major cladogenesis having taken place. If that is indeed the case, it is biologically most meaningful to designate the various taxa accepted here as subspecies, rather than full species, thus not implying more than genetically controlled fluctuation in size or proportion about an evolutionary mean. Only cranial or perhaps postcranial remains could clarify this point, as the dentition of macaques (and all other papionins) is of course highly stereotyped, but such fossils are rare. Revision is currently under way, but a provisional scheme may recognize six temporo-geographic forms, most tentatively ranked as sub-species of *M. sylvanus*.

The oldest of these is *Macaca libyca* (Stromer, 1920), represented only by a collection of dental remains, some in partial jaws, from the Late Miocene (ca. 6 MYA) locality of Wadi Natrun, in northern Egypt. As are almost all the European and north African fossils, these teeth are within the size range of *M. sylvanus* and may be referred to *Macaca* on the basis of geography and the (ancestral) condition of absence of mandibular corpus fossae on one well-preserved and several poorly preserved specimens. This taxon *may* predate the division of the genus into Asian and Mediterranean lineages, and thus is recognized as a full species. No facial regions are preserved, and in any event, it would be most difficult to distinguish a macaque of this age from the hypothetical common ancestor of the Papionini.

More interesting, if equally little-known, is the first macaque from Europe, probably the oldest cercopithecine outside Africa. *Macaca sylvanus prisca* (Gervais, 1859) is a slightly smaller form than *M. libyna*, known from a number of localities in southern Europe dating from the Ruscinian (5–4 MYA) and perhaps the early Villafranchian (4–3 MYA) mammal ages. The type locality is Montpellier, in southern France, where several teeth and partial jaws (see Fig. 2-5F) are found associated with the large terrestrial colobine *Dolichopithecus ruscinensis* and the smaller, perhaps more arboreal *Mesopithecus monspessulanus* in a moist forested habitat. A juvenile mandible of similar age from Hungary was named *M. praeinuus* by Kormos (1914), but it is probably referable to *M. s. prisca*. Additional fragmentary dental remains are known from definite or probable early Villafranchian local faunas in Italy, France, Hungary, Spain and Germany; these may also be tentatively referred to *M. s. prisca* as they appear somewhat smaller than the later Villafranchian *M. s. florentina*. These early specimens are most important in documenting the first extension of the cercopithecines out of Africa, probably by a crossing of, or circuit around, the Mediterranean during its desiccated phase at the end of the Miocene (see Azzaroli, 1975; Delson, 1975b).

A roughly contemporaneous but morphologically distinct population is known only from the cave breccia at Capo Figari, Sardinia, previously thought to be of "postglacial" age. From here come over 100 specimens, mostly dental, but including several partial crania and numerous postcranial elements (see Figs. 2-4C,D; 2-5A,D). Azzaroli (1946) described a few of these fossils as the "dwarf" species *M. majori*, named after the excavator Forsyth Major; it is here recognized tentatively as another subspecies of *M. sylvanus*. Preliminary comparisons suggest dental size differences from living *M. s. sylvanus* on the order of 5–10 percent (even less from *M. s. prisca*), statistically different at the 95 percent confidence level in most cases, but not as extreme as might be implied by the term dwarf (compared to other Pleistocene insular mammals). Moreover, the cheek teeth appear a bit more inflated or "puffy" than those of either *M. s. sylvanus* or *M. s. prisca*. Further studies of cranial and joint morphology may reveal additional diagnostic features to support a provisional suggestion that this is the most distinctive of the European macaque populations, potentially meriting status as a distinct species.

Seven localities or "fields" have yielded macaque fossils of middle and late Villafranchian age (later Pliocene, ca. 3–1.6 MYA); no material is definitively known from the Early Pleistocene, 1.6–1 MYA. The largest collection of remains, including some half-dozen mandibles, an equal number of partial toothrows or isolated teeth, and one partial ulna, derives from the Montevarchi Group of the Upper Val d'Arno, upstream of Florence, Italy. This unit spans a long time, perhaps the last 0.5 MY of the late Villafranchian, and many localities have yielded individual specimens. The first (and still most complete) specimen found was named *Macaca florentina* (Cocchi, 1872), which can be employed subspecifically for all populations of this age. These other samples are known mostly by isolated specimens from Spain, France, the Netherlands, Italy (earlier than Val d'Arno) and Yugoslavia. The Val d'Arno specimens are variable in both size and morphology, but essentially comparable to *M. s.*

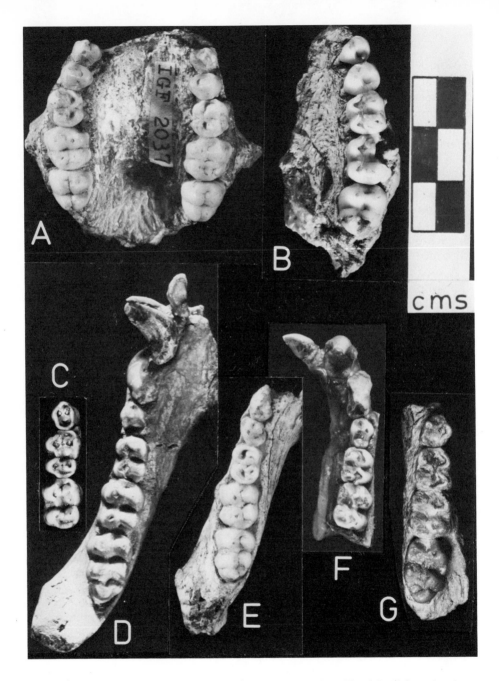

Fig. 2-5. Dentitions of fossil macaques. A: *M. ?sylvanus majori,* Capo Figari, Sardinia, male palate with right and left P³-M³; B: *M. sylvanus pliocena,* Heppenloch, Germany, left P³-M³; C: ?*Macaca* sp., Marceau, Algeria, composite left P₄-M₂ (isolated teeth); D: *M. sylvanus florentina,* Upper Val d'Arno, Italy, male left I₂-M₃; E: *M. ?s. majori,* male left P₃-M₃; F: *M. s. prisca,* Montpellier, France, male right I₂-M₂; G: ?*M. palaeindica,* Siwaliks, India, right P₄-M₃, sex unknown. All occlusal view.

sylvanus, larger than preceding *M. s. prisca*. The type mandible (Fig. 2-2D, 2-5D) of *M. s. florentina* possesses deep fossae on the corpus, but at least one other male specimen (Fig. 2-2C) does not, emphasizing the uncertainty of drawing morphological distinctions without even the cranium.

A number of sites across Europe have yielded Middle Pleistocene (ca. 1–0.125 MYA) macaque fossils, once again mostly fragmentary dentitions. The oldest name for these is *Macaca sylvanus pliocena* (Owen, 1845), for a single upper molar from an early warm phase of the "Riss" glacial stage at Grays Thurrock, near London. The nominal taxa *M. suevica* (Hedinger, 1891) and *M. tolosana* Harlé, 1892, from caves in Germany and France, are synonyms (see Fig. 2-5B). Over a dozen other localities have produced specimens, indicating a range from central Spain to East Anglia, Czechoslovakia, Italy, the Caucasus and Israel. All appear to be dated to warm phases, perhaps including the Last or Eemian Interglacial (125–75 thousand years ago). Although the local environments of these macaques varied between forest and steppe, they did seem to require a temperate climate to prosper; in the colder intervals, macaques may have retreated to the Mediterranean littoral (although they are not known from there except during warmer phases) or even out of Europe altogether, perhaps via the Near East. Morphologically, these specimens are nearly identical to the earlier *M. s. florentina*, but the cheek teeth may be slightly wider. Such refuges appear to have failed during the last glacial, if not earlier, as no monkeys occur in any of the numerous "postglacial" faunal assemblages from Europe or the Levant.

In the Pliocene and Pleistocene of northern Africa, a few fossil macaques are known, all of which can tentatively be termed *M. sylvanus* aff. *sylvanus*. In Tunisia, a humerus from Ichkeul is Ruscinian in age, while two molars from Ain Brimba are Villafranchian. An incorrect association of the latter with two incisors of a saber-tooth "cat" led Arambourg (1969–1970) to suggest the name "Anomalopithecus bicuspidatus" for this "aberrant" monkey, without formally describing the specimens or designating a type; the name is thus, happily, unavailable. Other sites are generally later Late Pleistocene (younger than 20,000 years old and thus clearly linked to the living Maghreb *M. sylvanus*), but Ain Mefta, Algeria, may be later Middle Pleistocene. From this site come a number of postcranial elements on which Pomel founded two species, *M. trarensis* and *M. proinuus*, in 1892 and 1896, respectively. It must be assumed that he simply changed his mind about the name, in ignorance of the practices of zoological nomenclature (not yet formalized in the Règles), as the later monograph does not even cite the first, short paper.

Asian Fossil Macaques

The potential proliferation of names for circum-Mediterranean *Macaca* should not confuse what is most probably a single closely knit lineage, with no evidence for more than one species at any given time, except perhaps for the apparently short-lived *M. ?sylvanus majori*. In Asia, on the other hand, there are today at least a dozen species in several distinct groups. In the past, the evidence for coexistence of two or more forms is unclear, but the record is even less complete than in Europe. As in the Mediterranean region, the earliest known Asian cercopithecids are colobine, a few dental remains being known from the Late Miocene (Dhok Pathan) of Pakistan. These specimens have been variously named *Macaca sivalensis* and *Cercopithecus* or *Semnopithecus asnoti*, but as Simons (1970) partly suggested, they are all referable to a single colobine species best termed *?Presbytis sivalensis.*

The oldest Asian cercopithecines, on the other hand, have erroneously been thought colobine by all previous workers who even considered them, until Delson (1975a) indicated that they were papionin and could be termed *?Macaca palaeindica* (Lydekker, 1884). The specimens involved are two mandibular fragments (see Fig. 2-5G) from the "Tatrot-equivalent" (Late Pliocene, ca. 3 MYA) of northern India, which are clearly cercopithecine in dental morphology, although the type specimen has a deep and narrow corpus. The teeth are similar in size to those of *M. sylvanus*, that is, rather large for Asian macaques. It is worth noting that there may be a still older specimen of macaque-like cercopithecid in Asia, as yet unpublished; this damaged mandible is said to be associated with a northeastern Chinese local fauna of Late Miocene age (see Delson, 1977).

All other Asian fossil macaques are Pleistocene in age, and a number of them can be referred to living species. A partial mandible of an old male from the ?Early Pleistocene Pinjor beds of northern India was originally termed *Cynocephalus* (= *Papio*) *falconeri*, but Jolly (1967) suggested it to be a large macaque. I now consider that it more probably represents *Procynocephalus*, known by two other jaws from the same horizon. In China, a male face (Figs. 2-1B, 2-4B,E) from the ?Early Pleistocene of Honan was named *M. anderssoni* by Schlosser (1924), while Young (1934) coined *M. robusta* for a series of mostly dental specimens (but see Fig. 2-4A) from the Middle Pleistocene of Choukoutien, near Peking. These two taxa are probably not distinguishable at the specific level. Moreover, *M. anderssoni* (sensu lato) is cranially similar to the modern *M. thibetana* and *M. arctoides* (placed in different species groups by Fooden, but see below), though all are rather different in shape and relative tooth size from the equally large-skulled *M. nemestrina* (see Fig. 2-3).

Additional, less complete fossil remains have been reported from Chinese sites ranging in age from later Early Pleistocene (the "*Gigantopithecus*-cave" at Liucheng, far to the southwest) through Late Pleistocene. The youngest of these may refer to *M. mulatta*, known in south and central China today, but few details have been published (for discussion, see Delson, 1977). Farther south, in the Indochinese peninsula, macaque remains have been recovered from a number of localities, most of which are probably later Middle Pleistocene in age (as are the majority of the Chinese sites). Until these are reported in more detail, their specific identity will remain uncertain. The most complete specimen is a partial skull from Tung-Lang, northern Vietnam, described by Jouffroy (1959) as *M. speciosa subfossilis*, after comparison especially with *M. speciosa* (sic) *thibetana*. The cranium does closely resemble *M. thibetana* and might belong to this species, or to *M. arctoides* or even to *M. anderssoni*, so similar is the morphology among these taxa.

Hooijer (1962) and Fooden (1975) discussed fossil macaques from Java ranging in age back into the Middle Pleistocene. These suggest that populations of *M. fascicularis* and/or *M. nemestrina* were present, but as cranial remains are lacking, precise identification is not yet possible. *Macaca* is also known in the fossil record of Japan, but here the age of the specimens is even less certain, none being clearly older than Late Pleistocene. Iwamoto (1975) has reviewed most of the previous work and described a new skull which he considered to share some features with *M. "robusta."* The Japanese fossils are most probably a distinct subspecies of the living *M. fuscata* (of the *fascicularis* species group), and thus it seems unlikely that they would be closely related to the early Chinese fossils, which may belong to the *sinica* species group (see below). Macaque fossils are also known from caves in South Korea.

PHYLOGENY

Review of Published Data

This brief review of the fossil record of *Macaca* does not directly lead to further inference as to the evolutionary relationships within the genus. But if the insightful studies of Fooden (see this volume for summary) on reproductive organ morphology and other external characters are combined with the fossils and with scattered additional comments on macaque relationships and Asian biogeography by Albrecht (1978), and Brandon-Jones and Hiiemae (unpublished ms) among others, a dim picture begins to emerge. A review of this nonpaleontological data leads to the formulation of a set of testable phylogenetic hypotheses, which may then be extrapolated further with the incorporation of time and adaptational factors.

Following Fooden, the male and female genital morphology seen in the *silenus-sylvanus* group of macaque species is ancestral for the genus. Female genital morphology and the presumably ancestral multiple-mount copulation pattern is shared with the *fascicularis* group, which differs in having a narrow, rather than a broad, glans penis. The *sinica* group is distinguished by a broad but acute or sagittate glans, hypertrophied uterine cervix and colliculi and a single-mount copulation pattern. Finally, *M. arctoides*, placed in its own species group by Fooden, is characterized by an acute, elongate (lanceolate) glans, lack of cervical colliculi but presence of a vestibular collicle and single-mount copulation with an increased number of intromissive thrusts. At least the male morphology and behavior pattern appear to form a three-pole morphocline with one trend from the ancestral (*silenus*) condition through the *sinica* to the *arctoides* (in which the female genital tract may show an evolutionary reversal), and a second to the *fascicularis* group's quite distinctive mosaic. Within each group, tail length, long thought to be a valuable taxonomic character, appears to have reduced convergently, as populations moved northward and body size increased. If Fooden's four groups are recognized, such convergent reduction would have occurred five times, or perhaps six if there were two episodes within the *fascicularis* group.

New Interpretations

Considering this evidence along with the early date of entry into Europe of macaques apparently related to *M. sylvanus*, it seems likely that this species is more distinctive than Fooden has previously admitted (as recognized also by Albrecht, 1978, p. 18). It is linked to *M. silenus* and its relatives (Fooden, 1975) essentially by retention of ancestral conditions, not by any shared derived states which would uphold its placement within the same species group. It may thus be tentatively recognized as a separate group, monotypic today but possibly containing one or more fossil species noted above (see also Cronin, *et al.*, this volume).

On the other hand, *M. arctoides* could well be placed within the *sinica* group. Although more derived than other members of that group, it appears to be on the same morphocline trend in penial morphology and copulation pattern. The female reproductive tract of *M. arctoides* is certainly specialized (reciprocally with that of the male) in a different way than is that of the rest of the *sinica* group. But Fooden (1967) suggested that this was an adaptation to ensure reproductive isolation from neighboring relatives—I suggest the *assamensis/thibetana* subgroup for this role. Hill and Bernstein (1969) also thought that *M. assamensis* was less derived in penial morphology and cranial hair-flow pattern than *M. sinica* or *M. radiata*, thus supporting the long separation of these subgroups. More detailed comparisons among all five species are needed to refute or confirm this hypothesis of their relationship. The cranial similarity of *M. arctoides* to *M. thibetana* (and *M. assamensis*) is another important link, as is the distribution pattern of the three species. *M. assamensis* is basically a smaller, slightly longer-tailed, more southerly version of *M. thibetana*, and the two are clearly closely related and parapatric. *M.*

arctoides is quite close in size (especially of skull) and relative tail length to *M. thibetana*, and it is widely distributed in southern Indochina, an area unoccupied by any other *sinica*-group member. However, *M. arctoides* also ranges northward to overlap fairly extensively with *M. assamensis* and slightly with *M. thibetana*; Albrecht (1978, p. 20) also considered that this distribution suggested "an ecological and/or evolutionary relationship between these" taxa. In sum, *Macaca arctoides* would appear to be related to the *assamensis/thibetana* pair, but to have developed more extreme (uniquely derived) conditions since its origin; it is thus included formally in the *sinica* group.

The hypothesis of relationships resulting from these inferences is presented in Fig. 2-6. It differs slightly from one which might reflect Fooden's published views, but the overall agreement is obvious. Some of the character states discussed above are shown on this cladogram (which has no simple time axis), as is relative tail length. According to this hypothesis, only one major tail reduction trend is required for each species group (four in all): *M. arctoides* and *M. thibetana* underwent continued reduction from an already short-tailed ancestor, and a similar canalization may have occurred in the *fascicularis* group (see below). Some fossils are tentatively indicated on the cladogram, although only one (*M. anderssoni*) is of any importance. An indication of the zoogeographic model presented below is also indicated. The seven taxa of macaques of Sulawesi (the Celebes) which Fooden consistently recognizes as species, are here indicated as a single unit. Albrecht (1978) basically accepted Fooden's ranking, but included comments by Groves (p. 87) which suggested some intergradation in the north. I imagine that perhaps four full species may eventually be recognized (see Groves, this volume), but any number from one to seven is feasible.

HISTORICAL ZOOGEOGRAPHY

A final step which can be taken here is to suggest an overall scenario of deployment for the genus, again based extensively on Fooden's work, but also on the phylogeny developed above. Fooden has suggested that the strongly disjunct distribution pattern of the *silenus-sylvanus* group(s) suggests early dispersal, followed by that of the moderately disjunct *sinica* group, after which the contiguously ranging *fascicularis* group may have dispersed. Each successive dispersal probably was closely linked to the disjunction of previous group(s), as well as Pleistocene climatic alternations.

After an origin in northern Africa, early macaques spread into Eurasia, probably via the Near East. One line entered southern Europe by the beginning of the Pliocene, here recognized as a succession of temporal subspecies of *Macaca sylvanus*. North African Late Miocene *M. libyca* and Sardinian later Pliocene *M. ?s. majori* might be distinct species, referred to the

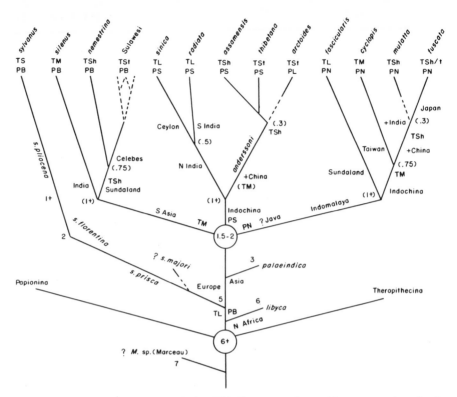

Fig. 2-6. Phylogram of macaque evolution. This diagram can be read in two ways. As a simple cladogram, the lines indicate relationships among the species of *Macaca* (all Sulawesi forms being treated as an indivisible monophyletic group). In a cladogram, there is no simple time axis; branching points along any lineage are more recent toward the top of the diagram, but nodes at a similar height along different lineages are not necessarily of similar age. All species are of the genus *Macaca,* and some fossil species are placed as independent lineages, as are the two other subtribes of the Papionini. Due to uncertainty as to their relationships, the two additional genera of the Macacina, *Procynocephalus* and *Paradolichopithecus,* are not included at all. The distribution and inferred history of two characters, tail length and penial glans shape, are also included. TL = tail long (≥ head + body length[HB], the inferred ancestral condition; TM = medium (between 1.0 and 0.5 HB); TSH = short (between 0.5 and 0.2 HB); and TSt = stump tail (less than 0.1 HB); TSh/t is intermediate between the last two. These classes are arbitrary, merely to indicate an approximate pattern, data from Fooden, this volume, PB = glans blunt and broad, the inferred ancestral pattern; PS = sagittate (acute and broad); PL = lanceolate (acute and elongate); PN-narrow (and blunt) (data from Fooden). The two large circles represent unresolved trichotomies. Note: TS at *M. sylvanus* should read TSt.

With further extrapolation, as described in the text, time, geography and character evolution may be added. Thus, the numbers indicate the ages of branching points and fossil taxa (those in parentheses being relatively less certain); nodes of similar age have been brought to equivalent heights. Changes from the inferred ancestral conditions of TL and PB are indicated by the newly derived condition on the segment involved, except that no changes are indicated for terminal segments. Several fossil species and subspecies are placed along segments which they might actually represent. Based on Fig. 2-7, an indication is given to the geographic dispersal of the various lineages: a " + " indicates the extension of a range; new terms following a node represent the subdivision of a range, and no term indicates continuity.

sylvanus species group on geographic grounds alone at present. *M. sylvanus* apparently flourished in Europe only during the warmer phases of the Pliocene and Pleistocene, eventually retreating from the continent entirely. Continuous occupation of the north African littoral, especially in the west, is likely.

A second line of macaques moved eastward, reaching India by the later Pliocene, if not earlier (Fig. 2-7). This group, like early *M. sylvanus*, would have been long-tailed, with *silenus*-type reproductive organs. The ancestral species presumably spread from southern India, along the coast (and perhaps inland) to Burma and into Malaya. As in Europe, climatic fluctuations affected southern Asia and "Sundaland" (a collective term for Indonesia and the neighboring margins of the Sunda shelf, see Medway, 1970). Brandon-Jones and Hiiemae (unpublished ms) suggest that cool/arid phases (linked to drops in sea-level and thus to island connnections permitting migration of suitably adapted species) led to restriction of forest in relict areas within this region (see also Eudey, this volume). Such phases would have alternated, several times during the past two million years and more rapidly after 0.5 MYA, with episodes of warmer and moister climate during which islands and their species were isolated from one another. The timing of these oscillations is quite unclear, but the relative sequence of evolutionary events postulated for *Macaca* is estimated to best agree with the underlying phylogeny, known fossils and rare biogeographic evidence (in part suggested by Brandon-Jones and Hiiemae). Fig. 2-7 depicts, in a very tentative manner, six stages in the deployment of Asian macaques.

An early drop in sea level permitted entry into Sundaland, after which I suggest that ancestors of the *fascicularis*-group became isolated in Indonesia (perhaps Java) and a proto-*sinica*-group species arose in or west of Burma; later, *M. nemestrina* or its ancestor speciated from the *silenus*-like stock in Malaya or Sumatra. This isolation is thought to have been younger both because of the overall closer relationship between *M. silenus* and *M. nemestrina* (mostly based on conservative retentions) and because these two show some tail reduction, suggesting that the trend had already commenced between the time the other two groups diverged and the point when these species became isolated.

With alternating climatic amelioration and decay, the several Asian macaque groups came into contact and competition. Long-tailed early *sinica*-group populations moved south and west into peninsular India, perhaps competing strongly with indigenous *M. silenus*. The former also moved north toward China, where they would have become larger and shorter-tailed, perhaps being represented by some of the earlier fossils of *M.* cf. *anderssoni*. Later, *M. silenus* may have been restricted to its present southwest Indian relict distribution, while the northern and southern members of the *sinica* group may have become (partly) genetically isolated. Ceylon may have been colonized by the southern form at this time (circa early Middle Pleistocene).

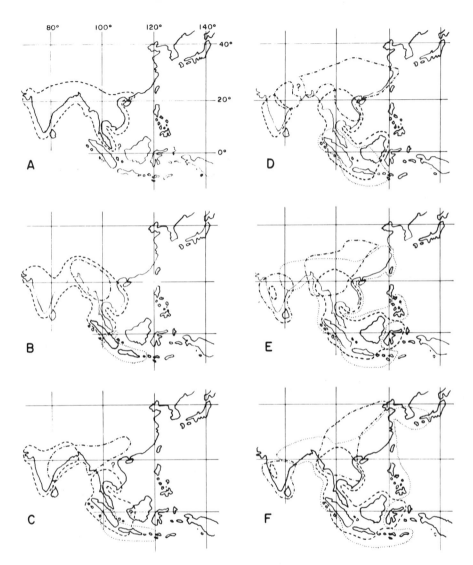

Fig. 2-7. Hypothetical dispersal pattern of Asian macaques. On the same base map used by Fooden (this volume) to present modern distributions, the inferred ranges of the three species groups are given for six time periods: A, later Pliocene (2.5–2.0 MYA); B, Early Pleistocene (1.5–1.3 MYA); C, late Early Pleistocene (ca. 1.1 MYA); D, early Middle Pleistocene (ca. 0.9 MYA); E, mid Middle Pleistocene (ca. 0.6 MYA); F, late Middle Pleistocene (ca. 0.25 MYA). ---------- *silenus* group (and ancestral dispersal pattern) .... *fascicularis* group _ _ _ _ _ _ _ _ _ *sinica* group. The boundaries are, of course, extremely questionable, but do include all known fossils (see text); the maps themselves are somewhat schematized.

In Sundaland, a population of *M.* cf. *nemestrina* may have entered Sulawesi from Borneo, while early *fascicularis* group populations began to spread onto various islands and up the Malayan peninsula. Succeeding mid-Pleistocene climatic fluctuations allowed the colonization of the Mentawai Islands by and divergence of a northern and southern major subspecies of *M. nemestrina* (Fooden, 1975); the diversification of the Sulawesi macaques from a *M. maura*-like ancestor (Fooden, 1969; Albrecht, 1978); the spread of *M. fascicularis* into the many islands of Indonesia and the Philippines; and the isolation, perhaps relatively early, of a northern member of the *fascicularis* group in Burma or eastern India (proto-rhesus).

The presence of *M. fascicularis* today in the Philippines and on small islands near the Mentawai group, but not on the latter (more distant from Sumatra) or on Sulawesi, suggests that *M. nemestrina* may have reached the last two areas early on, but that they were inaccessible by the time *M. fascicularis* pushed northward after colonizing the eastern islands (which lack *nemestrina* or relatives). Fossils of mid Pleistocene age on Java document the presence of at least one of these two species, but which one is yet uncertain; perhaps relative tooth size data may answer this question. Previously, Fooden has argued that the *fascicularis* group was the last to disperse, but as Albrecht (1978, p.20) noted, its distribution with respect to *M. nemestrina* in Sundaland is difficult to interpret on this premise; other authors (e.g., Medway, 1970) have suggested that *M. fascicularis* in fact arrived first. If it is postulated that *M. fascicularis* originated and diversified in southeastern Sundaland, but only entered the mainland at a later date, some of the opposing interpretations are reconciled.

Once a northern member of the *fascicularis* group invaded southern Asia, it may have spread rapidly both east and west. If tail length is considered, parsimoniously, to have reduced only once in the group, it must be suggested that *M. cyclopis* reached and became isolated on Taiwan at a relatively early stage, as its tail is longer than that of *M. mulatta*. On the other hand, being farther north and larger than the rhesus, *M. cyclopis* would be expected to have a shorter tail. Perhaps it penetrated into eastern China in a warm phase and then retreated to Taiwan during a glacial, without altering its tail/body proportions after subsequent isolation. Meanwhile, a population like modern *M. mulatta*, very ecologically adaptable, may have entered the Indian region and displaced *sinica*-group populations from the northern zones. *M. sinica* itself became isolated on Ceylon, while modern *M. radiata* developed in southern peninsular India, overlapping only partly with *M. silenus*. The presence of *M. mulatta* effectively completed the isolation of the southern and northern lines of the *sinica* group, whose territory it overlapped eastward. Japan was colonized by a form presumably like *M. mulatta* by the later Middle

Pleistocene. During the earlier Middle Pleistocene, *M. anderssoni* (= *M. robusta)* inhabited China, and its resemblance to *M. arctoides* and *M. thibetana* suggests that these species were not yet distinct. A search for derived cranial features among these taxa should prove useful. At some point, *M. arctoides* would have become isolated in southeastern Indochina (?), while *M. thibetana* and *M. assamensis* would have diverged even later in time, with tail length reducing in the first two of these.

Brandon-Jones and Hiiemae (unpublished ms) and Albrecht (1978) have noted that similar patterns of dispersal and distribution are known in various mammalian, avian and floral taxa over the same geographic range. Most interesting is the pattern seen by Brandon-Jones and Hiiemae for the Asian colobines. They argue that the south Indian-Ceylonese "*Kasi*" *senex* is closely related to the Indochinese-Indonesian *Trachypithecus* group; a possible sister of this taxon is "*Semnopithecus*" *entellus* of India. If this hypothesis were accepted, *Trachypithecus* might be recognized as a subgenus of *Semnopithecus*. In Sundaland, the smaller bodied species of *Presbytis* (sensu stricto) occur widely. There are also two groups of more distinctive colobines: *Pygathrix* [and *P. (Rhinopithecus)*] in southern China and eastern Indochina; and *Nasalis* [and *N. (Simias)*] on Borneo and the Mentawai Islands, respectively. The link between *S.* (*Trachypithecus*) of southern India and Indochina, separated by the highly adaptable and successful *S.* (*S.*) *entellus* in India recalls the pattern of *M. mulatta* separating members of both the *sinica* and *silenus* groups. Moreover, the distribution of *Pygathrix* parallels that of *M. arctoides*, compared to their respective "ancestral stocks" of *Semnopithecus* and *M. assamensis/M. thibetana*.

It is hoped that this set of hypotheses will be tested in the future through studies of morphologic and molecular systematics of macaque species (giving due consideration to the great ranges of variation within many species), as well as by finds of new fossils and studies of stereotyped behavior patterns and Pleistocene climatic shifts. This is a first step, built upon the groundwork of others and probably far too detailed, but it is offered in the hope of arousing interest and thus a desire to test and refute, among my colleagues.

ACKNOWLEDGMENTS

I thank Dr. Jack Fooden for many discussions of the morphology, distribution and systematic relationships of macaques, without which this paper would never have been possible. I also thank Drs. Colin Groves and Gene Albrecht and Messrs. Douglas Brandon-Jones and Alfred Rosenberger for discussions and/or the privilege of reading unpublished studies relevant to this article. Figures 2-6 and 2-7 were expertly drawn by Mr. Ray Gooris, Fig. 2-1 (A) by Ms. Biruta Akerbergs; the photos in Figs. 2-1, 2, 4, 5 were taken by the author. Finally, I am grateful to Dr. Donald Lindburg for requesting this contribution and persevering with publishing difficulties—I would not have considered writing such a paper without this instigation. The preparation of this article was financially supported, in part, by grants from the City University of New York Faculty Research Award Program: FRAP 11152, PSC-BHE 11785 and 12188.

REFERENCES

Albrecht, Gene H. The Craniofacial Morphology of the Sulawesi Macaques. Multivariate Approaches to Biological Problems. *Contributions to Primatology*, **13**: 1–151. Basel : S. Karger (1978).

Arambourg, Camille. Vertébrés continentaux du Miocène supérieur de l'Afrique du nord. *Publications Serv. Carte Geol. Algérie* (nouv. série), *Paléontologie*, **4**: 5–159 (1959).

_____. Les vertébrés du Pleistocène de l'Afrique du nord. *Arch. Mus. Nat. Hist. Nat.*, Paris, sér. 7, **10**: 1–126 (1969–1970).

Azzaroli, Augusto. La scimmia fossile della Sardegna. *Riv. Sci. Preist.*, **1**: 168–176 (1946).

_____. Late Miocene interchange of terrestrial faunas across the Mediterranean. In M. T. Alberdi and E. Aguirre (eds.), *Actas I Coloquio Internacional sobre Biostratigrafia Continental del Neogeno Superior y Cuaternario Inferior*, 1974. Madrid: Istituto Lucas Mallada (CSIC), pp. 67–72 (1975).

Blainville, H. M. D. de. *Ostéographie des Mammifères. I. Primates, Atlas*. Paris: Baillière (1839).

Brandon-Jones, Douglas and Hiiemae, Karen. Present primate distributions as a guide to Quaternary climatic change in Asia (Unpublished ms).

Cocchi, I. Su di due Scimmie fossili italiane. *Boll. Comitata Geol. Italia*, **3**: 59–71 (1872).

Delson, Eric. Evolutionary history of the Cercopithecidae. *Contributions to Primatology*, **5**: 167–217. Basel: S. Karger (1975a).

_____. Paleoecology and zoogeography of the Old World monkeys. In R. Tuttle (ed.), *Primate Functional Morphology and Evolution*. The Hague: Mouton, pp. 37–64 (1975b).

_____. Vertebrate paleontology, especially of non-human primates, in China. In W. W. Howells and P. J. Tsuchitani (eds.), *Paleoanthropology in the People's Republic of China*. Washington, D. C. : National Academy of Sciences, pp. 40–65 (1977).

Elliot, D. G. *A Review of the Primates*. Monographs, Amer. Mus. Nat. Hist., 3 vols., New York (1913).

Fooden, Jack. Complementary specialization of male and female reproductive structures in the bear macaque, *Macaca arctoides*. *Nature*, **214**: 939–941 (1967).

_____. Taxonomy and evolution of the monkeys of Celebes (Primates: Cercopithecidae). *Bibliotheca Primatologica*, **10**: 1–148, Basel: S. Karger (1969).

_____. Taxonomy and evolution of liontail and pigtail macaques (Primates: Cercopithecidae). *Fieldiana: Zoology*, **67**: 1–169 (1975).

_____. Provisional classification and key to living species of macaques (Primates: *Macaca*) *Folia primatol.*, **25**: 225–236 (1976).

Harlé, Edouard. Une mandibule de singe du repaire des hyènes de Montsaunès (Haute-Garonne). *Bull. Soc. Hist. Nat. Toulouse*, **26**: ix–xi (1892).

Hedinger, A. Uber den Pliocänen Affen des Heppenlochs. *N. Jb. Min., Geol., Paläont., I:* 169–177 (1891).

Hill, William Charles Osman and Bernstein, Irwin. On the morphology, behaviour and systematic status of the Assam macaque (*Macaca assamensis* McClelland, 1839). *Primates*, **10**: 1–17 (1969).

Hooijer, Dirk Albert. Quaternary langurs and macaques from the Malay archipelago. *Zool. Verhandelingen Mus. Leiden*, **55**: 3–64 (1962).

_____. Miocene Mammalia of Congo. *Ann. Mus. Roy. Afr. Centr.*, sér. in *8ᵛᵒ*, sci. *Géol.*, **46**: 1–71 (1963).

Jolly, Clifford J. The evolution of the baboons. In H. Vagtborg, (ed.), *The Baboon in Medical Research*, vol. 2. Austin: Univ. Texas Press, pp. 427–457 (1967).

Jouffroy, Françoise Kyou. Un crâne subfossile de Macaque du Pleistocène du Viet Nam. *Bull. Mus. Nat. Hist. Nat., Paris*, sér. 2, **31**: 209–216 (1959).

Kay, Richard F. Molar structure and diet in extant Cercopithecidae. In P. M. Butler and K. Joysey (eds.), *Studies in the development, function and evolution of teeth*. London: Academic Press, pp. 309–339 (1978).

Kormós, Theodor. Die phylogenetische und zoogeographische Bedeutung praeglazialer Faunen. *Verhandlungen K.-K. Zool.- bot. Gesell., Wien*, **64**: 218–238 (1914).

Lydekker, Richard. Rodents and new ruminants from the Siwaliks, and synopsis of Mammalia. *Mem. Geol. Survey India, Paleont. Indica*, ser. X, **3**: 105–134 (1884).

Medway, Lord. The monkeys of Sundaland: ecology and systematics of the cercopithecids of a humid equatorial environment. In J. R. Napier and P. H. Napier (eds.), *Old World Monkeys*. London: Academic Press, pp. 513–553 (1970).

Napier, John R. Paleoecology and catarrhine evolution. In J. R. Napier and P. H. Napier (eds.) *Old World Monkeys*. London: Academic Press, pp. 55–95 (1970).

Owen, Richard. Notice sur la découverte, faite en Angleterre, de restes fossiles d'un quadrumane du genre Macaque, dans une formation d'eau douce appartenant au nouveau pliocène. *C. R. Acad. Sci.* (Paris), **21**: 573–575 (1845).

Pomel, A. Sur un macaque fossile des phosphorites quaternaires de l'Algérie. *C. R. Acad. Sci.* (Paris), **115**: 157–160 (1892).

_____. Singe et Homme. *Publ. Serv. Carte Geol. Algérie, Paléont. Monogr.*, **11**: 1–32 (1896).

Schlosser, Max. Fossil primates from China. *Palaeont. Sinica*, **1** (2): 1–16 (1924).

Simons, Elwyn L. The deployment and history of Old World monkeys (Cercopithecidae, Primates). In J. R. Napier and P. H. Napier, (eds.), *Old World Monkeys*. London: Academic Press, pp. 97–137 (1970).

Stromer, Ernst. Mitteilungen über Wirbeltierreste aus dem Mittelpliocän des Natrontales (Aegypten). 5. Nachtrag zu 1. Affen. *Sitz.-ber. Bayerischen Akad. Wiss., Math.-phys. Kl.* 1920, pp. 345–370 (1920).

Szalay, Frederick S. and Delson, Eric. *Evolutionary History of the Primates*. New York: Academic Press (1979).

Walker, Phillip and Murray, Peter. An assessment of masticatory efficiency in a series of anthropoid primates with special reference to the Colobinae and Cercopithecinae. In R. H. Tuttle, (ed.), *Primate Functional Morphology and Evolution*. The Hague: Mouton, pp. 135–150 (1975).

Young, Chung-chien. On the Insectivora, Chiroptera, Rodentia and Primates other than *Sinanthropus* from locality 1 at Choukoutien. *Palaeont. Sinica*, ser. C, **8**(3): 1–160 (1934).

Chapter 3
Molecular Evolution and Systematics of the Genus Macaca

John E. Cronin,
Rebecca Cann
and *Vincent M. Sarich*

INTRODUCTION

The development of the evolutionary history and systematics of the macaques has proven to be a substantial problem for evolutionary biologists. Neither their positioning within the 42-chromosome clade of cercopithecoid monkeys (Papionini) nor the relationships among their various populations, subspecies, and species is at all clear. The lack of an adequate fossil record and the fact that the Asiatic macaques have undoubtedly undergone a recent (Late Pliocene–Pleistocene) and successful radiation with the speciation process continuing even now, have contributed to the problems involved in deciphering the origin and subsequent history of this very successful, morphologically diverse, and geographically widespread group.

PREVIOUS WORK IN MACAQUE SYSTEMATICS

Morphology and Paleontology

Traditionally, phylogenetic relationships, and by extension, taxonomic assignments, have been based on morphological comparisons of living and fossil forms. For example, Hill (1974) recognized two macaque genera containing some 14 species. Ellerman and Morrison-Scott (1951) recognized 11 species, Kellog (1945) 16, while Albrecht (1977) and Fooden (1969) see 7 species on Sulawesi alone.

While the macaque clade is certainly part of the 42-chromosome group of the Cercopithecoidea, there would appear to be no conclusive evidence linking that clade with any of the other genera in the tribe Papionini (*Papio, Theropithecus, Cercocebus, Mandrillus*).

Late Miocene fossils from Europe and North Africa have been referred to *M. sylvanus* by Delson (1975). Plio-Pleistocene macaques appear in some numbers in Asia and have been referred to living and extinct species (Delson, 1975). Delson (1975) and Delson and Andrews (1975) suggest that the macaque line is part of the primary adaptive radiation of the 42-chromosome clade beginning about 6–7 MYA and is not clearly associated with any other of the major lineages. Delson does suggest that the macaque cranial and dental pattern may be primitive for the Papionini as a whole. He points out: "Not enough is known of the early history of the dentally typical African Papionini to support more than a 'guesstimate' of genus-group relationships" (Delson, 1975, p. 210). In other words it becomes difficult to sort true phyletic macaques from primitive ancestral forms of the various Papionini lineages, and cladistic conclusions drawn from the currently very fragmentary early fossil record, whether at the genus or species level, ought to be considered very tentative ones.

Chromosomal Evidence

Karyotypic analysis of species in the genera *Papio, Macaca, Theropithecus,* and *Cercocebus* has revealed that they all possess the same diploid number (2N = 42, NF = 84), with 13 submetacentric pairs, 6 metacentrics, a large achromatic pair with a secondary constriction, and sex chromosomes (Chiarelli, 1971; Schmager, 1972), Chiarelli minimizes the importance of the small differences in the chromosomes of this group, citing large numbers of fertile interspecific hybrids as ample evidence of shared genetic homology.

One might expect that chromosome banding studies would provide increased resolution, but this has not thus far proved to be the case among the Papionini. Of eight individuals studied, *G* bands were identical between *M. fascicularis* and *P. sphinx*, except for one chromosome in the large submetacentric group possessing a small pericentric inversion (Rubio-Goday *et al.*, 1976). Within the macaques, *M. mulatta* and *M. fascicularis* show no major differences in G banding (DeVries *et al.*, 1975). C banding on these same animals gave identical results, revealing no differences in the placement of constituative heterochromatin between the two species. The karyotype of *M.*

mulatta has also been examined in detail with trypsin banding (Perticone *et al.*, 1974), giving some credence to the notion that chromosomes 6 and 13 are conservative in the Catarrhini (Egozcue, 1973).

Although banding techniques have been important in both elucidating the phylogenetic affinities and revealing chromosomal rearrangements among other closely related species, the relationships of animals within the 42-chromosome group still remain somewhat puzzling. Stock and Hsu (1973) have emphasized the homologies using G bands between *M. mulatta* and *C. aethiops* ($2N = 60$), and have reconstructed the banding pattern of human chromosomes from that of rhesus, assuming mostly large Robertsonian translocations (see Seth *et al.*, 1976 however for a contrasting viewpoint). In addition, they propose that the generation of *Macaca* chromosomes from an ancestral *Cercopithecus* involved the loss of a large amount of heterochromatin. They allow, however, that as in the case of *Mus,* it may be equally likely that new species evolve their own highly repeated DNA sequences after divergence. The detection of such sequences specific to the macaques, baboons, and mangabeys (Gillespie, 1977) may reflect such a case.

The shared marker achromatic chromosome, no. 9, has recently been shown by *in situ* hybridization to be the chromosome carrying genes for the production of ribosomes in *P. cynocephalus, P. hamadryas, T. gelada,* and *M. mulatta* (Henderson *et al.*, 1977; Henderson *et al.*, 1974), although there does seem to be variation in the number of RNA genes per chromatid. Henderson *et al.* (1976) estimate 140 cistrons are present in *M. mulatta,* versus 230 in *P. cynocephalus.* As more probes for discrete products become available through cloning techniques, it will be possible to study the variation in placement and number of different genes, detecting minute translocations and deletions only grossly hinted at by banding. Cell fusion for gene mapping has already been used to demonstrate the conservatism of primate chromosome no. 1 (Finaz *et al.*, 1977) in the retention of placement of three metabolic enzymes, indicating the importance of position effects in gene regulation.

Macromolecular Evidence

Recently the use of genetic information contained in macromolecules has proven exceedingly useful in elucidating the phylogenetic history of organisms, particularly primates. (For a recent review of this subject see *Molecular Anthropology,* 1976, edited by M. Goodman and R. Tashian). Within the Old World monkeys comparative information is fairly substantial. There exist data from immunodiffusion studies (Goodman and Moore, 1971; Dene *et al.*, 1976); protein sequences (as summarized in Goodman, 1976); immunological comparisons (Sarich, 1970; Cronin, 1975; Cronin and Sarich, 1976); DNA annealing (Kohne, 1975; Benveniste and Todaro, 1976, 1977; Gillespie, 1977);

and electrophoresis (Darga *et al.* 1975; Avise and Duvall, 1977; Bruce, 1977; Cronin and Sarich, 1976).

Nonetheless, major issues remain unsolved. Are macaques associated with any other of the Papionini lineages? Are macaques part of a true multilineate radiation, or are they the initial diverging lineage of this 42-chromosome clade? The molecular picture, at present, is one of a single primary adaptive radiation of the Papionini with, at the genetic level, only *Papio* and *Theropithecus* clearly sharing a substantive subsequent period of common ancestry.

Basically, with respect to positioning the macaques as a group, there are two choices. Data from immunodiffusion (Dene *et al.*, 1976), alpha hemoglobin sequences (Hewett-Emmett *et al.*, 1976) and microcomplement fixation comparisons of the albumins and transferrins (Cronin and Sarich, 1976) suggest, tentatively, that the macaque line represents the earliest separation within the 42-chromosome clade. Alternatively, the beta hemoglobin (Hewett-Emmitt, 1976), DNA (Gillespie, 1977; Benveniste and Todaro, 1976), and electrophoretic evidence (Cronin and Sarich, 1976; Bruce, 1977) does not clearly resolve the macaque line from other Papionini. On balance, then, we lean toward the former placement and have so indicated this in Fig. 3-1, though a realistic appraisal would set the odds here at no better than 2 or 3 to 1. The

MOLECULAR PHYLOGENY of the OLD WORLD MONKEYS

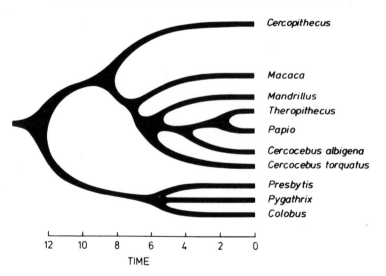

Fig. 3-1. A molecular phylogeny of the major Old World monkey lineages. Data from our laboratory and the literature have been analyzed cladistically in the sense of defining lineages on the basis of molecular changes which can be seen as having occurred along them. Groupings are thus defined on the basis of shared, derived amino acid or nucleotide substitutions.

quality of the available data is simply too poor to allow any clear-cut decisions as to the details of generic interrelationships among the Papionini. The molecular data are supportive of a radiation of the Papionini beginning some 5-7 MYA with the intragenetic distances among the extant genera being only slightly greater than those found among *Homo, Pan* and *Gorilla*.

MATERIALS AND METHODS

Immunological procedures

Albumin and transferrin were purified according to methods detailed in Cronin and Sarich (1975). The purified proteins were injected into rabbits and the antisera produced used for microcomplement fixation comparisons. Genetic differences are expressed in terms of immunological distance units where, for albumin and transferrin, one unit is approximately equal to one amino acid substitution (Champion *et al.*, 1974; Sarich and Cronin, 1976).

Electrophoretic Procedures

Electrophoresis of serum proteins was carried out in a vertical slab polyacrylamide gel in both continuous and discontinuous systems with various concentrations of acrylamide (Cronin and Sarich, 1976; Sarich, 1977). Electrophoresis of red cell enzymes and selected serum proteins was carried out by microzone electrophoresis on cellulose acetate membranes modified from the method of Gruenbaum (1977).

When adequate sample sizes were available, allele frequencies were determined. In general, though, this was not possible, and individuals were scored as identical when they shared the same allele or alleles, and different when they were homozygous for different alleles. Genetic similarities and distances were calculated according to Nei (1972). Nei's coefficient of similarity (I) can vary from 0 to 1.0, where 1.0 equals genetic identity. Genetic distance D ($-$ ln I) can vary from 0 to infinity and is the number of electrophoretically detectable substitutions per locus that have occurred since two populations diverged.

For our plasma protein electrophoreses we used a value S as a measure of genetic similarity. When two individuals are compared, this is simply the ratio of bands that have identical mobility on the gel to the total number of bands countable in both. Usually some 20-25 bands are resolvable, and presumably these represent a similar number of loci. Unfortunately, very few of these loci have been identified other than in humans, and we are currently involved in filling this gap in our knowledge. Specific plasma protein loci that have been identified in macaques are reported in this paper. Similarity values (*S*) are converted into genetic distance values ($D = -\ln S$) to yield a plasma protein electrophoretic (PPE) D value.

MOLECULAR APPROACH TO SYSTEMATICS

The study of evolution is, in essence, the study of adaptive speciation. This has been revolutionized in recent years by the advent of electrophoretic measures of genetic differentiation within and between populations. The genetic distances so measured have the potential of providing realistic cladistic and temporal frameworks within which evolutionary change must be structured and understood. Indeed, at the lower taxon levels of concern here, the data from the fossil record and comparative anatomy currently appear to be either totally lacking, or, when available, fail to provide sufficient resolving power towards delineating the desired phylogeny. The realization of the potential of the electrophoretic approach has also been delayed because of the concentration on easily scored loci—which generally meant those enzymes involved in important and complex metabolic pathways—and the failure to distinguish the sometimes differing goals of the population geneticist and systematist.

The success of any comparative effort in reconstructing phylogenies, be it based in molecules, anatomy, or behavior, cannot be better than its ultimate data base. This means that, not only must we be able to read the evolutionary message in the characters used, but there has to be a message to read. In other words, a lineage can have no reality to those of us who look for it, unless something which remains discernible to us today occurred along it. At the nucleic acid and protein levels especially, this means that the rates of evolution of the molecules chosen have to be commensurate with the demands of the systematic problem under study. In an electrophoretic effort, where, in effect, each can provide only one bit of information—identity or nonidentity—per pair of individuals compared, this becomes a critical and stringent requirement.

Thus our first concern is to gain some measure of the relative and absolute rates of electrophoretic differentiation at the various loci potentially usable. It has been shown that those rates are widely and bimodally distributed, with one group of loci accumulating electrophoretically detectable substitutions about 10 times more rapidly than the other (Sarich, 1977). This rapidly evolving group is made up, at present, of the various plasma proteins, nonspecific esterases, and others such as ribonuclease, lysozyme, and carbonic anhydrase. For the plasma proteins and nonspecific esterases the rate of change appears to be about one electrophoretically detectable substitution per lineage per 5 MY. This is consistent with the available immunological data on rates of evolution of two plasma proteins, albumin and transferrin, which appear to be typical of the entire set (Cronin, 1975). These show about one substitution per lineage per one MY, and if we assume that 20 percent of those substitutions to involve charge changes, we obtain our observed rate of electrophoretic change. As there are approximately 20 to 25 well-defined loci expressed clearly in fresh mammalian plasma samples, this would suggest that even if we limited ourselves to looking at only those, we would expect to see, on the average, one electrophoretically detectable substitution every 0.25 MY. Thus one might begin to detect fixed allele differences between populations separated for as little as 10^5 years. These figures are so small that, in a sense, they suggest the very real possibility of achieving a degree of phylogenetic resolving power sufficient to answer many questions not yet even asked—which means that we might be able to give away a little of that power for practical reasons and still obtain some very useful results. The obvious question which occurs in this context is that of sample size.

This has much more to do with the number and nature of the characters involved in the comparisons, rather then with the number of individual organisms. For example, by the usual morphological criteria, chimpanzees and gorillas are very similar to one another, and it takes a relatively sophisticated eye to sort them accurately. On the other hand, even though the genetic distances involved are quite similar, humans can be routinely sorted from both gorillas and chimps by even the most unpracticed eye. At the molecular level, comparing their albumins either immunologically or electrophoretically, we can readily sort all *individual* chimps, gorillas, and humans into three non-overlapping sets; while the fibrinopeptides prove to be quite nondiscriminating (Wooding and Doolittle, 1972). Thus, the degree to which individuals will be representative of the evolutionary units to which they belong for the purposes of systematics will very much depend on the features chosen, be they morphological or molecular, and on the rates of change of those features. The various proteins and nucleic acids have the advantage here in that their rates of change are minimally, if at all, lineage-dependent, and thus will do just as good (or poor) a job of sorting humans and chimps as gorillas and chimps, or, for all that goes, horses and zebras, or Indian and African elephants. What the molecules can ultimately give us are the cladistic and temporal dimensions of the phylogenies linking the organisms containing them—and this is all we can ever ask of them—thus the quality or resolving power of our result can be assessed in objective terms, as, for example, the number of changes which can be seen as having occurred along a particular lineage. The smaller the number, the less certain we can be that the lineage did have any existence in historical reality. It is just at this point that immunological comparisons of single proteins, DNA annealing procedures, amino acid sequencing, or the usual allozyme electrophoretic efforts begin to give us problems when dealing with evolutionary relationships extending over the last 5 MY or so. They provide us with too few differences to count reliably. Yet with the plasma proteins we can, on one gel, see 20 to 25 differences between a pair of individuals belonging to species separated for that long a period. So the differences are there to find, and one asks how they are to be counted most effectively. Basically this comes down to asking how much of our maximum possible resolving power is achieved using n individuals per evolutionary unit, where n starts at 1.

Clearly two individuals drawn from any polymorphic population will be genetically different from one another. The question is how different. Electrophoretically, at each variable locus with two alleles we might select two identical homozygotes, one heterozygote and one homozygote, two heterozygotes, or the two different homozygotes. If one now decides to compare only features, and, thus, by implication, alleles present in both individuals, then the only time one would count two individuals drawn from the same gene pool as different would be the last case; and the maximum probablility for such an occurence is 1 in 8 when these two alleles are present in equal amounts. This figure will, of course, increase as the number of alleles increases, but the presence of more than two alleles at relatively high frequency in a population is a relatively rare phenomenon. Thus even if one-third of our 20 to 25 loci were significantly polymorphic, it would be unlikely for two individuals to be counted as different at more than one locus. If one wishes to eliminate even this noise level, then four individuals can be compared two against two, and the 1 in 8 chance drops to 1 in 128. In theory then, one can be reasonably certain that where two randomly chosen individuals in one population have alleles at even one locus not present in two other individuals from a second, the two populations have been isolated for long periods from one another. Beyond this one gains nothing until a sufficiently large number of individuals has been looked at so that one can reliably speak of allele frequencies, and even then one does not obtain a gain in resolving power anywhere nearly commensurate with the enormous increase in time required to look at 30 individuals per population as compared to one or two. All one really loses by making single individuals represent populations is that component of genetic distance contributed by polymorphic loci. This is unlikely to be a very serious loss, as considerations of known heterozygosity levels lead one to the conclusion that even for rapidly evolving loci, the life span of a polymorphism is likely to be relatively brief in comparison to the time between successive substitutions. What eventuates, then, is a lag time before we can begin to reliably measure genetic differentiation between isolated populations.

The approach, then, is to subject plasma samples to polyacrylamide gel electrophoresis (see Fig. 3-2), thus providing maximum routinely available resolution. We use, as a measure of genetic identity, simply the ratio of the number of bands with the same mobility to the number counted in both. We already know that this measure is highly correlated in primates with the albumin plus transferrin immunological distances between the same species pairs (Sarich and Cronin, 1976). We also know, again using the immunological distance criterion, that one will never see a significant identity value (0.15 or so) when the albumin plus transferrin distance is more than 20 to 25 units; that is, where the indicated time of separation is more than 5–6 MY. We have therefore undertaken to apply this band counting approach as a preliminary effort to help sort out the very complicated systematics of the genus *Macaca*, where the divergence times among the various Asian species are unlikely to exceed 3 MY or so.

It is instructive to briefly dwell upon the difficulties in dealing with such short time spans molecularly. 3 MY or so would give us maximum DNA annealing differences of 1° or less within the genus, and such values are not very much larger than the errors associated with the measurements involved. One could compare the albumins and transferrins immunologically, but this would require the preparation of antisera to those proteins from several macaque species at an appreciable cost in time and money. Even then the distances obtained will be so small as to be of only marginal utility in deriving the cladistic relationships involved. The usual allozyme electrophoretic approach will be implemented with great difficulty for a number of the species, which simply aren't going to be available in the required numbers. So one turns logically to the plasma protein electrophoretic approach, and here a few gels involving trivial expenditures of time and money provide considerably more information than could have been available to anyone previously. The problems are by no means all solved, but many of them are, and the remaining ones are sharply delineated.

The major concerns one has above this "band-counting" approach are, of course, homologies and overcounting. The matter of comparing only homologous bands can be approached directly—most straightforwardly by immunofixation—but this requires having prepared antisera which will cross-react with the proteins in the taxa being studied, an impractical requirement if the approach is to be routinely applied. If, however, one can accept the idea of not being able to know everything—an idea basic to the efficient application of any molecular approach—and has something resembling a graded series of relationships with which to work, then the homologies are generally readily apparent using mobilities, band densities, and band appearances as homology criteria. The possibility of overcounting is perhaps best typified at the transferrin locus, where one will often see two or more iron-binding bands in all individuals of a population. This phenomenon, when investigated, has generally been attributed to variation in the carbohydrate (often sialic acid) portion of the transferrin molecule (Palmour and Sutton, 1971). Thus an amino acid substitution in the polypeptide portion will affect the mobility of each of the transferrin bands, and a naive comparison might count three to five differences where, in fact, only one substitution has occurred. By the same token, one could count three to five identities when in

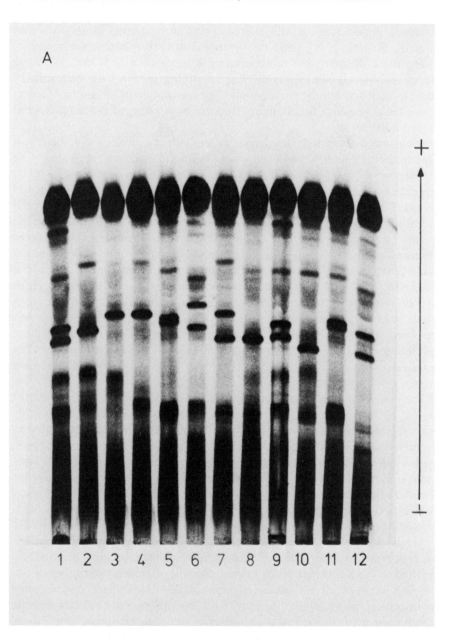

Fig. 3-2. Polyacrylamide gel electrophoresis of various macaque plasma samples. (A) 1. *mulatta*, 2. *sylvanus*, 3. *silenus*, 4. *nigra*, 5. *fuscata*, 6. *arctoides*, 7. *nemestrina*, 8. *fascicularis*, 9. *mulatta*, 10. *cyclopis*, 11. *fuscata*, 12. *radiata*. (B) 1. *mulatta*, 2. *fuscata*, 3. *nemestrina*, 4. *maura*, 5. *arctoides*, 6. *nemestrina*, 7. *fascicularis*, 8. *mulatta*, 9. *radiata*, 10. *mulatta*, 11. *radiata*, 12. *mulatta*, 13. *radiata*.

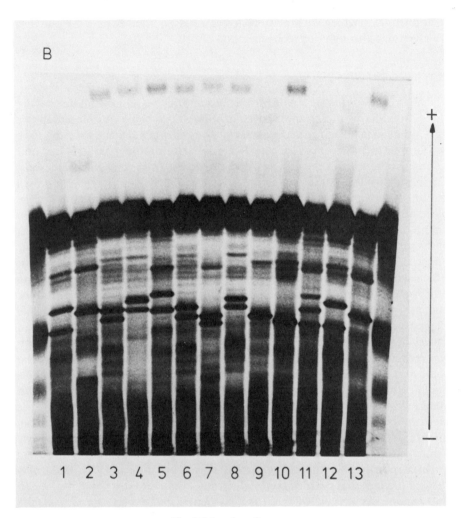

Fig. 3-2 (continued)

fact only one is justified. However, it is again clear, upon brief reflection, that this problem is directly identified when a series of bands is seen to shift mobilities as a unit. Because of the prevalence of glycoproteins in plasmas, however, it is a problem to which one must remain continually alert.

Our conclusions here are simple. Yes, there are problems with the approach, but they may be readily dealt with, even if not completely "solved," by exercising some intelligent judgment. The subjective element in that judgment is in large part dealt with by following the simple rules outlined above. We are left, however, with an issue not so readily solved. This is the matter of testing our conclusions. At this point we need to recognize that the evolutionary relationships in which we are interested can have no reality save the one that we give them through our reconstruction efforts.

Macaque Systematics

The phylogenetic dimension of work in systematics has as its ultimate goal the cladogram linking in time and space the populations being considered. In this light, we need to recognize at the outset that any efforts we can make at this time will be but a limited approach to this goal. The problems are several and soluble, but reaching the solution will not be an easy task for logistical reasons. The basic problem is that the samples with which we work represent only a portion of the variety of the genus. We can have appreciable confidence in the placements achieved for those individuals (and for the populations—often unknown—from which they are drawn), but we do not know what proportion of the taxonomic space occupied by *Macaca* in nature is accounted for by our material. This failing can only be resolved by concerted field efforts of the sort which are now routine for any effort in biochemical systematics among modern workers in the field, but clearly political and logistical considerations make this an operation more readily contemplated than accomplished. We suggest that excellent candidates for such an effort would be *mulatta* and *fascicularis* sampled over their entire ranges. The magnitude of the problem was recognized by Darga *et al.* (1975), who saw in their data a suggestion that some *mulatta* populations are more closely related to other *Macaca* species than to other *mulatta*. Although this material, collected some 10 years ago and generously made available to us by Dr. Goodman, has lost some of its electrophoretic integrity and so cannot provide as much information as we might like, we can see, for example, that Thai *mulatta* and *arctoides* have very similar plasma protein electrophoretic patterns ($S = 0.8$ or more), while our *mulatta* (presumably Indian) is appreciably more distant ($S = 0.65$). Assuming the material to be properly identified, these findings could well mean that *arctoides* is a recent offshoot of a particular *mulatta* lineage. *Assamensis* may well be another, and *cyclopis* probably stands in a similar position with respect to *fuscata*. While at this time we may be in no better position answering these questions than raising them, it is important to appreciate their presence. We do not wish to be seen as suggesting that a given placement for particular samples from our collection—especially for such geographically diverse forms such as *mulatta* and *fascicularis*—be necessarily representative of the entire species. As suggested above, it may well be that ultimately a cladogram of all *mulatta* will contain branches terminating in modern *arctoides* and *assamensis*, but we are at present in no position to strongly support or deny such a possibility—nor do we wish to get involved with the potential implications for formal taxonomy. We must begin by noting that, to the best of our knowledge, no biochemical studies of macaque systematics (except for *fuscata*, Nozawa *et al.*, 1975) have been carried out on *natural* populations. Intraspecific variation detectable by electrophoretic techniques of *M. nemestrina* laboratory raised or wild caught individuals have been carried out by Anderson and Giblett (1975). *M. mulatta* troops on Cayo Santiago have been surveyed by Buettner-Janusch and colleagues over a number of years (1974, 1977a, 1977b). Electrophoretic studies comparing some macaque species have been carried out by Darga *et al.* (1975), Avise and Duvall (1977), and by Bruce (1977). No primate study exists which can even begin to compare with what has for several years been routine in efforts on small American mammals. Our efforts at systematics are therefore truncated in a taxonomic sense in that we start with representatives of populations already differentiated at the species (or, in one case, subspecies) level. Our experiences among the primates in general, and Old World monkeys in particular, indicate that forms sufficiently distinct morphologically to merit subspecific or specific status (the distinction here is often a fuzzy one) show a plasma protein electrophoretic Nei *D* of at least 0.25 to 0.35; i.e., 5–6 band mobility differences of the 20 or so usually seen. This is typical of subspecific comparisons within the *Papio cynocephalus* complex, between East and West African *Cercopithecus aethiops*, or, to use a nonprimate example, between the domestic dog and coyote. In our *Macaca* sample we have only one well-documented pair of subspecies (*M. nemestrina nemestrina* and *M. n. leonina*), and they give a PPE *D* of 0.3. On the other hand, distances among the Asian macaques generally do not exceed 1 ($S = 0.37$; 12 to 13 band mobility differences out of 20), and thus our potential resolving power is severely restricted. Put in temporal terms, this means simply that the range of divergence times among the Asian macaques is quite narrow—at the species level, from 1 to 2–2.5 MYA. In other words, any lineage associating one or more macaque clades with one another would probably have existed for no more than 1 MY or so. This means that the best we can do moleculary without having actual populations to work with will be to see that MY or so with, on the average, the four electrophoretically detectable substitutions which might be expected in a sample of 20 or so rapidly evolving loci. The variance on four is of course large, and the range of selectively allowed mobilities for a given protein is generally narrow, leading to frequent parallelisms and convergences. These considerations should lead us to temper our conclusions with a good deal of caution, especially when divergence time differences of less than 0.5 MY or so between two potential nodes might be indicated.

As indicated in Table 3-1 we have available at least one sample from every extant macaque species with the exception of *thibetana* and some of the Sulawesi forms. Their electrophoretic integrity was, for various reasons, variable, and we feel least comfortable with any real attempts to place *sylvanus* and *silenus* on the basis of the information we currently have on them.

Table 3-1.   List of Macaque Species, Sample Sizes and Sources Available for this Study.

SPECIES	COMMON NAME	SAMPLE SIZE (N)	SOURCES
M. assamensis	Assamese macaque	1	1,2,3.
M. radiata	Bonnet macaque	5	1,2.
M. fuscata	Japanese macaque	5	7.
M. arctoides	Stump-tailed macaque	5	1,2,7.
M. fascicularis	Crab-eating macaque	5	2,7.
M. maura	Moor macaque	3	1,4,6.
M. nigra	Sulawesi Black ape	3	1,4.
M. tonkeanna	Tonkenese macaque	1	4.
M. cyclopis	Formosan macaque	2	1,7.
M. sinica	Toque macaque	3	1.
M. mulatta	Rhesus macaque	18	1,2,3.
M. n. nemestrina	Pig-tailed macaque	33	1,3,7.
M. n. leonina	Pig-tailed macaque	5	1,3.
M. silenus	Lion-tailed macaque	1	5.
M. sylvanus	Barbary Ape	2	5.

The following are the list of sources who generously provided us with samples.
1. Dr. M. Goodman—Wayne State University
2. Davis Regional Primate Center
3. Ames Research Center—Dr. C. Cann
4. Laboratory for Experimental Medicine and Surgery in Primates, New York
5. San Diego Zoo—Dr. O. Ryder
6. Houston Zoo—Dr. B. Whitlock
7. Dr. E. Bruce—University of California, Davis

Ideally we would prefer to carry out a cladistic analysis on the plasma protein data, but this would require a suitable reference species—and we are lacking suitable material from the only one that is potentially reasonable, *sylvanus*. Its albumin plus transferrin distance from two Asian species (*fascicularis* and *nigra*) averages 14 units, which is more than the distance to any Asian form; and such plasma protein electrophoretic comparisons as we have been able to make suggest a D of at least 1.2. These data, plus the fact of its geographical separation from any other macaque species, make the probability that all Asian macaques form a clade relative to it a strong one. When this is shown to be the case, then we will be in a position to use *sylvanus* to judge the relative amounts of change along each of the Asian macaque lineages. Failing this at present, we are forced to go to the far less suitable phenetic cluster analysis—to which we can add a small amount of cladistic information deriving from the allozyme data involving other catarrhine forms.

Table 3-2 gives the data for 25 proteins coded for by 29 loci in 11 species. These 29 loci include both slowly and rapidly evolving loci (or Group I plus Group II and III loci) (Selander, 1976; Johnson, 1976; Sarich, 1977). The data are those of Darga *et al.* (1975) and Bruce (1977) as well as our own. The most

Table 3-2. Genetic Similarities[1] (Below Diagonal) and Distances[2] (Above Diagonal) Based on 25 Proteins Encoded For By 29[3] Loci For 11 Species of Macaque[5].

SPECIES	1	2	3	4	5	6	7	8	9	10	11
1. *M. cyclopis*	—	0.65	0.54	1.25	0.63	0.74	0.49	0.49	0.41	0.37	0.76
2. *M. fuscata*	0.52	—	0.83	0.85	0.76	0.56	0.36	0.65	0.56	0.25	0.76
3. *M. radiata*	0.58	0.44	—	0.34	0.51	0.85	0.54	0.18	0.48	0.41	1.32
4. *M. sinica*[4]	0.29	0.43	0.71	—	0.14	1.95	0.85	0.56	0.56	0.56	0.00
5. *M. maura*	0.53	0.47	0.60	0.14	—	0.51	0.31	0.63	0.51	0.31	1.10
6. *M. nigra*	0.48	0.57	0.43	0.14	0.60	—	0.27	0.74	1.10	0.41	0.41
7. *M. nemestrina*	0.62	0.70	0.58	0.43	0.73	0.76	—	0.55	0.34	0.20	0.51
8. *M. arctoides*	0.62	0.52	0.83	0.57	0.53	0.48	0.58	—	0.74	0.43	0.76
9. *M. mulatta*	0.67	0.57	0.62	0.57	0.60	0.33	0.71	0.48	—	0.21	1.10
10. *M. fascicularis*	0.69	0.78	0.67	0.57	0.73	0.67	0.82	0.65	0.81	—	0.76
11. *M. tonkeanna*	0.47	0.47	0.27	0.00	0.33	0.67	0.60	0.47	0.33	0.47	—

[1] *S* values calculated as described in the text. The most frequent allele at each locus within each species was chosen to calculate *D* and *S* values. Sample sizes are small so genetic distance values are tentative.

[2] Nei's Index.

[3] The following loci were screened by Darga *et al.*, (1975). 1. TBPA, 2. Tf, 3. CA I, 4. 6PGD, 5. Hb alpha, 6. Hb beta.
The following loci were screened by Bruce, (1977). 1. AK, 2. Fum, 3. G6PD, 4. 6PGD, 5. GOT-S, 6. IDH, 7. LDH alpha, 8. LDH beta, 9. MDH, 10. NADH-dia, 11. PGI, 12. PGM 1, 13. PGM 2, 14. ALB, 15. Acp, 16. Cer, 17. EST A, 18. EST B, 19 Hpt alpha, 20. Hpt beta, 21. LAP.
The following loci were screened in our laboratory. 1. ALB, 2. GC, 3. G6PD, 4. Tf, 5. G6PD, 6. 6PGD, 7. PGI, 8. PGM 2, 10. AK, 11. ADA, 12. EST D, 13. Acp.

[4] Only 7 loci were screened, all of which are highly polymorphic loci among primates. Thus, *D* values involving *M. sinica* are higher than other species pair comparisons where more conservative, less polymorphic loci were included.

[5] The allozyme data was not converted into a dendrogram; but instead analyzed in a cladistic manner where we used shared derived electrophoretic changes along lineages to define clades.

Table 3-3. Genetic Similarities[1] (Below Diagonal) and Distances[2] (Above Diagonal) Based on Plasma Protein Loci in Nine Species of *Macaca*.

	1	2	3	4	5	6	7	8	9
1. *M. radiata*	---	0.80	0.69	1.05	0.82	0.97	0.89	0.97	N.D.
2. *M. mulatta*	0.45	---	0.40	0.63	1.05	0.87	0.80	0.94	N.D.
3. *M. arctoides*	0.50	0.67	---	0.64	0.69	0.51	0.82	0.89	0.92
4. *M. fuscata*	0.35	0.53	0.53	---	0.69	0.76	0.76	0.80	0.60
5. *M. fascicularis*	0.44	0.35	0.50	0.50	---	0.69	0.84	1.02	0.87
6. *M. nemestrina*	0.38	0.42	0.60	0.47	0.50	---	0.60	0.55	0.43
7. *M. nigra*	0.41	0.45	0.44	0.47	0.43	0.55	---	0.48	0.40
8. *M. maura*	0.38	0.39	0.41	0.45	0.36	0.58	0.62	---	0.40
9. *M. silenus*[3]	N.D.	N.D.	0.40	0.55	0.42	0.65	0.67	0.67	---

[1] Plasma Protein S value.
[2] Nei's Index.
[3] Some comparisons involving *M. silenus* were not done (N.D.).

frequent allele at each locus for each species was used in the comparisons and S and D values calculated. There were 8 monomorphic loci and a further 5 with only one species showing a variant allele; thus these 13 contribute little to the measured genetic distances and nothing to resolving the cladistics. By far the largest portion of the genetic distances obtained is contributed by the plasma proteins and certain erythrocyte proteins such as hemoglobin and 6-PGD. All six of the identified plasma proteins screened showed at least two alleles, with Gc having 3 and transferrin 7. Table 3-3 gives the plasma protein S and D values for comparisons among 9 species. Figure 3-3 is the phenogram for the macaques derived from the data of Tables 3-2 and 3-3.

Phenetic clustering begins by producing some clear-cut associations where the S values approximate 0.7 ($D<0.4$). These include *sinica* with *radiata*, *tonkeanna-maura*, *assamensis-mulatta*, and *cyclopis-fuscata*. In each of these pairs one of the species is, because of the quality of the samples available, preferable for use in further clustering. These are *radiata*, *maura*, *mulatta*, and *fuscata*.

The data of Bruce (1977) clearly indicate at least three derived changes along a common *radiata-arctoides* lineage involving substitutions at the GOT, PGM_2 and Hp loci. There is also a possible fourth derived change at the ceruloplasmin locus, and we have evidence of a fifth at the ADA locus. The data of Darga *et al.* (1975) indicate an *arctoides-mulatta* association and our comparisons support that view. These comparisons then can be combined to produce our first lineage cluster including *radiata*, *sinica*, *arctoides*, *assamensis*, and *mulatta*.

Similarly, we find that *maura-tonkeanna*, *nigra*, *nemestrina*, and *silenus* comparisons consistently give smaller D values than between any of these and a member of the first cluster such as *mulatta*, *radiata*, or *arctoides*.

This leaves, among the Asian forms, *fuscata-cyclopis* and *fascicularis* to be

MACAQUE MOLECULAR PHYLOGENY

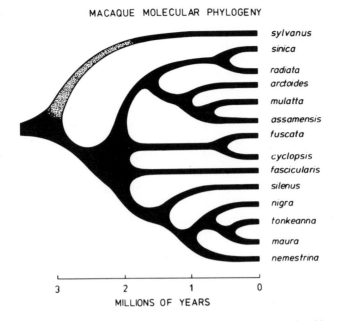

Fig. 3-3. A phenogram of the various macaque species based on electrophoretic evidence. We have used the available plasma protein and allozyme data for cladistic analyses whenever possible. Beyond this, the phenogram is to be seen as a phylogeny to the extent that electrophoretically detectable substitutions have occurred in similar numbers along the various lines since the beginning of the macaque adaptive radiation, here seen as about 3MYA. As discussed in the text, as soon as good *sylvanus* material becomes available, a complete cladistic analysis becomes possible. The time scale is based on the calibration in Sarich (1977), where a PPE *D* of 1 is reached in about 2.5MY.

placed. We do not find the available data (either ours or those of Darga *et al.* and Bruce) sufficiently convincing to reliably associate this pair with one another or with one of our two large clusters. We are thus left with extant lines going back to the primary adaptive radiation among Asian macaques. As we have already pointed out, *sylvanus* appears phenetically to be significantly more distant from the Asian macaques than they are from one another. We have therefore pictured it as a separate clade. We know of no other molecular studies on this species, and so this positioning must be considered a very tentative one pending the availability of better samples and further comparisons with them.

DISCUSSION

These conclusions, while broadly similar to groupings recently proposed by systematists using more traditional lines of evidence, do differ significantly in three respects: our association of *assamensis* with *mulatta*, placement of *arctoides*, and separation of *sylvanus* from the *nemestrina-silenus*-Sulawesi cluster. In the past, characters most heavily weighted in drawing taxonomic conclusions have been tail length, head and neck pelage, and baculum morphology (Fooden, 1976). More recently, Albrecht (1978) has compared craniofacial morphologies using multivariate techniques with the aim of sorting out the Sulawesi forms.Considering the variety of features used, it is certainly the strong congruency of the conclusions reached by different workers that ought to be emphasized—with any disagreements being easily attributable to the inadequacy of the current data base and lack of cladistic analysis of any of the morphological, and much of the molecular, data. Phenetic clustering, while a useful first approximation whether in the molecular or morphological realms, is no substitute for the intelligent sorting out of primitive from derived features, and cladistic allocation of the latter—whatever the data base. It should also be noted that cladistic analysis will also allow one to make intelligent decisions as to the suitability of various characters for evolutionary reconstructive efforts—simply stated, the most suitable ones will be those which lend themselves most readily to cladistic analysis and produce the smallest number of possible phylogenies consistent with the data.

In the absence of cladistically generated conclusions concerning macaque systematics, then, we believe it inappropriate to engage in any lengthy discussions of such phenetic noncongruences as those mentioned above. The reason is that phenograms do not purport to represent an actual history, and thus are not working hypotheses in any evolutionary sense. For example, the unique bacular morphology of *arctoides* could well be a recent, highly derived feature entirely consistent with the phylogeny implied in Figure 3-3. Similarly, the similarities between *sylvanus* and the *nemestrina-silenus*-Sulawesi cluster might well represent the retention of features primitive for the macaques as a whole, and again such an interpretation is entirely consistent with Figure 3-3 viewed as an actual phylogenetic statement. We need to structure our comparisons of conclusions drawn from different data sets, and it seems to us that the only available structure is the cladistic one. Any taxonomic decisions must come after the cladistic and temporal structuring of the phylogeny is well in hand, and we are a long way from that goal in the morphological realm. We are closer with the molecules, but much critical work remains to be done. In particular, as mentioned previously and reemphasized here, we require studies of macaque populations in something approximating a natural habitat. Without them, it is impossible to adequately dispel the aura of unreality which necessarily accompanies studies limited to captive animals and museum specimens.

SUMMARY

In conclusion, macaques appear to be the most primitive of the 42-chromosome monkey clade cranially and dentally. They also appear to be among the first diverging lineage within this clade, although this latter conclusion is not certain, as some molecular evidence suggests macaques are part of an adaptive radiation of the major lineages of the tribe.

Within the macaques electrophoretic evidence suggests there are five major lineages. The first clade is that leading to *M. sylvanus*. The second lineage cluster is that involving *M. sinica* and *M. radiata* clustering with *M. arctoides—M. mulatta* with *M. assamensis* being extremely close to *M. mulatta*. A third major lineage is that leading to *M. fuscata* and *M. cyclopis*. *M. fascicularis* is a fourth monotypic clade. The fifth and last cluster involves *M. maura—M. tonkeanna* clustering with *M. nigra*. The Celebes forms there join sequentially to lineages leading to *M. nemestrina* and *M. silenus*. These results are compared with data from other studies, and it is suggested that populational data from the field, with animals of known localities, will substantially improve our knowledge of macaque evolution at the behavioral, anatomical and molecular levels.

ACKNOWLEDGMENTS

We would like to thank the individuals and institutions listed in Table 3-1 who provided us with material. We would also like to thank Dr. A. Wilson of the Department of Biochemistry and Dr. R. Palmour of the Department of Genetics at the University of California, Berkeley, in whose laboratories the work was done. We would like to thank Anne Childs for excellent technical help and Carla Simmons for the excellent illustrations.

REFERENCES

Albrecht, G. H. Methodological approaches to morphological variation in primate populations: the Celebesian Macaques. *Yearbook of Phys. Anthrop.*, **20**: 290–308 (1977).

_____. The craniofacial morphology of the Sulawesi macaques. *Contrib. to Primatol.*, **13**: 1–151 (1978).

Anderson, J. E., and Giblett, E. R. Interspecific red cell enzyme variation in the pigtailed macaque (*Macaca nemestrina*). *Biochem. Genet.*, **13**: 189–212 (1975).

Avise, J. C., and Duvall, S. W. Allelic expression and genetic distance in hybrid macaque monkeys. *J. Hered.*, **68**: 23–30 (1977).

Benveniste, R. E., and Todaro, G. J. Evolution of Type C viral genes: evidence for an Asian origin of man. *Nature*, **261**: 101–108 (1976).

_____. Evolution of primate oncoviruses: An endogenous virus from langurs (*Presbytis* spp.)

with related virogene sequences in other Old World monkeys. *Proc. Nat. Acad. Sci.,* **74:** 4557–4561 (1977).

Bruce, E. J. A Study of the Molecular Evolution of Primates, Using the Techniques of Amino Acid Sequencing and Electrophoresis. Ph.D. Dissertation, University of California, Davis (1977).

Buettner-Janusch, J., Dave, L., Mason, G. A., and Sade, D. S. Primate red cell enzymes: glucose 6 phosphate dehydrogenase and 6-phosphogluconate dehydrogenase. *Amer. J. Phys. Anthrop.,* **41:** 4–14 (1974).

Buettner-Janusch, J., Mason, G., Dave, L., Buettner-Janusch, V., and Sade, D. Genetic studies of serum transferrins of free ranging rhesus macaques of Cayo Santiago, *Macaca mulatta* (Zimmerman 1780). *Amer. J. Phys. Anthrop.,* **41:** 217–232 (1974).

Buettner-Janusch, J., and Sockol, M. Genetic studies of free ranging macaques of Cayo Santiago. I. Description of the population and some nonpolymorphic red cell enzymes. *Amer. J. Phys. Anthrop.* **47:** 371–374 (1977a).

_____. Genetic studies of free ranging macaques of Cayo Santiago. II. 6-phosphogluconate dehydrogenase and NADH-methemoglobin reductase (NADH-diaphorase). *Amer. J. Phys. Anthrop.,* **47:** 375–380 (1977b).

Champion, A.C., Prager, E. M., Wachter, D. and Wilson, A. C. Microcomplement fixation. In C. A Wright (ed.), *Biochemical and Immunological Taxonomy of Animals.* London: Academic Press (1974).

Chiarelli, A. B. Comparative cytogenetics in primates and its relevance for human cytogenetics. In B. Chiarelli (ed.) *Comparative Genetics in Monkeys, Apes and Men.* pp. 273–308, London: Academic Press (1971).

Cronin, J. E. Molecular systematics of the order primates. Ph.D. Dissertation, University of California, Berkeley (1975).

Cronin, J. E. and Sarich, V. M. Molecular systematics of the New World Monkeys. *J. Hum. Evol.,* **4:** 357–375 (1975).

_____. Dual origin of mangabeys among Old World monkeys. *Nature,* **260:** 700–702 (1976).

Darga, L. L., Goodman, M., Weiss, M. L., Moore, G. W., Prychodko, W., Dene, H., Tashian, R. and Koen, A. Molecular systematics and clinal variation in macaques. In C. L. Markert (ed.) *Isozymes, Genetics and Evolution.* New York: Academic Press (1975).

Delson, E. Evolutionary history of the Cercopithecidae. In F. Szalay (ed.) *Approaches to Primate Paleobiology, Contrib. to Primatol.,* **5:** 167–217 (1975).

Delson, E., and Andrews, P. Evolution and interrelationships of the Catarrhine Primates. In W. P. Luckett and F. Szalay (eds.) *Phylogeny of the Primates.* New York: Plenum Press (1975).

Dene, H. T., Goodman, M., and Prychodko, W. Immunodiffusion evidence on the phylogeny of the Primates. In M. Goodman and R. Tashian (eds.) *Molecular Anthropology,* New York: Plenum Press (1976).

DeVries, G. F., DeFrance, H. F., and Schvriers, J. A. M. Identical gimsa banding patterns of two *Macaca* species: *Macaca mulatta* and *M. fascicularis. Cytogenet. Cell. Genet.,* **14:** 26–33 (1975).

Egozcue, J. Constancy of Banding Patterns in Primate Chromosomes. V°Conv. Annuale di Citotassonomia dei Cordati, Roma. *Bolletino di Zoologia, Unione Zoologia Italiana.* (1973).

Ellerman, J. R. and Morrison-Scott, T. C. S. *Checklist of Paleoartic and Indian mammals, 1758–1946.* London: British Museum of Natural History (1951).

Finaz, C., Van Cong, N., Cothet, C., Frezal, J., and DeGrouchy, J. Fifty-million-year evolution of chromosome 1 in the primates. *Cytogenet. Cell. Genet.,* **18:** 160–164 (1977).

Fooden, J. Taxonomy and evolution of the monkeys of Celebes. *Biblio. Primatol.,* **10:** 1–148 (1969).

_____. Provisional classification and key to the living species of macaques (Primates: *Macaca*). *Folia Primatol.,* **25:** 225–236 (1976).

Gillespie, D. Newly evolved repeated DNA sequences in Primates. *Science,* **196:** 889–891 (1977).

Goodman, M. Toward a genealogical description of the Primates. In M. Goodman and R. Tashian (eds.), *Molecular Anthropology.* New York: Plenum Press (1976).

Goodman, M. and Moore, W. Immunodiffusion systematics of the Primates. I. The Catarrhines. *Syst. Zool.*, **20:** 19–62 (1971).

Gruenbaum, B. A microanalytic electrophoresis technique for the determination of polymorphic blood proteins for medical and forensic applications. *Microclinica acta*, **2:** 339–352 (1977).

Henderson, A. S., Atwood, K. C., and Warburton, D. Chromosomal distribution of rDNA in *Pan paniscus, Gorilla gorilla berengei,* and *Symphalangus syndactylus*: comparison to related primates. *Chromosoma*, **59:** 147–155 (1976).

Henderson, A. S., Warburton, D., and Atwood, K. C. Localization of rDNA in the chromosome complement of the rhesus (*Macaca mulatta*). *Chromosoma*, **44:** 367–370 (1974).

Henderson, A. S., Warburton, D., Magraw-Ripley, S. and Atwood, K. C. The chromosomal location of rDNA in selected lower primates. *Cytogenet. Cell. Genet.*, **19:** 281–302 (1977).

Hewett-Emmett, D., Cook, C. N. and Barnicot, N. A. Old World monkey hemoglobins: deciphering phylogeny from complex patterns of molecular evolution. In M. Goodman and R. Tashian (eds.), *Molecular anthropology.* New York: Plenum (1976).

Hill, W. C. O. *Primates: Comparative Anatomy and Taxonomy,* Vol. *VII*: *Cynopithecinae.* Edinburgh: Edinburgh University Press (1974).

Johnson, G. B. Genetic polymorphism and enzyme function. In F. Ayala (ed.) *Molecular Evolution,* Sunderland: Sinauer (1976).

Kellog, L. Macaques. In D. Aberle (ed.), *Primate malaria*, Office of Medical Information (1945).

Kohne, D. DNA evolution data and its relevance to mammalian phylogeny. In W. P. Luckett and F. Szalay (eds.), *Phylogeny of the Primates,* New York: Plenum Press (1975).

Nei, M. Genetic distances between populations. *Amer. Nat.*, **106:** 385–398 (1972).

Nozawa, K., Shotake, T., Ohkura, Y., Kitajima, M. and Tanabe, Y. Genetic variation within and between troops of *Macaca fuscata fuscata. Contemporary Primatol.*, **5:** 75–89 (1975).

Palmour, R. and Sutton, H. E. Vertebrate transferrins, molecular weights, chemical compositions, and iron binding studies. *Biochem.*, **10:** 4026–4032 (1971).

Perticone, P., Rizzoni, M., Palitti, F., and diChiara, P. Banding patterns of the chromosomes of the rhesus monkey (*Macaca mulatta*). *J. Hum. Evol.*, **3:** 291–295 (1974).

Romero-Herrera, A. E., Lehmann, H., Joysey, K. A. and Friday, A. E. Evolution of myoglobin amino acid sequences in primates and other vertebrates. In M. Goodman and R. Tashian (eds.), *Molecular Anthropology* New York: Plenum Press (1976).

Rubio-Goday, A., Caballin, M. R., Garcia-Caldes, M., and Egozcue, J. Comparative study of the banding patterns of the chromosomes of Cercopithecidae. *Folia Primatol.*, **26:** 306–309 (1976).

Sarich, V. M. Rates, sample sizes and the neutrality hypothesis for electrophoresis in evolutionary studies. *Nature*, **265:** 24–28 (1977).

Sarich, V. M. and Cronin, J. E. Molecular systematics of the primates. In M. Goodman and R. Tashian (eds.) *Molecular anthropology,* New York: Plenum Press (1976).

Schmager, J. Cytotaxonomy and geographical distribution of the Papinae. *J. Hum. Evol.*, **1:** 477–485 (1972).

Selander, R. K. Genetic variation in natural populations. In F. Ayala (ed.), *Molecular evolution,* Sunderland: Sinauer (1976).

Seth, P. K., DeBoer, L.E. M., Saxena, M. B. and Seth, S. Comparison of human and nonhuman primate chromosomes using the fluorescent benzimidazol and other banding techniques. In P. L. Pearson and K. R. Lewis (eds.) *Chromosomes Today,* Vol. 5, pp. 315–322. New York: John Wiley and Sons (1976).

Stock, A. D. and Hsu, T. C. Evolutionary conservatism in arrangement of genetic material. *Chromosoma*, **43:** 211–224 (1973).

Wooding, G., and Doolittle, R. Primate fibrinopeptides: Evolutionary significance. *J. Hum. Evol.*, **1:** 555–563 (1972).

Chapter 4
Pleistocene Glacial Phenomena and the Evolution of Asian Macaques

Ardith A. Eudey

SPECIES GROUPS AND DISTRIBUTIONS

That species of macaques fall into several "natural groups," for which generic or subgeneric names are proposed, was recognized by a number of taxonomists earlier in this century, e.g., Allen (1916), Pocock (1921, 1926, 1939), and Ellerman and Morrison-Scott (1966). These groups are recognized or described by morphologic characters, mainly the length of the tail, the direction of hair growth on the head, and the structure of the external genitalia in the male. As early as 1921, Pocock referred to tail length and arrangement of crown hairs as "unsatisfactory characters of very little systematic importance" and grouped together macaques on the data available on the structure of the glans penis. The taxon *Macaca* is remarkable for asymmetry and wide variation in the form of the glans penis in different populations (Hill 1958, 1974) and, apparently, complementary structures of the female reproductive tract (Fooden 1967, 1971a, 1975, 1976; and Lemmon and Oakes 1967).

Recently, Fooden, (1976, this volume) assigns extant macaques, of which he recognizes 19 species, to four species groups: the *silenus-sylvanus* group; *sinica* group; *arctoides* group, which is represented only by *Macaca arctoides*; and *fascicularis* group. Allocations are based primarily on glans penis morphology and secondarily on apparent uterine cervix complementarity, female sexual skin, and copulatory patterns. Although the geographic ranges of all four species groups are partly sympatric, the ranges of species within each species group are basically allopatric. I have contacted sympatric groups of *M. mulatta* and *M. fascicularis*, both of which are members of the *fascicularis* group, in the Dawna range of western Thailand, an area which appears to have been of special importance in the evolution of macaques during the Pleistocene epoch, as discussed below.

In contrast to Fooden, Delson (this volume), based on a different interpretation of the morphological evidence, differentiates *Macaca sylvanus* from Asian macaques and assigns it to a separate *sylvanus* group and assigns *M. arctoides* to the *sinica* group. On the basis of electrophoretic examination of plasma proteins, Cronin et al. (this volume) likewise (tentatively) recognize *M. sylvanus* as a separate lineage or clade. The fact that *M. sylvanus* exhibits a single-mount ejaculatory pattern (Taub, this volume), not a multiple-mount pattern as previously reported, appears to provide further support for this differentiation. The single-mount ejaculatory pattern may be the ancestral condition for *Macaca*, and a different ejaculatory pattern, or specialization of the genitalia, may have evolved to reduce successful copulations between members of different species (and species groups) in areas of sympatry. In addition to the lineage leading to *M. sylvanus*, Cronin *et al.* (this volume) recognize four other lineages or clades consisting of the following species: (1) *M. sinica, M. radiata, M. arctoides, M. mulatta,* and *M. assamensis*; (2) *M. fuscata* and *M. cyclopis*; (3) *M. maura, M. tonkeana, M. nigra, M. nemestrina,* and *M. silenus*; and (4) *M. fascicularis*.

Fooden (1976, this volume) infers from disjunctions and continuities in the present distribution of species groups that the dispersal of the *silenus-sylvanus* group probably occurred earliest and was followed by that of the *sinica* group, with which the distribution of the *arctoides* group may be historically or ecologically related. The *fascicularis* group appears to have dispersed relatively recently.

According to Fooden (1976, this volume), competition between older and younger species groups may be responsible for the apparent reduction and disjunction of the ranges of the *silenus-sylvanus* group and *sinica* group respectively. Earlier, Hill and Bernstein (1969) offered a similar explanation for disjunctions in the distributions of related species. That the present ranges of the different species groups are a response to climatic changes in South and Southeast Asia associated with Quaternary glacial phenomena seems a more plausible explanation to me. Fooden (1975) does suggest that the dispersal of the *silenus-sylvanus* group from the Indian peninsula to the Indochinese Peninsula probably occurred during a pluvial period and the disjunction in the distribution of the species group was caused by a succeeding period of aridity, but he does not employ consistently a paleoclimatic approach in attempting to explain the dispersal of species groups. Climatic changes during the Pleistocene also may have produced the conditions necessary for the differentiation of macaque species themselves.

QUATERNARY CLIMATIC EVENTS

Accumulated data suggest considerable climatic and environmental change during the Quaternary in Southeast Asia (Verstappen, 1975), comparable to that of other humid tropical areas in Africa (Moreau, 1966) and South America (Haffer, 1969). Verstappen (1975) identifies three factors exerting considerable influence on Quaternary climatic conditions in Southeast Asia, especially during glacials: (1) the position of the Intertropical Convergence Zone (ITC)), with which is associated most of the precipitation in the area; (2) the world-wide drop in air and marine temperatures which caused a lowering of the snowline and forest line and affected the altitudinal zonation of vegetation in the area; and (3) the emergence of both the Sunda and Sahul shelves during glacials due to lowering of sea level. During glacials these three factors appear to have interacted to produce a continental climate with lower precipitation and a longer dry season, and a corresponding reduction of rain forest. Cores or refuges of rain forest probably alternated with, or were substantially replaced by, monsoon forest and tree savanna.

Fairbridge (1976) traces the supposed correlation between glacials and increased precipitation (pluvials) to the influence of Biblical (Noachian) tradition in the nineteenth century. Although some support for this correlation still persists, notably that of Flint (1971, 1976), Fairbridge (1976) cites evidence for glacial aridity in such diverse areas as Australia, India, central Asia, South America and North America and suggests as much as a 50 percent drop in continental precipitation during glacials.

In a major reconstruction of late Quaternary environments and climatic change for Australia and New Guinea, based on evidence derived from palynology, sedimentology, and geomorphology, Bowler et al. (1976) conclude that throughout Australia cold periods seem generally associated with less precipitation than that at present. Difficulties in interpreting the effectiveness of precipitation, or the relationship between rainfall and evaporation, are cited as contributing to contrary climatic interpretations. Bowler, et al. (1976) demonstrate that high lake levels during colder phases are the consequence of reduced or unchanged precipitation associated with depressed evaporation rates rather than increased rainfall.

Gross simulation of climate based on foraminiferal changes in deep-sea sediments likewise suggests a substantially cooler and drier climate over unglaciated continental areas at the end of the last glacial about 18,000 years ago (Gates 1976) and the expansion of more arid types of vegetation at the

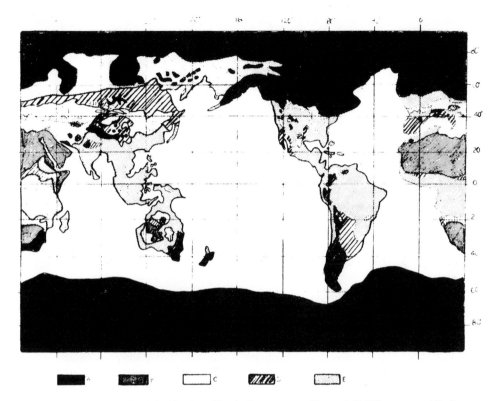

Fig. 4-1. Continental albedo for Northern Hemisphere summer (August) 18,000 years ago. Albedo values are given by the key (A) Snow and ice; albedo over 40 percent. (B) Sandy deserts, patchy snow, and snow-covered dense coniferous forests; albedo between 30 and 39 percent. (C) Loess, steppes, and semideserts; albedo between 25 and 29 percent. (D) Savannas and dry grasslands; albedo between 20 and 24 percent. (E) Forested and thickly vegetated land; albedo below 20 percent (mostly 15 to 18 percent). Sea-surface temperatures, ice extent, and ice elevation have been eliminated. (From CLIMAP Project Members, 1976, The surface of the ice-age earth. *Science.* 191: 131–137. American Association for the Advancement of Science, Washington, D.C.)

expense of forest (Fig. 4-1, as redrawn with simplifications from CLIMAP Project Members, 1976). Two such periods of aridity may be the most plausible explanation for the apparent zoogeographical anomalies in Southeast Asia, including the disjunct distributions of primates (Brandon-Jones, 1978).

At least nine times in the past million years, massive ice sheets have mantled much of Europe and North America. Approximately 120,000 years ago the climate of the earth appears to have been warm. An abrupt drop in temperature occurred about 90,000 years ago, and the last glacial advance began about 70,000 years ago. Although climatic fluctuations with at least four minima in temperature are subsequently recognized, the climate was basically cold until about 10,000 years ago (Matthews, 1976). Oxygen-18 and fo-raminifera analyses of deep-sea sediments suggest that ice-volume and general climatic changes may have occurred in a cycle near 100,000, 40,000, and 20,000 years (Hays *et al.*, 1976). If glacially induced periods of aridity did, as seems likely, influence the dispersal and differentiation of macaque species and species groups, the problem remains of identifying the climatic events which may have been most significant.

FOSSIL EVIDENCE

The earliest unquestionable fossil macaque, *Macaca libyca*, is represented by numerous dental remains found in association with *Libypithecus markgrafi*, an African-type colobine, at Wadi Natrun, Egypt, in latest Turolian deposits (Late Miocene) dating to about 6 MYA. A somewhat earlier fossil cer-copithecine population from Marceau, Algeria is macaque-like but not clearly referable to *Macaca*. Fossils of a comparable age from Oran, Algeria, which were assigned to *"Macaca" flandrini* have proved to be colobine (Delson, 1975a, b, this volume).

The first cercopithecines outside of Africa are found in Early Ruscinian local faunas of less than 6 MY and appear to be identifiably *Macaca* (Delson, 1975b, this volume). Palynological investigations at the Middle Ruscinian locality of Celleneuve/Montpellier, at which both *Macaca* and colobines have been recovered, indicate monsoon or seasonal forest conditions (Delson, 1975b).

Although colobines dominated European faunas in the Ruscinian, *Macaca* became the most common and subsequently the only cercopithecid in Europe during the Villafranchian and Pleistocene, persisting at least until the last interglacial. Most, if not all, European and later North African *Macaca* appear to be phyletically close to modern populations of *M. sylvanus* (Delson, 1975a,b, this volume). Delson (this volume) discusses in more detail fossil macaques of the circum-Mediterranean region.

During the Late Villafranchian, macaques appear to be associated with warm climate and deciduous forest which alternated with open-country parkland or steppe (Delson, 1975b). During the Pleistocene, *Macaca* is widespread in Europe in relatively warm times, extending from England to the Caucasus, but the genus is never present in faunas of distinctly cold aspect. The other mammals associated with *Macaca* are of forest or more commonly steppe type (Delson, 1975b).

The European faunal assemblages suggest that *Macaca* originally may have been adapted to seasonal or deciduous forest or even tree savanna, and that adaptation to evergreen forest is a derived character. Macroenvironmental conditions in the Dawna range in western Thailand may approximate those which prevailed during periods of extensive deployment of *Macaca* in Europe.

At present the fossil evidence which relates to the evolution and dispersal of Asian macaques is not plentiful. Two partial mandibles recovered from the Siwalik Tatrot formation, which is somewhat earlier than 3 MY, were thought to be colobine but are the earliest cercopithecines known in Asia and are assigned to *?Macaca palaeindica* (Delson, 1975b, this volume). No environmental reconstruction of the Tatrot or succeeding Pinjor formation is possible at this time.

Paradolicopithecus and *Procynocephalus* are younger populations, dating to about 2 MY, and represent large cercopithecines from Eurasia and Asia respectively, which may have evolved in parallel from *Macaca* or may be specially related to the taxon (Simons, 1970, 1972; Delson, 1975b). One or more large species of *Macaca* appears to be represented by populations dating to as recent as about 1 million years which are closely similar in morphology and geographically widespread. These populations include *'Cynocephalus' falconeri* (Siwalik Pinjor formation), *M. anderssoni* (Honan Villafranchian), *M. robusta* (north China), and *M. speciosa subfossilis* (Tung-Lang, north Vietnam). These specimens probably are related or "linked" to the modern *sinica* group (Delson, 1975a, b, this volume).

Delson (this volume) feels that the fossil record documents the differentiation of modern species of *Macaca* no later than 1.0 to 0.3 million years ago. Nozawa *et al.* (1977) likewise infer the average time of divergence among modern taxa of Asian macaques to be about 700,000 years based on genetic distance extrapolated from the results of electrophoretic examination of blood proteins from 13 populations representing seven species. From a different set of electrophoretic data, Sarich (1977, personal communication, Cronin and Sarich, 1976) concludes that differentiation among extant species of macaques must have occurred 3 to 2 million years ago, although recently Sarich and Cronin (Cronin *et al.*, this volume) suggest that divergence among Asian species of macaques occurred from 1 (or even less) to 2-2.5 million years ago. As a consequence of the above dating, it may be necessary to differentiate in some cases between the events of speciation and the dispersal of macaques which resulted in modern distributional patterns. Confronted with a limited fossil record, which is subject to alternative interpretations as discussed below, and uncertainty over the chronological significance of biomolecular differences observed between mammal populations (e.g., Uzzell and Pilbeam, 1971; Goodman *et al.*, 1971; Sarich, 1977), it also may be more productive to concentrate at this time on the problem of dispersal and relegate that of the ultimate differentiation of species groups (and sometimes species) to secondary consideration.

The study of dispersal patterns is, in itself, complicated by problems of interpreting the extent of eustatic changes in sea level and plate tectonic activity during the Quaternary. Brandon-Jones (1978) feels that climatic change basically is sufficient to account for the biogeography of the Asian region without recourse to topographic barriers, but fluctuations and changes in land mass as a result of tectonic activity and glacial eustacy may contribute to these climatic changes (cf. Beaty, 1978).

Fluctuations in climate and habitat during the Pleistocene do not appear to have produced among macaques the conditions necessary for sampling error or founders effect to result in significant chromosomal evolution, although environmental bottlenecks may have produced gene change. All species of the genus *Macaca* exhibit a diploid chromosome number of 2n = 42. Chiarelli (1962) considers that no gross morphologic differences can be detected among the chromosome sets of the different species. Butzer (1977) suggests that during the Middle Pleistocene, periods of high resource productivity may have resulted in the multiplication and dispersal of groups of hominids into marginal areas with the opportunity for random loss or fixation of genes to occur while periods of environmental stress may have exerted a centripetal effect by drawing these temporarily isolated groups into larger regional aggregates with the opportunity for gene flow to occur. Periods of aridity and deforestation during Pleistocene glacials may have had an effect similar to the latter on species of *Macaca*, causing groups to aggregate and exchange genes in refugia from which, with climatic amelioration, genetically similar populations dispersed. Provisional reconstructions of the dispersal (and differentiation) of species groups of macaques are presented below.

DISPERSAL OF SPECIES GROUPS IN ASIA

silenus-sylvanus Group

The disjunct distribution of extant species which extends from the Atlas mountains of North Africa eastward to Sulawesi suggests that the *silenus-sylvanus* group of macaques experienced the earliest dispersal throughout Asia (see Fooden, this volume). Electrophoretic data analyzed by Nozawa *et al.* (1977, Fig. 5) also suggest that the *silenus-sylvanus* group, specifically *Macaca nemestrina* in Thailand, may be considerably older than other species groups in Asia.

Fooden (1975) infers primarily from tail length and secondarily from skull morphology that *Macaca silenus* is morphologically most similar to the ancestral population of the Asian sector of the *silenus-sylvanus* group. Today *M. silenus* is reduced to a relic population with perhaps no more than 800 individuals in the South India High Range (Mohnot, 1976). Fooden considers that the *silenus* subgroup originated in peninsular India from where it dispersed northward to reach continental and subsequently insular Southeast Asia. According to Fooden (1975) the initial dispersal northward may have provided the impetus for tail reduction which expresses itself in a west to east gradient in the subgroup.

At present a gap of about 2,000 kilometers separates *Macaca silenus* in peninsular India from *M. nemestrina leonina* in Assam and Burma. Fooden (1975) recognizes that the dispersal of the *silenus* subgroup across this area probably took place when "more continuous rainfall" supported evergreen forest, while the subsequent disjunction between the Indian and eastern populations of the subgroup occurred during a succeeding arid interval. CLIMAP Project Members (1976, Fig. 1) reconstruct northern India as being an area of savanna and dry grassland during the most recent glacial about 18,000 years ago, and comparable desiccation and deforestation may have occurred during earlier glacials. A simplification of the CLIMAP reconstruction is presented in Fig. 4-1. Although the disjunction in the distribution of the *silenus* subgroup probably was initiated by glacially induced aridity and deforestation, this disjunction would have been reinforced, with amelioration of climate, by the expansion of the *sinica* group (cf. Fooden, this volume). *M. silenus* may have been driven into a forest refugium in south India from which it did not recover due to the subsequent expansion of *M. radiata* in peninsular India. Reduction in rain forest during a glacial may have caused populations of *M. nemestrina leonina* to aggregate in refugia in such areas as the Khasi hills near the foothills of the eastern Himalaya, Dawna range, and Annamitic cordillera (Fig. 4-2), from which they subsequently dispersed.

The dispersal of the *silenus* subgroup to the major islands of Indonesia, the Mentawai islands, and Sulawesi appears to be intimately associated with the effects of glacial phenomena on Sundaland. The Sunda shelf or Sundaland (Fig. 3, as redrawn from Verstappen, 1975, and Haile, 1969) is that area of the present sea floor in the western part of the East Indian archipelago which emerged from the sea during Pleistocene glacial maxima. Haile (1969) suggests that the 180-meter (100-fathom) contour of the present sea should be taken as the limit of the shelf. The resulting land area would be about 2,200,000 kilometers square. The greater part of the shelf is flat and smooth, but the northern area near Borneo is markedly indented with three major submarine valley systems, the Anambas, North Sunda, and Proto-Lupar systems (Fig 4-3), which probably are largely the result of tectonic activity (Haile, 1969). The northwest to southeast orientation of these Pleistocene river valleys suggests that they facilitated the dispersal of plants and animals between Borneo and Malaya and Sumatra.

A population like *Macaca nemestrina leonina* may represent the ancestral stock which penetrated Sundaland during a Pleistocene glacial. Fossil teeth from a number of Middle Pleistocene deposits in eastern Java (Bangle, Punung, Sangiram, Saradan, and Trinil) are among the oldest remains of macaques recovered in Southeast Asia. Interpretations of the affinities of these

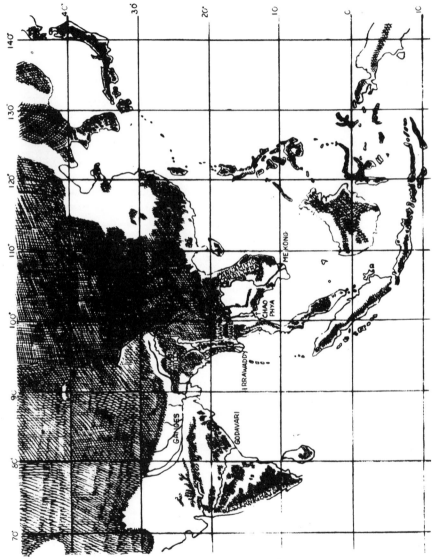

Fig. 4-2. Major mountain chains and rivers of Asia.

fossils are problematic and seem to depend on the significance attached to the relative breadth of premolars and molars (Hooijer, 1962a; Fooden, 1975). Dental measurements recorded for 14 of these specimens are compared by Fooden (1975: Tables 16 and 17) with measurements for subfossil and living *M. fascicularis mordax* from Java, and extant *M. n. nemestrina* and *M. n. leonina*. The premolar/molar breadth ratio of the fossils is similar to that of extant *M. n. leonina* (Fooden 1975). The ratio is less than that of living *M. n. nemestrina* and greater than that of *M. fascicularis mordax* (Hooijer, 1962a).

Two plausible interpretations for the fossil evidence have been offered. Hooijer (1962a) concludes that the fossils represent a population which experienced reduction in the premolar/molar breadth ratio as it evolved into *Macaca fascicularis mordax*. This interpretation implies the absence of the *silenus* subgroup from Java during the Quaternary. In contrast, Fooden (1975) suggests that the relatively broad premolars in living *M. n. nemestrina* may be an evolved condition related to allometric growth in the rostrum and anterior part of the dental arch from that in an ancestral population resembling *M. n. leonina* and, therefore, the fossils may represent a population ancestral to *M. n. nemestrina*. This interpretation implies that the *silenus* subgroup inhabited Java during the Pleistocene and became locally extinct.

During the Pleistocene the fauna of Sundaland was more evenly distributed over Java, Sumatra, and Borneo than at present. Java experienced the greatest reduction in "well-known" mammals, including the local extinction of both siamang and orang-utan, which are known from Middle Pleistocene faunas (Hooijer, 1975). Java and the Lesser Sunda Islands also are floristically distinct from Borneo, the Malay Peninsula, and Sumatra, which appear to form a phytogeographical unit (Keng, 1970). Verstappen (1975) reviews evidence for the existence of an increasingly dry and seasonal climate in a corridor or wider zone extending from lower Burma to eastern Java with the exposure of Sundaland during glacials. The same drought plants of Asiatic and/or Australian type are found in lower Burma and eastern Java separated by the core of humid evergreen forest that covers the western part of the Malay Peninsula. Very few of the drought plants are endemic to either area, which suggests that periods of arid climate were of short duration, and the exchange of drought plants must have been rather recent since almost all species are the same as those found in Asia and Australia. Glacial phenomena may have reduced temporarily evergreen forest in Java, resulting in the local extinction of some primates including *Macaca nemestrina*.

Fooden (1975) establishes the boundary between the subspecies *Macaca nemestrina leonina* and *M. n. nemestrina* as coinciding almost exactly with the Khlong Marui fault, the larger of two transcurrent faults which are responsible for the angular elbow-like feature in the Phuket area of peninsular Thailand which juts into the Andaman Sea (Garson and Mitchell, 1970). Biogeographic

considerations rather than geologic evidence suggest that a marine trans-
gression of either eustatic or tectonic origin may be responsible for the
differentiation of both plants and animals that appears to be centered on this
fault zone (Keng, 1970; Fooden, 1975). An equally plausible explanation for
the observed differentiation may be that the zone of dry and seasonal climate
connecting lower Burma with eastern Java during the emergence of Sundaland extended through this area. Secondary contact between *M. n. leonina*
and *M. n. nemestrina* resulting in limited gene exchange, appears to have
occurred at 8° to 9° N latitude in peninsular Thailand (Fooden, 1975)
subsequent to amelioration of the climate.

Populations of *Macaca nemestrina nemestrina* may have taken refuge in
montane forest during periods of increasing aridity and seasonality. The
principal habitat of this subspecies appears to be evergreen forest (Crockett
and Wilson, this volume; Rijksen, 1978), and specimens have been collected
from elevations up to 1700 meters in Borneo (Fooden, 1975). As a consequence
of glacial activity to the north, the snowline appears to have lowered as much
as 1000 meters, resulting in vertical compression of the climatological zonation
in mountains (Verstappen, 1975). In comparison with *M. n. leonina, M. n.
nemestrina* exhibits an increase in external size (head and body length) which
may reflect adaptation to colder conditions, and the largest specimens collected of *M. n. nemestrina* exhibit reduction in tail length (Fooden, 1975).
Eustatic lowering of sea level during the last glacial would have permitted gene
flow among the populations of *M. n nemestrina* in Sundaland, to which Fooden
(1975) attributes the morphologic similarity of the now disjunct insular and
peninsular populations of the subspecies.

Before 170,000 years ago the sea level was lowered by glacial eustacy to
about -200 meters (Jongsma, 1970) and maximal exposure of Sundaland may
have occurred. Fooden (1975) hypothesizes that a population ancestral to
Macaca nemestrina pagensis probably reached the Mentawai Islands west of
Sumatra by a transitory land connection with Sundaland. The minimal depth
of the strait between the Sunda shelf and the Mentawai Islands is about 200
meters. A land connection is postulated because of the presence in the
Mentawai fauna of *Hylobates klossi*, to which water might have posed a
formidable barrier. Recently, Brandon-Jones (1978) has postulated on morphologic evidence that within the primate fauna of the Mentawai Islands can
be found the progenitors of the *Presbytis* group of colobine monkeys, most, if
not all, extant *Hylobates*, and perhaps *M. nemestrina* (?*M. n. nemestrina*).

Based on the absence of gibbons and the general impoverished character of
the mammal fauna on Sulawesi, Fooden (1969, 1975) argues that the dispersal
of the *silenus* subgroup from Sundaland to this island was across a water gap
during eustatic lowering of sea level and probably by rafting. Biomolecular
data (Darga *et al.,* 1975) support the inclusion of the Sulawesi macaques in the

silenus subgroup. The minimum depth of the Makassar strait between Borneo and Sulawesi is about 600 meters. Hooijer (1975) likewise suggests that large mammals could have reached Sulawesi by island hopping around Makassar Strait from Palawan northward to, for example, Luzon and Mindanao and south via the Sangihe Islands to the northern peninsula of Sulawesi. Brandon-Jones (1978) finds sea barriers to have played a minor part in limiting recent colobine dispersal, and the effectiveness of such barriers in limiting the dispersal of macaques would have been even more negligible.

Groves (this volume), in contrast, suggests that the ancestral mammal fauna reached Sulawesi from Java by a Pliocene landbridge which was submerged subsequently by tectonic activity. Sartono (1973), however, contends that during the Pleistocene Borneo and Sulawesi were connected by a landbridge. He argues that tectonic activity during the Plio-Pleistocene may have uplifted the central part of Borneo causing a regression of the sea from the immediate area and adjacent areas including southeast Borneo and south Sulawesi. The last two areas are thought to have been occupied by a Neogene basin, and the uplift left the intervening Makassar Strait as a lowland which became a landbridge with subsequent eustatic lowering of sea level. Island chains north and south of Makassar Strait may be relics of this lowland. According to Sartono (1973), both of these island chains, and the area between them, were broken by block faulting at the end of the Pleistocene and submerged by postglacial rise in sea level resulting in the formation of the deep Makassar trough. Verstappen (1975) suggests that thick rhyolite ash deposit in West Malaysia may be associated with the renewed tectonic activity assumed by Sartono.

Fooden (1969, 1975) attributes the differentiation among Sulawesi macaques to the dispersal of the ancestral *silenus* population to peripheral islands in an hypothetical Pleistocene archipelago. Sartono (1973) suggests that southern Sulawesi experienced ongoing uplift until "very recent times," probably as an adjustment to the downward movements of the Makassar and Bone troughs. However, the climate of Sulawesi also may have fluctuated during the Pleistocene because of the island's proximity to Sundaland. Cold-loving mountain plants from temperate regions to the north migrated into at least southwest Sulawesi some time during the Pleistocene (Verstappen, 1975: Figure 5). Populations of the *silenus* subgroup may have taken refuge in a number of mountainous areas as evergreen forest retreated during a Pleistocene glacial. The robust body build and reduction in tail length of Sulawesi macaques suggest an adaptation to cold climatic conditions (cf. Wilson, 1972).

The explanations for such apparent anomalies in the distribution of the *silenus* subgroup as the absence of *Macaca nemestrina leonina* from evergreen forest in central and northern Laos and northern Vietnam, as raised by Fooden (1975), must wait upon detailed paleoclimatic reconstructions for these areas.

sinica Group

Disjunct distribution from peninsular India including Sri Lanka north-eastward to Szechwan and south China suggests that the *sinica* group probably dispersed after the *silenus-sylvanus* group but before the *fascicularis* group (Fooden, this volume). The *sinica* group is composed of allopatric populations which constitute a graded ecological series with respect to body size and tail proportions (Fooden, 1976). Following the logic of Fooden (1975) that a long tail in macaques is a primitive character, *Macaca sinica* in Sri Lanka and *M. radiata* in peninsular India probably are most closely related to the ancestral population of the *sinica* group. Brandon-Jones (1978) speculates that the *Semnopithecus* group of colobine monkeys (*Presbytis entellus*) may have differentiated in the highland area in south Sri Lanka, and this area also warrants consideration as that in which the *sinica* group may have arisen. Biomolecular data examined by Darga *et al.* (1975) suggest that *M. sinica* may have undergone slightly more evolutionary change than *M. radiata*. The strait separating Sri Lanka from the mainland would appear to be negotiable with any lowering of sea level, since it is about 64 kilometers wide and no more than 11 to 12 meters deep in some areas. Reduction in tail length in this species group also may be associated with a northward expansion in which colder night and winter temperatures were encountered at higher latitudes. However, as discussed above, a group of morphologically similar fossils widespread at more northern latitudes in Asia about 1 MYA is considered to be related to the *sinica* group by Delson.

A discontinuity of approximately 1,000 kilometers separates *Macaca radiata*, whose northernmost distribution in peninsular India extends from 19° N latitude at the Godavari River in the east to at least 18° N latitude at Satara in the west, from *M. assamensis* in the southern foothills of the Himalaya in Uttar Pradesh and Assam. A vegetational reconstruction 18,000 years ago suggests that northern India was an area of savanna and dry grassland (CLIMAP Project Members, 1976, Figure 1), and this period of aridity associated with the last glacial may have initiated the major disjunction in the distribution of the *sinica* group. With climatic amelioration, the recovery and expansion of the *sinica* group in the area of disjunction appears to have been checked by the invasion of *M. mulatta* (cf. Fooden, 1976). *M. radiata* exhibits sufficient ecological diversity to have occupied the area in the absence of competition.

The increasing seasonality and deforestation associated with the last glacial may have caused populations of *Macaca assamensis* to aggregate, with the opportunity for gene flow to occur, in forest refuges in the Khasi hills near the foothills of the eastern Himalaya, Dawna range, and Annamitic cordillera

(Fig. 4-2), from which they subsequently dispersed with improvement of climatic conditions.

The present distribution of little known *Macaca thibetana* extends from the high mountainous area of eastern Tibet and Szechwan eastward to Kwangtun and Fukien (Fooden, 1971a). The robust body build and marked reduction in length of tail of *M. thibetana* suggest that it is adapted to the extreme cold found in the montane areas of its distribution.

arctoides Group

Macaca arctoides is assigned to a separate species group on the basis of the morphology of the glans penis and cervix (Fooden, this volume). The species also appears to be unique from other macaques in its alleles for transferrin, 6-PGD, and CA 1 (Weiss *et al.*, 1973; Darga *et al.*, 1975; Nozawa *et al.*, 1977). The geographic range of *M. arctoides* generally conforms to that of the eastern branch of the *sinica* group except that it extends somewhat farther south. Pocock (1939) comments on the similarity between the skull of *M. arctoides* and that of *M. assamensis* in spite of differences in the structure of the glans penis in the two species, and Nozawa *et al.* (1977: Fig. 5) consider that *M. arctoides* is closest to *M. radiata* of the Asian macaques whose blood proteins they have examined by electrophoresis. On the basis of morphological characters, Delson (this volume) infers a special relationship between *M. arctoides* and *M. assamensis/M. thibetana* and includes the former in the *sinica* group.

The distribution of *Macaca arctoides* as established by Fooden (this volume) in itself appears to be disjunct and conforms to a number of major and minor mountain ranges in Southeast Asia while extending to between 20° to 30° N latitude in east Asia. The westernmost extension of the species includes the Chittagong Hills and foothills of the eastern Himalaya. In Burma and Thailand the species is known from the Dawna range and a smaller range in peninsular Thailand adjacent to 100° E longitude. In Laos and Vietnam, *M. arctoides* has been collected at sites in the Annamitic cordillera. With the spread of monsoon forest and savanna during a glacial, populations of *M. arctoides* may have taken refuge in a number of montane forests from which they did not or were not able to disperse widely, perhaps because of competition from populations of other species groups. The robust body and reduction in tail length of *M. arctoides* suggest adaptation to cold climatic conditions.

Extant populations of *Macaca arctoides* appear to be relict populations, analogous to *M. silenus*, isolated from one another basically in mountainous areas. A comparative study of these populations would seem to offer an opportunity to observe the effects of random change in small groups. Brandon-Jones (1978) remarks on a comparable situation in which Asian colobines on at least three instances have undergone prolonged and widespread isolation without significant morphologic differentiation.

fascicularis Group

Widespread and continuous distribution throughout much of South and East Asia suggests that the *fascicularis* group is the youngest of the species groups of *Macaca* (Fooden, this volume). The *fascicularis* group is composed of basically allopatric populations which, with the exception of *M. cyclopis*, constitute a graded ecological series with respect to body size and tail proportions (Fooden, 1971b; 1976). *M. cyclopis*, the Taiwanese macaque, is rather robust in body build, but the species is characterized by a stout and bushy tail at least one-half to two-thirds head-body length (Napier and Napier, 1967; Fooden, 1971b, Table 2). The monkey resembles *M. fuscata* but with gray pelage and a long tail (personal observation). Cronin *et al.* (this volume) suggest a special relationship between *M. cyclopis* and *M. fuscata* on the basis of electrophoretic data. Reports that *M. cyclopis* prefers coastal cliffs and rocks to forest can be attributed to Swinhoe (1862). All substantiated occurrences of the monkey have been in montane forest (C. Shuttleworth, personal communication) which, at higher elevations, are subject to seasonal snows. From the standpoint of body morphology, *M. cyclopis* may resemble the ancestral population of the *fascicularis* group, although molecular similarities between *M. fuscata* and *M. mulatta* suggest that the latter may be closest to the ancestral population (cf. Weiss *et al.*, 1973; Nozawa *et al.*, 1977). A more traditional interpretation would be that the relatively small body and long tail of *M. fascicularis* nominate the species as a candidate for the ancestor of the group which bears its name (cf. Delson, this volume).

The differentiation of, and initial divergence within, the *fascicularis* group may have occurred more than 500,000 years ago. Fragmentary fossil evidence suggests that a population ancestral to *Macaca fuscata* may have reached the Japanese islands by a landbridge connecting Japan with the mainland at Korea during the Middle Pleistocene (Iwamoto and Hasegawa, 1972), although Delson (this volume) considers that no fossil specimens recovered from Japan are clearly older than Late Pleistocene. The robust body build and reduction in tail length of *M. fuscata* suggest an adaptation to cold climatic conditions. On the basis of molecular comparisons (Weiss *et al.*, 1973; Darga *et al.*, 1975), the relationship of *M. fuscata* to other members of the *fascicularis* group appears somewhat problematic, although there is a suggestion of an early relationship between *M. fuscata* and *M. mulatta*. According to the electrophoretic data of Nozawa *et al.* (1977), *M. mulatta* is the taxon genetically closest to *M. fuscata*, and the time of divergence between them is estimated as being 538,000 years.

Subsequent differentiation and/or dispersal of the *fascicularis* group may have been held in check until the most recent glacials caused retreat of the *silenus* subgroup and *sinica* group to forest refugia. The combined adaptability of *Macaca mulatta* and *M. fascicularis* to more arid environments and riverine, disturbed, and secondary forests may reflect the dispersal of these species before climatic amelioration had occurred in southern Asia by about 10,000 years ago. Although the data are inadequate to draw any final conclusions about centers of dispersal, the protein studies by Nozawa *et al.* (1977) indicate that both *M. mulatta* in China and *M. cyclopis* are closer biomolecularly to *M. fascicularis* in Indonesia (provenance unspecified) than either is to Malaysian or, even more so, Philippine populations of this taxon.

Nozawa *et al.* (1977, Figure 5) appear to suggest that the time of divergence between *Macaca mulatta* and *M. fascicularis* is greater than 250,000 years. The fact that *M. mulatta* and *M. fascicularis* exhibit similar amounts of lengthening of the X chromosome (Chiarelli, 1962) and cannot be differentiated chromosomally by the G (Giemsa) banding technique (Paszter, 1976) may argue against such an early differentiation. Likewise, Weiss *et al.* (1973) remark on the high degree of similarity in proteins between the two species. Fooden (1975) argues convincingly that the failure of *M. fascicularis* to reach the Mentawai Islands and Sulawesi means that the species was absent from Sundaland at the time that maximal eustatic lowering of sea level, which may have occurred about 170,000 years ago, made possible the dispersal of the *silenus* subgroup to these peripheral islands.

Macaca fascicularis or an ancestral population, may have penetrated Sundaland somewhat prior to 40,000 years ago when the continental shelf of eastern Asia was exposed because of eustatic lowering of sea level. A series of strata in Niah Cave, northern Sarawak, dated from approximately 40,000 to 2,000 years ago, has yielded 76 identifiable subfossil macaque teeth and jaws of which the preponderance of material (83 to 100 percent) from each stratum is assigned to *M. fascicularis* while the remainder is assigned to *M. nemestrina* (Hooijer, 1962b). All of these specimens apparently are human food remains (Hooijer, 1962b).

About 20,000 years ago during the last glacial, *Macaca fascicularis* probably dispersed widely throughout Sundaland, reaching continental Southeast Asia and the Philippines by the northeastward extension of the Sunda shelf through Palawan. On the basis of protein differences, Nozawa *et al.* (1977), however, estimate the time of divergence among populations of *M. fascicularis* in Indonesia, Malaysia, and the Philippines to be about 75,000 years. Riparian forests (Fig. 4-3) may have facilitated the dispersal of *M. fascicularis* at a time when increasing seasonality and aridity characterized Southeast Asia. The length of the tail in *M. fascicularis* may be a derived character associated with its adaptation to warm humid environments. Rodman (n.d.) suggests that river systems may have provided the most significant refugia at times of climatic cooling during the Pleistocene, and the adaptation of *M. fascicularis* to riverine, disturbed, and secondary forests suggests the dispersal of the species prior to general improvement of the climate.

Napier and Napier (1967) recognize 21 subspecies of *Macaca fascicularis*, of which the majority are insular populations which probably became isolated from one another with eustatic rise in sea level accompanying the retreat of the last glacial about 10,000 years ago. The presence of *M. f. fascicularis* in the Malay Peninsula, Borneo, and Sumatra (and a number of lesser islands) strengthens the contention that these regions constitute a natural floristic and faunal unit. Pleistocene river valleys may have facilitated extensive gene flow among the populations of *M. fascicularis* in these regions of Sundaland.

Three cave deposits of unknown origin in eastern Java dated from 20,000 to 10,000 years ago, have yielded 28 fragments of jaws and teeth in which tooth dimensions closely correspond to those in *Macaca fascicularis mordax*

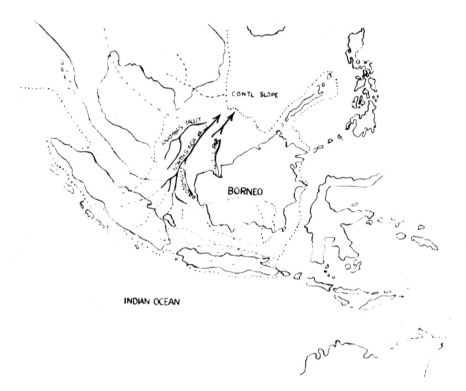

Fig. 4-3. Sundaland (dotted lines) as it was exposed during Pleistocene glacials. Major drainage patterns then existing on the Sunda shelf are indicated by dashed lines. (From Verstappen, 1975, and Haile, 1969)

(Hooijer, 1962a). River systems between southern Borneo and eastern Java (Fig. 4-3) may have made possible the dispersal of *M. fascicularis* into this area about 20,000 years ago. The dispersal may have taken place during a period of desiccation, with the result that *M. fascicularis mordax* became isolated from other populations of the species; or subsequent to the dispersal of the species a zone of drought extending from lower Burma to eastern Java, followed eventually by eustatic rise in sea level, may have isolated the Javanese population from others.

The northernmost distribution of *Macaca fascicularis* appears to be about 20° N. latitude, at approximately the latitude of the headwaters of the Chao Phya River and the area where the Mekong River begins to enter mountainous terrain (Fig. 4-2). The deltas of these rivers may have formed the basin of the South China Sea during the exposure of Sundaland (Fig. 4-3, Sartono, 1973). *M. fascicularis* may have dispersed into continental Asia by following forests adjacent to the rivers. More northward expansion of the species may have been checked by the westward dispersal of *M. mulatta*.

During a 14-month "census/survey" of primates on the island of Sumatra, Crockett and Wilson (this volume) recorded information on the habitat preference of *Macaca fascicularis*. Over 70 percent of contacts with the species were made in riparian habitats, including river banks, lake shores, and along the sea coast, and only 15 percent of contacts occurred "within the forest." Field studies in both the Kutai Reserve, East Kalimantan, in Indonesian Borneo by Fittinghoff and Lindburg (this volume) and the Ketambe area of the Gunung Leuser Reserve in Aceh, north Sumatra by Rijksen (1978) established that groups of *M. fascicularis* are mainly arboreal and confined to riverbank habitat. In addition, *M. fascicularis* appears to partition the riverine and adjacent habitat among conspecific groups through the establishment of somewhat overlapping ranges with specific sleeping sites or roosts from which the members of a group rhythmically disperse and return. In the Ketambe area such ranging usually takes groups no more than 300 meters inland from the rivers. This preference for, or restriction to, riverine habitat appears to date back to at least the late Pleistocene dispersal of *M. fascicularis* described above, and may be under phylogenetic constraints.

Biomolecular data suggest a degree of divergence between *Macaca cyclopis* and *M. mulatta* comparable to that exhibited by different populations of *M. fascicularis* (Weiss *et al.*, 1973; Darga *et al.*, 1975; Nozawa *et al.*, 1977). Nozawa *et al.*, (1977) estimate the time of divergence between the two taxa to be about 67,000 years and suggest that *M. cyclopis* should be considered a local population of *M. mulatta* isolated on Taiwan (Formosa).

Palynological research has established fluctuating glacial conditions (the Tali glacial) on Taiwan from before 60,000 to 10,000 years ago (Tsukada, 1967). During this glacial Taiwan was a part of the Asian mainland (Tsukada,

1967) and appears to have been on the route of vertebrate migration from eastern Asia to the Philippines and Sulawesi (Sartono, 1973). The maximal stage of the glacial occurred some time subsequent to 60,000 to 50,000 years ago as temperatures decreased about 8° to 11° C to support boreal conifers and pine. Although cool to temperate conditions prevailed from about 50,000 to 10,000 years ago, an increase in temperature, although still colder than at present, occurred between 45,000 to 40,000 years ago and may have been significant in the invasion of Taiwan by *M. cyclopis*. The limited differentiation between *M. cyclopis* and *M. mulatta* suggests that the ancestral population of the former did not reach Taiwan during an earlier glacial such as that associated with maximal sea regression prior to 170,000 years ago.

The expansion of *Macaca mulatta* may have been held in check until glacial phenomena caused the retreat of populations of the *silenus* subgroup and *sinica* group into forest refugia. Reduction in tail length, although exhibiting variation, suggests the exposure of *M. mulatta* to colder climatic conditions. From the results of electrophoretic examination of blood proteins obtained from samples collected throughout the distribution of *M. mulatta*, Nozawa *et al.* (1977) estimate the time of divergence among different populations of the taxon to about 23,000 years. This time period approximates to the date of the last glacial about 20,000 years ago. The westward dispersal of *M. mulatta* occurred at latitudes north of those which *M. fascicularis* was invading in Southeast Asia and through the arid (or formerly arid) zone in northern India which had disrupted the distribution on the *sinica* group. Biomolecular similarities between populations of *M. mulatta* and *M. fascicularis* in Thailand (Weiss *et al.*, 1973) may in part be a consequence of limited hybridization at the border of the two species (cf. Fooden, 1964). Southward penetration of the Indian subcontinent by *M. mulatta* appears to have been halted by the recovery and northward expansion of *M. radiata* with climatic amelioration. The importance of competition between species groups in determining the distribution of macaques can be better understood by considering some of the events apparently surrounding the dispersal of *Presbytis entellus,* the common langur of India. The same glacial events which disrupted the distributions of the *silenus* subgroup and the *sinica* group probably reduced species of Indian colobines such as *P. johnii* to refugia populations, thereby making possible the rapid spread of *P. entellus* throughout the diverse environments of the Indian subcontinent during, or subsequent to, the last glacial. Sugiyama (1976) even considers *P. entellus* to be second only to *Homo sapiens* in the diversity of the microenvironments which it exploits.

The reconstructions presented above, although provisional, establish the importance of climatic change and forest refugia during the Quaternary in the differentiation and dispersal of species groups of *Macaca*. A series of paleo-climatic investigations must be undertaken throughout Asia before more detailed reconstructions can be attempted. The present distribution of macaques suggests, however, that three refugia (Fig. 4-2) probably were of most importance in the evolution of Asian forms: the Khasi hills, Annamitic cordillera, and Dawna range.

SOME PRELIMINARY OBSERVATIONS ON MACAQUES IN THE DAWNA RANGE

During the periods October 1973 to March 1974 and January to May 1975, I initiated surveys for macaques (*Macaca* spp.) in a known area of sympatry in Huay Kha Khaeng Game Sanctuary in western Thailand. Actual residence in the field was discontinuous. During July and August 1977, I returned to the sanctuary to assist in the development of an ecological research station. Additional observations were made on macaques at this time.

Description of the Dawna Range

Huay Kha Khaeng Game Sanctuary encompasses 165,014 hectares in the western provinces (*changwat*) of Uthaithani and Tak, and is the second largest wildlife sanctuary in Thailand. The sanctuary lies between 15°10' – 15°50' N latitude and 99°00' – 99°22.5' E longitude (Fig. 4-4). The area belongs to the Burmese-Malayan folded mountains, specifically the Dawna range, which consists of a series of Paleozoic limestones which underwent magmatic invasion with granitic intrusion during the late Triassic to early Jurassic (Kobayashi, 1964). In the southwest, mountains of limestone formation rise up to over 1,500 meters while the west is girded by a mountain chain of similar formation over 1,200 meters. Dry evergreen, mixed deciduous, and dry dipterocarp forests intergrade with one another, at least to an elevation of 1,000 meters, creating a patchy environment.

The area has three seasons: a six-month monsoon or rainy season (May through October); a cool, dry winter with infrequent rains (November through January); and a hot, dry summer with occasional, but increasing rains (February through April). Table 4-1 presents monthly rainfall for the period 1965 to 1975 recorded at two meteorological stations in *changwat* Uthaithani. Although a few large rivers (*huay*) such as *huay* Kha Khaeng flow throughout the year, most water courses are dependent on the monsoon rains, and dry up or are reduced to pools of standing water during the hot, dry season. In mountainous areas some streams may be fed by springs.

The mammal fauna in Huay Kha Khaeng Game Sanctuary is rich and diverse. The combined distribution of the different faunal assemblages represented in the sanctuary encompasses at least southern China, the

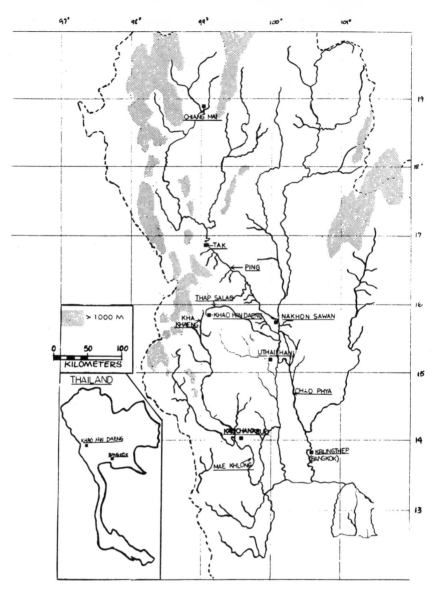

Fig. 4-4. Khao Hin Daeng Ranger Station, Huay Kha Khaeng Game Sanctuary, and adjacent areas in west-central Thailand.

Philippines, Indochina, the Malay Peninsula, Borneo, Sumatra, Java, the southern Himalaya, Burma, Bangladesh, India, and Sri Lanka. The mammals include the following: barking deer (*Muntiacus muntjak*), sambar deer (*Cervus unicolor equinus*), gaur (*Bos gaurus*), banteng (*Bos banteng*), wild buffalo (*Babalus bubalis*), tapir (*Tapirus indicus*), elephant (*Elephas maximus*), "wild

Table 4-1. Monthly Rainfall in Millimeters for the Period 1965—1975, Recorded at Nong Chang and Ban Rai, *changwat* Uthaithani.

NONG CHANG	1965*	1966*	1967*	1968	1969	1970	1971	1972	1973	1974	1975+
January				0.0	75.6	11.8	32.5	0.0	0.0	0.0	17.0
February				0.0	0.0	0.0	18.8	3.4	0.0	13.8	0.0
March				0.0	17.4	57.3	35.4	4.7	12.2	5.8	72.1
April				103.2	52.3	134.0	67.3	107.5	28.3	96.9	67.4
May				154.9	133.1	485.9	97.2	92.3	300.4	301.9	125.8+
June				75.3	209.9	365.2	133.9	156.4	128.0	149.3	
July				217.0	282.6	176.7	159.0	106.4	101.7	206.7	
August				100.3	36.7	355.5	365.6	166.2	241.0	216.0	
September				235.5	593.2	301.6	266.7	448.6	415.2	277.9	
October				153.4	297.8	221.5	161.5	170.3	103.4	279.7	
November				0.0	66.1	46.7	T	104.5	39.7	56.5	
December				0.0	0.0	13.9	T	30.4	0.0	0.0	

BAN RAI	1965	1966	1967	1968	1969	1970*	1971*	1972*	1973*	1974	1975+
January	0.0	42.5	0.0	0.0	54.8					0.0	63.9
February	30.7	30.0	0.0	32.2	0.0					0.0	29.0
March	40.5	40.2	0.0	0.0	92.7					127.1	83.3
April	5.5	44.0	167.7	115.9	0.0					114.4	169.8
May	202.0	207.9	119.4	140.6	80.3					109.0	45.0+
June	117.0	75.2	107.3	88.9	27.8					36.0	
July	56.8	0.0	99.0	172.6	25.6					224.8	
August	172.3	227.0	65.6	35.8	39.8					97.9	
September	299.8	206.0	170.8	168.7	259.6					498.7	
October	73.5	339.9	249.3	0.0	219.1					401.4	
November	41.5	56.0	34.4	0.0	0.0					79.8	
December	0.0	40.0	0.0	0.0	0.0					0.0	

* no record   + incomplete record

73

dog" *(Cuon alpinus)*, Malayan sun bear *(Helarctos malayanus)*, mustelids such as yellow-throated marten (*Martes flavigula*) and river otter (*Lutra lutra*), many felids including tiger (*Panthera tigris*) and leopard (*Panthera pardus*), and many civets and squirrels. The primates are the white-handed gibbon (*Hylobates lar*), macaques (*Macaca* spp.), colobine monkeys (*Presbytis* spp.), and slow loris (*Nycticebus coucang*).

Contacts with Macaques

During the three field sessions I established contact with groups of *Macaca fascicularis, M. nemestrina leonina, M. assamensis,* and *M. mulatta* numbering approximately 5–10 to 30 monkeys and perhaps as few as only three individuals of *M. arctoides*, although several kinds or subspecies of the latter are reported for the area. Some assignments to *M. assamensis* and *M. mulatta* are provisional, and the monkeys in question may represent short-tail *M. fascicularis* or morphologic intermediates between *M. mulatta* and *M. fascicularis* (cf. Fooden, 1964). One group of monkeys exhibited such morphologic variability that it may have been a mixed species and/or hybrid group. The details of these contacts are summarized in Table 4-2. Contact also was established with groups of *Presbytis phayrei* and *P. cristata*, both of which are members of the *Trachypithecus* group, although all contacts with the latter were limited to the 1973–1974 field session.

A total of 130 survey days was completed from ranger stations at Khao Hin Daeng (15°34′ N latitude, 99°19.5′ E longitude), SapFarPa (15°31′ N latitude, 99°17.5′ E longitude), and Khao Nang Rum (15°29′ N latitude, 99°17.5′ E longitude) and from a camp on *huay* Chang Tai to the west of Khao Nang Rum. Contacts with macaques were both limited and sporadic. Contact was established with macaques on 25 of the survey days. Successive contacts in a given area can be attributed to the clumping of preferred foods (fruits) in time and space. Different groups of macaques were contacted in spatial and temporal proximity to each other on four of the 25 contact days. On only one of the four days was it apparent without question that two different groups of monkeys were members of the same species. In this instance contact may have been established twice with the same group (or foraging subgroups of the same group).

Habitat Preference

Macaques in the areas surveyed in Huay Kha Khaeng Game Sanctuary are primarily frugivorous, and seasonality occurs in most of the fruits preferred by macaques (and other primates). Table 4-3 identifies fruits and other vegetation consumed by macaques during the three field sessions. All macaques in the

Table 4-2. Contacts with *Macaca* spp. in Huay Kha Khaeng Sanctuary During 1973-74, 1975, and 1977.

COORDINATES	ELEVATION	SPECIES	NUMBER OF ANIMALS	FOREST TYPE[1]	TIME OF CONTACT ACTIVITY[2]	DURATION OF CONTACT IN MINUTES	DATE
15° 30' N 99° 16' E	400 m	*M. arctoides*	1	E	13:30 (?) S	1.0	12 October 1973
15° 36' N 99° 19' E	220 m	*M. fascicularis*	<10	D, B	15:30 ST	3.0	29 November 1973
15° 30' N 99° 17' E	340 m	*M. mulatta* (or short-tailed *M. fascicularis*)	>20	D, B	10:30 F	90	3 December 1973
15° 30' N 99° 16' E	460 m	unknown "large, brown" jumped to ground to escape	unknown	E	14:00 F	2.0	4 December 1973
15° 28' N 99° 15' E	500 m	unknown (?) escape in trees	unknown	D	15:00 F	15	20 December 1973
15° 27' N 99° 15' E	440 m	unknown escape on ground	unknown	D	10:00 F	5.0	21 December 1973
15° 27' N 99° 15' E	500 m	*M. mulatta* (or short-tailed *M. fascicularis*)	2	D	12:15 FG	2.0	21 December 1973
15° 27' N 99° 15' E	(?) 500 m	unknown	unknown	D	12:00 W	1.0	19 January 1974
15° 27' N 99° 15' E	400 m	*M. assamensis*	>20	D	10:15 P	315	23 January 1974
15° 30' N 99° 18' E	500 m	*M. assamensis* (long tail)	>20	D	15:25 F	25	13 February 1974
15° 30' N 99° 17' E	400 m	unknown "large, red, with x red face"	2	D, B	11:00 F	1.0	26 February 1974

Table 4-2. (continued)

COORDINATES	ELEVATION	SPECIES	NUMBER OF ANIMALS	FOREST TYPE[1]	TIME OF CONTACT ACTIVITY[2]	DURATION OF CONTACT IN MINUTES	DATE
15° 28' N 99° 19' E	660 m	*M. assamensis*	>20	E	11:40 (?) F	5.0	18 March 1974
15° 28' N 99° 19' E	600 m	unknown morphologic variability	unknown	E	13:05 (?) P	3.0	18 March 1974
15° 28' N 99° 19' E	600 m	*M. assamensis* morphologic variability	<10	E	13:20 F, R, PL	40	19 March 1974
15° 30' N 99° 18' E	280 m	unknown jumped to ground	<10	E, B	16:15 ST	90	20 January 1975
15° 35' N 99° 19' E	280 m	*M. fascicularis*	>20	D, B	16:10 P, (?) to ST	40	25 January 1975
15° 30' N 99° 18' E	320 m	*M. mulatta* (or short-tailed *M. fascicularis*)	>10	B	11:30 F	60	3 March 1975
15° 28' N 99° 18' E	580 m	*M. n. leonina*	>20	E	10:25 F	45	4 March 1975
15° 28' N 99° 18' E	600 m	unknown darker than above	unknown	E	11:05 (?)	2.0	4 March 1975
15° 28' N 99° 18' E	640 m	*M. n. leonina*	>20	E	13:00 F	55	4 March 1975
15° 28' N 99° 18' E	580 to 800 m	*M. assamensis*	>20	E to B	11:05 (?) P	111	6 March 1975
15° 28' N 99° 18' E	680 m	unknown "large, very red"	unknown	D	12:20 (?) F	1.0	23 April 1975
15° 38' N 99° 18' E	200 m	*M. fascicularis*	<10	D, B	10:40 (?)	2.0	15 July 1977

Coordinates	Elevation	Species	Group size	Forest type	Activity / Time		Date
15° 30' N 99° 17' E	460 m	*M. assamensis*	>20	D, B	10:05 P, F	65	22 July 1977
15° 30' N 99° 17' E	460 m	*M. assamensis*	>20	D, B	9:45 F, P	136	23 July 1977
15° 30' N 99° 17' E	460 m	unknown	unknown	D, B	9:40 P	5.0	24 July 1977
15° 30' N 99° 16' E	540 m	*M. mulatta*	(?) 6	D, B	11:45 (?) P, (?) I	10	24 July 1977
15° 30' N 99° 16' E	480 m	*M. assamensis*	>20	D, B	11:55 (?) I, P	55	24 July 1977
15° 28' N 99° 19' E	540 m	*M. n. leonina*	>30	D	10:30 F	5.0	30 July 1977
15° 28' N 99° 19' E	580 m	*M. arctoides*	(?) 2	E	9:30 (?)	2.0	31 July 1977
15° 30' N 99° 17' E	480 m	unknown	unknown	D, B	9:45 P	15	7 August 1977

forest type
B = bamboo
D = mixed deciduous
E = dry evergreen

[2] activity
F= feeding
FG= feeding on ground
P= progression
PL= playing
R= resting
S= sentinel
ST= sleeping trees
W= responded to whistle
I= intergroup encounter

Table 4-3.   Seasonal Fruits and Other Foods Eaten by Macaques in Huay Kha Khaeng Game Sanctuary, Recorded by Month of Observation.

	Oct	Nov	Dec	Jan	Feb	Mar	Apr	May	Jun	Jul	Aug	Sep
Afzelia xylocarpa	M	M	M							E	E	
Zizyphus oenoplia		M										
Ardisia sp.?			M									
Lagerstroemia sp.			M	M								
Phyllanthus emblica			M									
Zingiberaceae roots			M									
Globba sp.										M		
Spondias pinnata			M	M	M?							
Albizzia sp.					M							
Aglaia sp.					M							
Ficus cf. *glabella*					M							
Baccaurea ramiflora					I	M?						
Euphoria longana					I							
Artocarpus dadah uniq.					IM							
Garcinia xanthochymus					M							
Ficus sp.					M							
Bambusa arundinacea seeds						M*						
Bambusa sp. seeds						M*						
Ficus racemosa						M						
Platymitra siamensis						M						
Mangifera cf. *sylvatica*						I						
Schleichera oleosa							I			M	M	
Artocarpus gomezianus							M					
Alangium salviifolium							M					
Aradirachta sp. ?							M					
Baccaurea sapida										M		
Uvaria sp.										M	M	
Sumbairopsis albicans										M	M	
Vitex glabrata											M	
Willughbia sp.											M	
Artidesima sp.											M	
Bouea oppositifolia											M	
Mangifera sp.											M	
Bambusa sp. shoots											M	

M   mature fruit
I   immature fruit, too green to be eaten
E   immature fruit, edible
*   aseasonal

areas surveyed to date in the Dawna range appear to exploit large spatio-temporal ranges which are altitudinally graded, thereby encompassing lower elevations and mountainous regions, in order to sample seasonally available fruits and other vegetation which occur in different microenvironments. This ranging behavior appears to necessitate that groups travel through various intergrading forest types.

No gross differences in habitat utilization were apparent among the different macaques although *Macaca fascicularis* may be less far-ranging than other species. During 1973-1974 and 1975, all contact with *M. fascicularis*, conceivably the same group of monkeys, by me, as well as personnel of the Royal Forest Department, occurred at what appear to be night resting sites in an area of mixed deciduous-bamboo forest along the road connecting Khao Hin Daeng and SapFarPa. The shortest distance from this area to a permanent water course is at least 2,000 meters. In 1977 contact was established with a group of *M. fascicularis* adjacent to *huay* Thap Salao. The patchy distribution of preferred foods (fruits) and the presence of a potential predator in the form of the leopard (*Panthera pardus*) in the roadside area, may result in ranging patterns which necessitate the repeated use of a non-riverine sleeping area or site by *M. fascicularis*. For this species much of the Dawna range may constitute marginal habitat in which phylogentic constraints are tested by environmental variables.

Comparative Observations

The only area with a patchy environment approaching the complexity of that of Huay Kha Khaeng Game Sanctuary in which a systematic field study has been undertaken on macaques is the Asarori forest at 30°21′ N latitude, 78°0′ E longitude in Uttar Pradesh, north India. During 1964 to 1965, Lindburg (1971, 1977) accomplished 527 hours of observation in the Asarori forest on *Macaca mulatta*, one of the species of macaques in the Dawna range.

The Asarori study area encompasses a band of tropical moist deciduous forest running parallel to the crest on the northern slopes of the Siwalik Hills at an elevation increasing from 400 to 1,000 meters. The area experiences three seasons: a warm, wet monsoon; a cool season with light rains; and a hot, dry season. In contrast to the Dawna range, the majority of the area is covered with immature *Shorea robusta*, a deciduous hardwood commonly known as *sal*. But dry and heavily eroded stream beds fringed by grassy meadows and scattered stands of shrubs (mixed inferior forest) dissect the forest and contribute to the patchiness of the environment (Lindburg 1971, 1977).

The ranging behavior of groups of *Macaca mulatta* in the Asarori forest appears to be based upon a foraging strategy that involves the exploitation of food patches and movement between such patches with night resting occurring at feeding sites. Fruits constituted 65 to 70 percent of the diet of the monkeys, and were the only foods on which an entire group fed for all or part of an observation day. Leaves made up the second largest quantity of foods, and stems, flowers, shoots, and fungi also were consumed. These foods were exploited when no single item was dominant, or as supplemental or incidental foods.

Lindburg (1977) considers the spatio-temporal distribution of food and a limited water supply in the dry season, to be the two primary constraints on ranging behavior, and the clumping of food and water within the Asarori forest appears to result in considerable range overlap of groups of *Macaca mulatta*. Contacts may be minimized at feeding sites through temporal variations in the use of these resources, but physiological demands during the dry season may account for a high incidence of encounters at water. In contrast, low population densities and far-ranging over difficult and complex terrain to exploit seasonal fruits, appear to minimize contacts between groups of the same and different species of macaques in the Dawna range, in spite of seasonal reduction in water and similarities in ecological adaptation.

CONCLUSIONS

The dispersal, and probably the differentiation, of species groups of *Macaca* in Asia appears to be intimately associated with climatic events of the Pleistocene epoch. Periods of increasing aridity and seasonality associated with Pleistocene glacials probably are responsible for the disjunctions in the present distributions of species groups. Competition between different species groups may have reinforced these disjunctions. The sympatry of both *Macaca* spp. and *Presbytis* spp. as well as the diversity of mammals in Huay Kha Khaeng Game Sanctuary suggests that the Dawna range was an area of refuge forest during periods of glacial aridity. *M. nemestrina leonina* probably is the earliest of the macaques to have invaded the Dawna range, followed by *M. assamensis* and *M. arctoides*. Competition between the latter two species, which appear to exhibit a special relationship to each other, may have contributed to specializations of the genitalia of *M. arctoides* and the limited representation of the species in the sanctuary and elsewhere. The *fascicularis* group would have entered the area most recently. *M. fascicularis* probably dispersed westward from the Chao Phya River along the rivers and streams such as *huay* Thap Salao which flow from the Dawna range into the central plain of Thailand. Further exploration may reveal the presence of larger concentrations of *M. fascicularis* adjacent to the few major rivers such as *huay* Kha Khaeng in the interior of the range. Some southward penetration by *M. mulatta* has occurred in the area, but the expansion of the *fascicularis* group appears to have been checked by both intragroup and intergroup competition.

The retreat of macaques into major and other refugia during the Pleistocene would have provided an opportunity for gene flow to occur among populations of the same species and some hybridization between different species, including populations of different species groups. Similarities in the electrophoretic patterns of plasma proteins between *Macaca mulatta* and *M. arctoides* in Thailand, and to a lesser extent *M. mulatta* and *M. assamensis*, as reported by Cronin *et al.* (this volume) may be a consequence of such hybridization. The morphological complexity of some groups of macaques in the Dawna range may reflect such gene exchange between different species.

Recent paleoclimatic reconstructions suggest that environmental conditions similar to those in Huay Kha Khaeng Game Sanctuary may have been more widespread in southern Asia as a consequence of glacially induced aridity during the Pleistocene.

ACKNOWLEDGMENTS

Field work was supported by the Wenner-Gren Foundation and Sigma Xi in 1973–1974 and the National Science Foundation, the National Institute of Mental Health, and Sigma Xi in 1975. My participation in the development of an ecological research station in Huay Kha Khaeng Game Sanctuary in 1977 was made possible by awards from the New York Zoological Society and the Fauna Preservation Society. I wish to thank the National Research Council and the Wildlife Conservation Division, Royal Forest Department, for sponsoring my research in Thailand.

REFERENCES

Allen, J.A. The proper generic name of the macaques. *Bull. Amer. Mus. Nat. Hist.*, **35**: 49–52 (1916).

Beaty, C.B. The causes of glaciation. *Amer. Scientist*, **66**: 452–459 (1978).

Bowler, J.M., Hope, G.S., Jennings, J.N., Singh, G., and Walker, D. Late Quaternary climates of Australia and New Guinea. *Quaternary Res.*, **6**: 359–394 (1976).

Brandon-Jones, D. The evolution of recent Asian Colobinae. In D.J. Chivers and K.A. Joysey, (eds.) *Recent Advances in Primatology*, Vol. 3, *Evolution*. London: Academic Press pp. 323–325 (1978).

Butzer, K.W. Environment, culture, and human evolution. *Amer. Scientist*, **65**: 572–584 (1977).

Chiarelli, B. Comparative morphometric analysis of primate chromosomes. II. The chromosomes of the genera *Macaca, Papio, Theropithecus* and *Cercocebus. Caryologia*, **5**: 401–420 (1962).

CLIMAP Project Members. The surface of the ice-age earth. *Science*, **191**: 131–137 (1976).

Cronin, J.E., and Sarich, V.M. Molecular evidence for dual origin of mangabeys among Old World monkeys. *Nature*, **260**: 700–702 (1976).

Darga, L., Goodman, M., Weiss, M.L., Moore, G.W., Prychodko, W., Dene, H., Tashian, R. and Koen, A. Molecular systematics and clinal variation in macaques. In C.L. Markert, (ed.). *Isozymes IV: Genetics and Evolution.* New York: Academic Press (1975).

Delson, E. Evolutionary history of the Cercopithecidae. In F. Szalay, (ed.), *Contributions to Primatology.* Basel: S. Karger (1975a).

———. Paleoecology and zoogeography of the Old World monkeys. In R.H. Tuttle, (ed.), *Primate Functional Morphology and Evolution.* The Hague: Mouton (1975b).

Ellerman, J.T. and Morrison-Scott, T.C.S. *Checklist of Palaeoarctic and Indian mammals, 1758–1946.* London: British Museum of Natural History (1966).

Fairbridge, R.W. Effects of Holocene climatic change on some tropical geomorphic processes. *Quaternary Res.*, **6:** 529–556 (1976).

Flint, R.F. *Glacial and Quaternary Geology.* New York: John Wiley (1971).

_____. Physical evidence of Quaternary climatic change. *Quaternary Res.*, **6:** 519–528 (1976).

Fooden, J. Rhesus and crab-eating macaques: intergradation in Thailand. *Science*, **143:** 363–365 (1964).

_____. Complementary specialization of male and female reproductive structures in the bear macaque, *Macaca arctoides. Nature,* **214:** 939–941 (1967).

_____. Taxonomy and evolution of the monkeys of Celebes (Primates: Cercopithecidae). *Bibl. Primat.,* **10:** 1–148 (1969).

_____. Female genitalia and taxonomic relationships of *Macaca assamensis. Primates,* **12:** 63–73 (1971a).

_____. Male external genitalia and systematic relationships of the Japanese macaque (*Macaca fuscata* Blyth, 1875). *Primates,* **12:** 305–311 (1971b).

_____. Taxonomy and evolution of liontail and pigtail macaques (Primates: Cercopithecidae). *Fieldiana: Zoology,* **67:** 1–169 (1975).

_____. Provisional classification and key to living species of macaques (Primates: *Macaca*). *Folia Primat.,* **25:** 225–236 (1976).

Garson, M.S. and Mitchell, A.H.G. Transform faulting in the Thai peninsula. *Nature,* **228:** 45–47 (1970).

Gates, W.L. Modeling the ice-age climate. *Science,* **191:** 1138–1144 (1976).

Goodman, M., Barnabas, J., Matsuda, G. and Moore, G.W. Molecular evolution in the descent of man. *Nature,* **233:** 604–613 (1971).

Haffer, J. Speciation in Amazonian forest birds. *Science,* **165:** 131–137 (1969).

Haile, N.S. Quaternary deposits and geomorphology of the Sunda shelf off Malaysian shores. *Proceedings INQUA Congress, Paris,* (1969).

Hays, J.D., Imbrie, J. and Shackleton, N. J. Variations in the earth's orbit: pacemaker of the ice ages. *Science,* **194:** 1121–1132 (1976).

Hill, W.C.O. B. External genitalia. In H. Hofer, A.H. Schultz, and D. Stark, (eds.), *Primatologia* III (1). Basel: S. Karger, (1958).

_____. *Primates: Comparative Anatomy and Taxonomy.* **VII.** *Cynopithecinae: Cercocebus, Macaca, Cynopithecus.* New York: John Wiley, (1974).

_____ and Bernstein, I.S. On the morphology, behaviour, and systematic status of the Assam macaque (*Macaca assamensis* McClelland, 1839). *Primates,* **10:** 1–17 (1969).

Hooijer, D.A. Quaternary langurs and macaques from the Malay archipelago. *Zool. Verh. Rijksmus. Nat. Hist. Leiden,* **55:** 1–64 (1962a).

_____. Prehistoric bone: the gibbons and monkeys of Niah great cave. *Sarawak Mus. J.* (n.s.), **11:** 428–449 (1962b).

_____. Quaternary mammals west and east of Wallace's line. In G.-J.J. Barstra, and W.A. Casparie, (eds.). *Modern Quaternary Research in Southeast Asia.* Rotterdam: A.A. Balkema. (1975).

Iwamoto, M. and Hasegawa, Y. Two macaque fossil teeth from the Japanese Pleistocene. *Primates,* **13:** 77–81 (1972).

Jongsma, D. Eustatic sea level changes in the Arafura sea. *Nature,* **228:** 150–151 (1970).

Keng, H. Size and affinities of the flora of the Malay peninsula. *J. Trop. Geog.,* **31:** 43–56 (1970).

Kobayashi, T. Geology of Thailand. In T. Kobayashi, (ed.). *Geology and Palaeontology of Southeast Asia.* Tokyo: University of Tokyo (1964).

Lemmon, W.B., and Oakes, E. "Tieing" between stump-tailed macaques during mating. *Lab. Primat. Newslet.,* **6(1):** 14–15 (1967).

Lindburg, D.G. The rhesus monkey in north India: an ecological and behavioral study. In L.A. Rosenblum, (ed.), *Primate Behavior: Developments in Field and Laboratory Research,* vol. 2. New York: Academic Press (1971).

_____. Feeding behaviour and diet of rhesus monkeys (*Macaca mulatta*) in a Siwalik forest in north India. In T.H. Clutton-Brock, (ed.). *Primate Ecology: Studies of Feeding and Ranging Behaviour in Lemurs, Monkeys and Apes*. New York: Academic Press (1977).

Matthews, S.W. What's happening to our climate? *Nat. Geog.*, **150**: 576–615 (1976).

Mohnot, S.M. The conservation of non-human primates of India. *Abstracts*, Sixth Congress of the International Primatological Society, Cambridge, (1976).

Moreau, R.E. *The Bird Faunas of Africa and its Islands*. New York: Academic Press (1966).

Napier, J.R. and Napier, P.H. *A Handbook of Living Primates*. New York: Academic Press (1967).

Nozawa, K., Shotake, T., Ohkura, Y. and Tanabe, Y. Genetic variations within and between species of Asian macaques. *Japan. J. Genet.*, **52**: 15–30 (1977).

Pasztor, L.M. Species identification by chromosome analysis. *Primate News*, **14(7)**: 3–7 (1976).

Pocock, R.I. The systematic value of the glans penis in macaque monkeys. *Ann. Mag. Nat. Hist.*, **9**: 224–229 (1921).

_____. The external characters of the catarrhine monkeys and apes. *Proc. Zool. Soc. London* 1925: 1479–1579 (1926).

_____. *Mammalia. 1. Primates and Carnivora in Fauna of British India, including Ceylon and Burma*. London: Taylor and Francis (1939).

Rijksen, H.D. *A Field Study on Sumatran Orang Utans (Pongo pygmaeus abelli* Lesson 1827): ecology, behaviour and conservation. Wageningen: H. Veenman and Zonen B.V. (1978).

Rodman, P.S. Pleistocene climates in the Congo: fluctuation and subspeciation in the genus *Cercopithecus*, unpub. man., n.d.

Sarich, V.M. Rates, sample sizes, and the neutrality hypothesis for electrophoresis in evolutionary studies. *Nature*, **265**: 24–28 (1977).

Sartono, S. On Pleistocene migration routes of vertebrate fauna in Southeas Asia. *Bull. Geol. Soc. Malaysia*, **6**: 273–286 (1973).

Simons, E.L. The deployment and history of Old World monkeys. In J.R. Napier, and P.H. Napier, (eds.), *Old World Monkeys; Evolution, Systematics, and behavior*. New York: Academic Press (1970).

_____. *Primate Evolution*. New York: Macmillan (1972).

Sugiyama, Y. Characteristics of the ecology of the Himalayan langurs. *J. Human Evol.*, **5**: 249–277 (1976).

Swinhoe, R. On the mammals of the island of Formosa (China). *Proc. Zool. Soc., London* 1862: 347–365 (1862).

Tsukada, M. Vegetation in subtropical Formosa during the Pleistocene glaciations and the Holocene. *Palaeogeog. Palaeoclimat. Palaeoecol.*, **3**: 49–64 (1967).

Uzzell, T. and Pilbeam, D. Phyletic divergence dates of hominoid primates: a comparison of fossil and molecular dates. *Evolution*, **25**: 615–635 (1971).

Verstappen, H.T. On Palaeo climates and landform development in Malesia. In G.-J. Bartstra, and W.A. Casparie, (eds.), *Modern Quaternary Research in Southeast Asia*. Rotterdam: A.A. Balkema, (1975).

Weiss, M.L., Goodman, M., Prychodko, W., Moore, G.W. and Tanaka, T. An analysis of macaque systematics using gene frequency data. *J. Human Evol.*, **2**: 213–226 (1973).

Wilson, D.R. Tail reduction in *Macaca*. In R. Tuttle, (ed.). *The Functional and Evolutionary Biology of Primates*. Chicago: Aldine-Atherton, (1972).

Chapter 5
Speciation in MACACA: The View From Sulawesi

Colin P. Groves

The taxonomic revision of the monkeys of Celebes (Sulawesi) by Fooden (1969) was a timely piece of work in many respects. The "received doctrine" on these monkeys (Forbes, 1894; Pocock, 1926; Carter *et al.*, 1945; Laurie and Hill, 1954; Napier and Napier, 1967) placed them in two separate genera: the widespread *Macaca* was said to have only one species on Sulawesi, *M. maura*, while the only other species of the island was allocated to a separate, monotypic genus *Cynopithecus (C. niger)*. In general the former, known as the Moor macaque, was supposed to range over the southern and central parts of the island, while the latter, the Celebes Black ape, inhabited the long northern peninsula.

Already in 1917 Büttikofer had attempted a modification of this system; for him not merely the Black ape, but all the Sulawesi monkeys were referable to *Cynopithecus*, and there were not two species but many; the precise number could not be determined because of the paucity of material, but provisionally eight were recognized. In spite of this admitted defect, Büttikofer's arrangement made considerable sense: the magnitude of the differences between most of the species, and the consistent geographic pattern they demonstrated, spoke for itself. Notwithstanding this careful study, the two-species theory with its *Macaca/Cynopithecus* dichotomy continued to hold sway with a few exceptions, generally overlooked by primatologists as Büttikofer had been (Sody, 1949; Khajuria, 1953).

And so things remained until the publication of Fooden's monograph. Based on plentiful material, Fooden was able to substantiate Büttikofer's conclusions and to add to them; he preferred to return the Sulawesi monkeys to *Macaca*, on the grounds that the differences between them and other macaques seemed to be no greater than those between other species allocated to *Macaca*. One of Büttikofer's species, the so-called *Cynopithecus tonsus* (Matschie, 1901), was synonymized with *Macaca tonkeana* (Meyer, 1899); the other seven were upheld. The distributions were fully mapped, and the differences between the species in skull, external characters, baculum etc., were fully analyzed. The seven, with their range limits as deducible from available material, were as follows:

1. *Macaca maura* (Schinz, 1825). Southern peninsula; northern boundary record—Parepare (4.01 S, 119.38 E).
2. *Macaca brunnescens* (Matschie, 1901). Islands of Buton and Muna.
3. *Macaca ochreata* (Ogilby, 1841). Southeastern peninsula. Nothern boundary record—Wawo (3.41 S, 121.02 E).
4. *Macaca tonkeana* (Meyer, 1899). Central mass of Celebes; southern boundary record—Uru (3.30 S, 119.53 E); southeastern—Palopo (3.01 S, 120.13 E) and Tonkean (1.24 S, 122.30 E); northern—Labua Sore (0.27 S, 120.03 E), parhaps atypical specimens, and Parigi (0.48 S, 120.10 E), typical example.
5. *Macaca hecki* (Matschie, 1901). Proximal part of northern peninsula; southern boundary record—Kampungbaru (1.02 N, 120. 48 E); eastern boundary record—Kwandang (0.50 N, 122.53 E).
6. *Macaca nigrescens* (Temminck, 1849). Middle part of northern peninsula; western boundary record—Gorontalo (0.33 N, 123.03 E); eastern—Sungei Onggak Dumoga, between Doloduo and Molibagu (0.30 N, 123.54 E).
7. *Macaca nigra* (Desmarest, 1822). Distal end of northern peninsula; western boundary record—Modayag (0.45 N, 124.25 E).

The advisability of referring all these forms to the genus *Macaca* will of course depend on a final revision of the Asiatic Cercopithecines with a comparison of the breadth of morphological variation within this group, with the variation within other primate genera. At such a time it could turn out that all species will remain with a single genus, *Macaca*, or else several genera will be recognized.

Since Fooden's revision, the macaque volume in Osman Hill's monograph on Primates has appeared (1974). This continues to place the Celebes macaques in two genera as before, but recognizes, with certain unexplained modifications, Fooden's species, of which *maura, ochreata* and *brunnescens* are placed in *Macaca* (subgenus *Gymnopyga*), the remainder in *Cynopithecus*. Hill unreservedly accepts Fooden's demonstration of monophyly in the group, keeping *Cynopithecus* solely on the basis of its "extreme specializations" (although some of his definition of this genus would not apply to two of its species, *C. tonkeanus* and *C. hecki*), so that the basis of the difference is a solely phenetic, nonphylogenetic, concept of the genus category.

The question of the number of species in the Celebes group is a much more controversial one. Fooden stated that he had kept all seven as full species because there was no clear evidence of intergradation at that time; future research might well result in reducing some or all of them to subspecific status. Some commentators (Thorington and Groves, 1970) noting that the chain of species formed to some extent a morphocline, hypothesized that such a situation would be consistent with a single-species solution. Indeed, Fooden does on pp. 110–111 of his monograph offer evidence that marginal interbreeding between two of his species may occur, on the basis of three juveniles from Labua Sore, the northernmost locality of *tonkeana*, which showed features tending towards *hecki*.

METHOD

I therefore decided to examine the situation at the border areas between some of Fooden's species, to see whether there was reproductive isolation or interbreeding in such places. In the time available to me, it would probably not be possible to visit all the border areas, so it seemed advisable to take them in geographic sequence modified by consideration of the accessibility of the area. As far as possible, a visit to a border area was preceded by observation of at least one of the species concerned in a known typical habitat outside the border zone.

The island of Sulawesi (Celebes) is divided among four Indonesian provinces (Fig. 5-1): Sulawesi Selatan ("Sulsel," south Sulawesi; capital, Ujung Pandang, formerly Macassar), Sulawesi Tenggara ("Sultra", southeast Sulawesi; capital, Kendari), Sulawesi Tengah ("Sulteng", central Sulawesi; capital, Palu), and Sulawesi Utara ("Sulut", north Sulawesi; capital, Manado). As it turned out, no visit was made to Sulawesi Tenggara, but all of the other three provinces were visited. All of them contain nature reserves, of varying sizes and security, although none of them are adequately staffed or patrolled in spite of their enthusiastic promotion by the local Nature Protection officials.

Ujung Pandang, Palu and Manado are all served by airports, as is the town of Gorontalo in the northern province (and certain other towns not visited by me). From Ujung Pandang (south Sulawesi) a road, with frequent bus services, goes north as far as Palopo, and in the southern mountains and the Tempe Depression there are networks of drivable roads and tracks. From Manado (north Sulawesi) a network of roads spreads out through Minahasa (the northernmost Kabupaten, or Regency, of the province), and a road extends into the Kabupaten of Bolaang-Mongondow to its main town, Kotamobagu, and beyond down the valley of the Sungei (= river) Onggak Dumoga to Imandi and Doloduo. Maps mark a road along the northern coast of the peninsula as far as Kwandang, but this is in a fact no more than a cart-track,

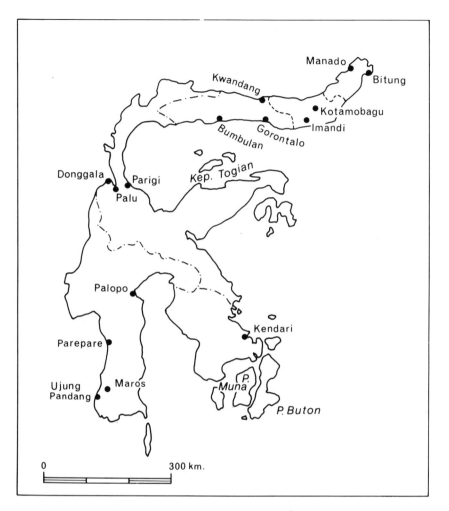

Fig. 5-1. Map of Sulawesi (Celebes) showing political divisions mentioned in text.

traversable by a motorcycle but not by larger motor vehicles; so that essentially the Kabupaten of Gorontalo is not accessible by road from the Mina-hasa/Bolaang-Mongondow network. From Gorontalo, roads go east for some distance along the Sungei Bone, north to Kwandang, and east to Bumbulan and Marisa. From Palu (C. Sulawesi) roads go north for some distance on either side of the peninsula, across the peninsula to Parigi, northwest to Donggala, and south to Kulawi. This network is separate alike from those of the northern and southern provinces.

In three out of four cases, observation of wild specimens proved unsatisfactory by itself, and the fascination with monkeys shown by many Sulawesi people who commonly keep them as pets proved decisive. The existence of so many captive monkeys, though not a great help to the species' conservation, and in most cases totally abhorrent from a humanitarian point of view, was an indispensible adjunct to the study. In most instances the animals were known to have been captured nearby, and in a few cases they were stated to have been obtained from a more distant region. Whenever there was independent corroboration, the information was nearly always accurate. If, for example, I had previously seen free-living monkeys in the same region as a certain pet animal was said to have come from, they were always of the same taxon. A few people proved very knowledgeable about monkeys. A man in Palu owned an adult male *M. nigra* which he had admitted he had bought from a trader so that he could not be sure of its provenance; but, he opined, it surely was caught in Minahasa because of the characteristic shape of its ischial callosities.

THE GEOGRAPHICAL BACKGROUND

Sulawesi is a spidery-shaped island, consisting of a central land mass and four peninsulas: southern, southeastern, eastern and northern. Broadly, the southern peninsula is the habitat of *maura*, the southeastern of *ochreata*, the central landmass and eastern peninsula (though this latter is in fact not confirmed) of *tonkeana*, and the northern of *nigra, nigrescens* and *hecki*.

The historical geology of the island is the subject of controversy. For Audley-Charles *et al.* (1973) the southern and northern arms were part of Sundaland, the eastern and southeastern arms being the leading edge of the Australian continent which became glued onto the other half as the Australian plate drifted northwestwards to collide with the Southeast Asian plate. For Katili (1974) on the other hand the eastern and southeastern arms are a product of the collision itself: a section of Oceanic crust which was uplifted and folded when the Australian plate margin in that region was subducted beneath the Southeast Asian plate. The difference between the two hypotheses, from a zoogeographic point of view, is that under the Audley-Charles model Australian land fauna could have been carried to Sulawesi, while under the Katili view it could not have been. Under both theories the western half of Sulawesi is agreed to have been a part of Sundaland, and largely above sea-level during most or all of the Quaternary.

Vegetationally, Sulawesi contrasts strongly with the Sundaland region. Although, unlike the Lesser Sunda Islands, it has a rainfall level broadly comparable with that of Sundaland, the near absence of the dominant family Dipterocarpaceae gives the forests a totally different aspect. Broadly speaking, we have moist deciduous forest in the southern half of the southern peninsula and in parts of the southeastern, and mixed rain forest elsewhere: lowland up to 1500 meters, changing gradually to montane above this level. Dipterocarps, mainly *Shorea*, occur most commonly along the north coast of the northern peninsula between Kwandang and Paleleh; a few grow in the Malili district, at the "root" of the southern peninsula, and again in the Maros region further south. Apart from that, mixed forests predominate, with single-dominant associations being nearly confined to the drier country of the western half of the northern peninsula (west from Gorontalo) and on the eastern arm. In these regions, *Agathis celebica* (Araucariaceae), the biggest tree in Sulawesi, is dominant, with *Palaquium* spp. (Sapotaceae), and species of Verbenaceae and Ebenaceae. Throughout Sulawesi, *Koordersiodendron pinnatum* (Anacardiaceae) is a common tree, especially on the northern peninsula east of Gorontalo; many species have distributions restricted to one part of the island or another (teak, for example, is reported to grow only in Muna and Buton). Table 5-1 lists some of these species of restricted distribution according to the "macaque area" each occurs in.

It should also be remarked that a large number of trees of Sulawesi are undescribed. Areas to be designated Nature Reserves are surveyed botanically beforehand. In the published report on Karaenta (South Sulawesi) 54 percent of 126 tree species are as yet undescribed (Zakaria *et al.*, 1975), while in the Gunung Tangkoko Batuangus Reserve the figure (from a list kindly supplied by the PPA office, Manado) is 37 percent of 89 species. In the former case, 10 undescribed tree species are, together with *Ligustrum* sp., *Arenga pinnata, Palaquium bataanense* and *Laportea* sp., among the most dominant in the reserve, with an area in each case of over 0.5 m^2 per hectare.

Rainfall is responsible for some of the intraisland diversity, but not for all by any means. It can be seen from Table 5-2 that in the south the rainfall is high, above 2,000 millimeters p.a., on the coast and in the southerly mountainous region; in the Tempe depression (Anabanua, Amparita, Rappang) it falls below this level, and rises again in the Toraja highlands. In the north there is high rainfall in the eastern half of the northern peninsula, falling steeply to the west, beginning somewhat east of Gorontalo. Palu and Talisse have the lowest rainfall in Indonesia, but these towns lie in a marked rain-shadow, and rainfall in the nearby mountains is much greater.

Crude annual rainfall figures do not, of course, give a complete picture. Seasonality is rather more marked on the southern peninsula than elsewhere; thus Ujung Pandang has for two months of the year a rainfall below 50 centimeters, and so do most other stations on the southern peninsula, whereas such a marked dry season does not occur even in the overall low-rainfall part of the northern peninsula (only at Gorontalo and Limboto is there even a single month whose figures fall as low as 50 millimeters).

A useful way of assessing seasonality is by calculating the coefficient of variation for monthly rainfall figures. As seen in Table 5-2, these levels

Table 5-1. Distribution of Important Tree Species in Sulawesi.

SPECIES	FAMILY	MACAQUE SPECIES-AREA					
		MAURA	TONKEANA	OCHREATA	HECKI	NIGRESCENS	NIGRA
Koordersiodendron pinnatum	Anacardiaceae	x			x	x	x
Agathis celebica	Araucariaceae	x			x	x	x
Hopea dolosa	Dipterocarpaceae			x			
Diospyros buxifolia	Ebenaceae	x	x	x		x	
" celebica	"	x	x		x		
" ferrea	"			x	x		
" hebecarpa	"	x				x	x
" pilosanthera	"			(Muna)		x	x
" rumphii	"						x
Homalium foetidum	Flacourtiaceae				x		
Elmerrillia spp.	Magnoliaceae		x	x	x	x	x
Metrosideros petiolata	Myrtaceae		x	x	x	x	
Ligustrum sp.	Oleaceae	x	x				
Arenga pinnata	Palmae	x					
Pometia pinnata	Sapindaceae		x	x	x		
Manilkara kaukioides	Sapotaceae		x				
Palaquium bataanense	"	x	x		x	x	
Laportea sp.	Urticaceae	x					x
Avicennia marina	Verbenaceae				x		
Vitex cofossus	"	x			x	x	
" quinata	"				x	x	x

Compiled from Prawira et al., 1972 (a and b); Zakaria et al., 1975; I Made Taman (pers. comm.); and information supplied by U bus Wardyu Maskar and the staff of Kantor PPA, Manado.

correspond well with taxonomic differentiation in macaques. Seasonality is relatively low in *hecki* and *tonkeana* areas (C.V. in most cases aroung 25-35), higher in *nigra* and *ochreata* areas (approx. 35-45), higher still for *nigrescens* and *brunnescens* (50-60) and extreme for *maura* (more than 60). It is perhaps significant that shifts in seasonality separate neighboring taxa, except in the case of *hecki* and *tonkeana* where the absolute rainfall level itself differs.

The consequence is that the vegetation on the southern peninsula is a mosaic of forest and open grassland, so that *M. maura* perforce comes to the ground more often than other forms, is able to forage in the open, and is easier to observe for such reasons. It is only on the southern peninsula—with the exception of a few, easily irrigated areas elsewhere—that fairly dense human settlement based on wet rice cultivation has occurred. The range of *M. maura* has therefore been fragmented more than that of any other form.

The other mammals of Sulawesi have been treated in Groves (1976), but a much more detailed study is in preparation by G.G. Musser. Internal speciation has, strikingly, been largely altitudinal, neither highland nor low-land species show much tendency to vary geographically. Where there is geographic variation, however, differentiation from central taxa tends to occur (1) on the southern peninsula, south from the Tempe Depression; (2) on the northern peninsula, east of Gorontalo. Stresemann (1928) finds a similar pattern in birds. The sudden break at the depression has obvious isolation potential in the south. In the north the rather steep rainfall change may have helped to maintain speciation processes initiated by the marine incursion invoked by Fooden (1969) to explain *hecki/nigrescens* separation.

The macaques would therefore seem to be unusual in their failure to speciate vertically (though they occur from sea level to 7,500 feet—G.G. Musser, pers. comm.); fairly standard in their observance of boundaries at Gorontalo and in the Tempe Depression; and perhaps unique in their observance of additional boundaries in the regions round Palu and Lake Matana.

THE MAURA/TONKEANA BOUNDARY

The differences between *M. maura* and *M. tonkeana* which proved to be more or less observable in the field were as follows:

1. Form and color of rump patch. In *maura*, the rump patch is not highly conspicuous: it is brownish-gray, considerably lighter in tone than the brown or black-brown of the rest of the body, but not sharply set-off, not very broad, and not extending down the thighs to the knees. In *tonkeana*, however, it is markedly paler than the glossy black tone of the rest of the hair, and in addition is broader, extends down below the knee, and is set off laterally by a crest, giving it a bushy appearance. This is much the easiest way to differentiate the two at a glance in the field.

Table 5-2. Annual Rainfall At Various Locations in Sulawesi.

STATION	PROVINCE	KABUPATEN	MACAQUE	ALTITUDE	MEAN ANNUAL RAINFALL (MM)	COEFFICIENT OF VARIATION	MONTHS WITH <50 MM
Mapangat	Sulut	Minahasa	nigra	30	3318	33	0
Kema	"	"	"	1	1349	42	2
Manado	"	"	"	4	2783	47	0
Tondano	"	"	"	640	2033	30	0
Noongan	"	"	"	800	2284	25	0
Amurang	"	"	"	1	2120	48	0
Boyong	"	"	"	400	3819	28	0
Poigar	"	"	"	6	2793	42	0
Kotamobagu	"	Bolaang-Mongondow	"	610	2060	25	0
Modayog	"	"	"	766	2497	20	0
Malibagu	"	"	nigrescens	1	2703	67	0
Lolak	"	"	"	2	2654	50	0
Gorontalo	"	Gorontalo	hecki	1	1264	24	0
Limboto	"	"	"	20	1338	28	1
Kwandang	"	"	"	9	2890	26	0
Tilamuta	"	"	"	1	1556	31	0
Paleleh	Sulteng	Buol-Tolitoli	"	1	3487	31	0
Buol	"	"	"	1	1823	34	0
Tolitoli	"	"	"	1	2404	23	0
Tinombo	"	Donggala	"	1	1247	48	1
Dampelas	"	"	"	5	1962	28	0
Sirenja	"	"	"	5	1995	31	0
Talisse	"	"	"	3	604	31	5
Palu	"	"	"	1	736	28	7

Name	Region	Species	District				
Bora	"	"		30	1109	30	0
Parigi	"	tonkeana		6	1667	46	0
Kapiru	"	"		525	1773	17	0
Kulawi	"	"		735	2408	22	0
Tomado	"	"		1000	2407	26	0
Poso	"	"	Poso	1	2368	27	2
Luwuk	"	"	"	2	1051	46	0
Kolonedale	"	"	"	1	3670	29	1
Kendari	Sultra	ochreata	Kendari	10	1944	42	0
Wawotebi	"	"	"	35	1792	39	0
Kolaka	"	"	Kolaka	1	1773	32	2
Raha	"	brunnescens	Muna	5	1593	54	3
Baubau	"	"	Buton	10	2021	57	0
Rantepao	Sulsel	tonkeana	Tator	700	4015	42	1
Rappang	"	tonkeana & maura	Sidrap	28	2003	41	0
Maiwa	"	"	"	71	2818	37	1
Amparita	"	maura	"	10	1619	43	0
Anabanua	"	"	Wajo	39	1859	61	0
Paria	"	maura	Wajo	80	2442	60	1
Parepare	"	"	Parepare	2	2362	61	1
Maros	"	"	Maros	5	3241	89	1
Loka	"	"	Gawa	1160	2636	53	2
Bulukumba	"	"	Bulukumba	1	1440	83	3

Sources: Lembaga Meteorologi Dan Geofisika, 1969 and 1974.

2. Shape of the head. The short muzzle and flat crown hair of *maura* are usually quite easily recognizable from the long muzzle (especially in the male), somewhat upstanding crown hair, and bushy cheeks of *tonkeana*.
3. Color of cheeks. The cheeks have a marked pale hue in *tonkeana*, emphasized by the bushiness of the cheek whiskers; this effect is absent from *maura*, where the cheeks are dark and the hair lies flat.
4. The ventral surface is often noticeably paler in *maura* than in *tonkeana*.

In addition, *tonkeana* is always black; *maura* occurs in two phases, one brown and one virtually black, which can be distinguished readily in field observations.

In and around the Bina Mulya Ternak ranch at Maiwa (southern peninsula, approx. 3.40 to 3.50 S, 119.20 to 119.45 E), both *M. maura* and *M. tonkeana* can be observed, living in separate troops without any evidence of hybridization in the form of animals showing intermediate characteristics. A walk along the southwestern boundary of the ranch (Fig. 5-2) during the late afternoon, when troops would leave the gallery forests and emerge onto the grassy ridges to catch the dying sun, encountered six troops of *M. maura* and one of *M. tonkeana*. The latter was encountered at 5:50 p.m. on 9.21.75, scattered over about 75 meters, its most dispersed members being only 150 meters from a troop of *M. maura*. The differential characters of (1) strongly marked, bushy rump-patch in *tonkeana*; and (2) color polymorphism among *maura* individuals, were easily visible to the naked eye. The pale cheeks of *tonkeana*, together with the long face and characteristic circumfacial hair pattern, could be seen through binoculars.

A second troop of *tonkeana* was observed on the northern boundary of the ranch, about 2 kilometers north of the nearest *maura* troop. No other troops of monkeys were seen along that transect.

Two captive specimens, both juveniles, were examined: a brown-phase *maura* in Salokarajae village, (Fig. 5-3) and a *tonkeana* in Malino. Both were stated to have been captured in the vicinity of the respective villages.

The southernmost record for *tonkeana* in Fooden's study is Uru; evidently not the village itself (which lies some 3 kilometers to the northeast of Maiwa at the foot of the escarpment on the north edge of the Tempe Depression) but at the top of the escarpment overlooking the village, as Stresemann (1940) gives an altitude of 800 meters and a latitude 10 minutes further north than the village itself. Uru was visited twice by me, but no monkeys were seen; the forest of the slopes above the village gives way directly to cultivated land round the village itself, so that undisturbed open grassland, which around the ranch itself makes for good visibility, is absent.

It is clear, in any case, that *M. maura* and *M. tonkeana* are good species, being marginally sympatric—probably, parapatric—at Maiwa; and that the range of

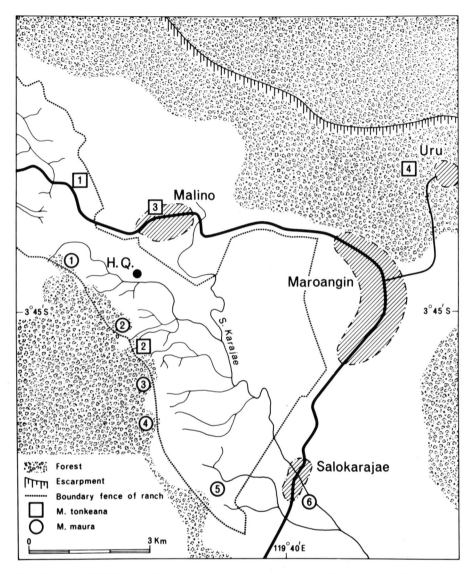

Fig. 5-2. Map of BMT Ranch, Maiwa, S. Sulawesi. Symbols show locations where wild troops of *M. maura* and *M. tonkeana* were encountered, except as follows: ⑥- *M. maura* captive juvenile in Salokarajae; ③- *M. tonkeana* captive juvenile in Malino; ④- Southernmost record (Fooden, 1969) of *M. tonkeana*. More probably this record should be on top of the escarpment rather than at its foot as depicted here.

Fig. 5-3. Juvenile *Macaca maura*, kept at Salokarajae, near Maiwa.

the former extends north across the Tempe Depression to the foot of the Toraja highlands. If, as Fooden suggests, the formation of the Depression was the geographic isolating factor of the two taxa, then it is clear that something in the environment south of the Depression, most likely the more seasonal climate and consequent interlacing of forest with natural grassland, has since enabled *maura* to out-compete *tonkeana* by reinvading the Depression and spreading north to the foot of the mountains which are the homeland of *tonkeana*.

Although actual counts could not be made, a few observations seem worth recording. First was the definite impression that at Maiwa, where marginally sympatric with the all-black *tonkeana*, troops of *M. maura* consisted predominantly of brown animals. One *maura* troop, seen on two evenings in exactly the same place, consisted of about two dozen individuals—two incomplete counts totaled 21 and 19—among which only one blackish monkey was seen; the northernmost *maura* troop, which consisted of over 40 individuals, had only three or four blackish ones. Among all the other troops, none seen in anything like the degree of completeness of these two, only one blackish animal was observed. By contrast, in each of three troops seen in the Maros region (two in Karaenta, one near Tompok Balang) there seemed to be more nearly equal proportions of blackish and brown individuals. This impression can certainly not, as it stands, be used to promote a theory of character displacement, but is certainly susceptible of disproof or substantiation by future observers.

A second observation concerns age-graying and albinism. Fooden (1969) reports a single instance of age-graying in *tonkeana*; in the present study one individual in the more southerly of the two *tonkeana* troops appeared to be iron-gray all over (its long face and the crest on the crown identified it as *tonkeana*, in addition to its association with normally colored monkeys of the same species). In the *maura* troops, possible cases of age-graying were noted, but it was usually difficult to be sure that simple lighting effects were not the cause of the gray patches. Cases of albinism, or rather, partial albinism were much more clear-cut: four definite examples, all in *maura*, were seen although not closely. One troop was followed into a patch of low forest where the animals seemed more confident than in the open. During a 1½-hour period of continuous observation, individuals approached to within 20 meters on occasion. An adult male was completely gray-white except for brown patches on shoulders, upper arms, rump and thighs, and a smaller monkey had only the anterior half of the body white. In a second troop (the northernmost *maura* troop) one monkey was snowy-white on the head, the rest of the body being brown; a second, walking beside the first, was white on both head and shoulders and blackish elsewhere.

Fooden suggests (p.87) that depigmentation in this species is related to conditions of captivity, as all the partial or total albinos seen by him were zoo specimens. These records of its occurrence in the wild disprove Fooden's surmise. The restriction of albinism to just two out of the six *maura* troops, with two affected monkeys in each, makes a genetic basis seem more likely, as indeed is normal in mammals. "Kera putih" (White Monkeys) are also reported in Karaenta. A forestry official told me he thought that they might be monkeys turned white with age, a view which seemed plausible to me until I actually saw the ones at Maiwa. A troop living near Malino (5.16 S, 119.51 E—not the Malino near Maiwa) is said by Dr. W.J. Meyer (pers. comm.) to be led by a formidable and aggressive white monkey known in Bahasa Makassar as "Karaeng Darre kebo" (the White Monkey King). But none of the white monkeys at Maiwa appeared to be in a control role.

THE TONKEANA/HECKI BOUNDARY

The main differences between *tonkeana* and *hecki* are as follows:

1. In *hecki* the shanks of the hindlegs are light brownish, and the forearms are dark brown; in *tonkeana* they are black like the body.
2. The underside of *hecki* is distinctly paler than the upperside: not so in *tonkeana*.
3. The cheek whiskers of *hecki* lack the pale grayish or brownish tint of

Fig. 5-4. Adult male *Macaca tonkeana tonkeana:* Landskron castle, Alsace (near Bale); troop kept by M. Nicolas Herrenschmidt of University of Strasbourg.

Fig. 5-5. Ischial callosities of juvenile male *M. t. hecki* (in some respects this animal approaches *tonkeana,* but callosity shape is typical for *hecki*).

tonkeana (Fig. 5-4), having at most a brown tinge to the tips of the hairs.

4. The rump patch, though somewhat paler than the dorsum, is not sharply marked in *hecki* (Fig. 5-5), and not pale and bushy as in *tonkeana*.

5. The ischial callosities of *tonkeana* are oval and orange-colored (as in *maura* and *ochreata*), while in *hecki* they are gray-yellow and "reniform" (Fig. 5-5), that is to say, basically with an upper and lower lobe separated medially by a transverse crease. It should be noted that the callosities in *hecki* are not precisely the same as those of *M. nigra* which are also reniform: in the latter the callosities are very bright orange-pink, more completely divided, the upper lobe shifted somewhat laterally with respect to the lower, and both lobes rounded, instead of flat as in *hecki*. But in *hecki*, as in *nigra* (Hill, 1974), and also *nigrescens*, the callosities are given extra prominence by the shape of the underlying ischium, which has a longer shank and more expanded tuberosity than in *tonkeana, maura* etc. (Fig. 5-6). This results in a situation where the anus lies in a depression anterior, as well as superior, to the callosities.

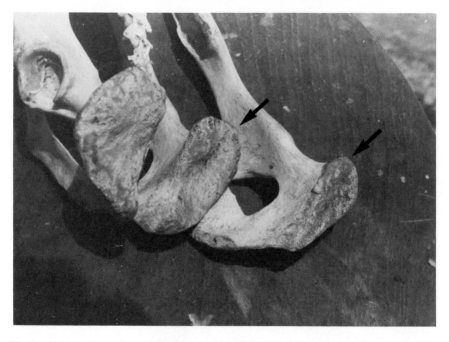

Fig. 5-6. Ischial tuberosities of (left) *M. t. hecki*, MZB 21339; (right) *M. t. tonkeana*, MZB 6543, both adult males. Note the broadly expanded tuberosities of the former.

6. The tail is tiny and stublike in *hecki*: another similarity to *nigra* and *nigrescens*. In *tonkeana* as is *maura* it is much longer (though still extremely short as monkey tails go!).
7. In *hecki* the "gluteal fields," the bare or thinly haired areas on either side of the tail, so conspicuous in *tonkeana* and *maura*, are virtually absent.

Fooden records typical *tonkeana* from Parigi (Fig. 5-7), at the base of the north-south isthmus that joins the northern peninsula onto the rest of the island. *Hecki* is not recorded on the isthmus, but two out of three juvenile skins studied by him from Labua Sore (0.27 S, 120 E), halfway up the peninsula, showed features tending towards *hecki*. He remarks, "The somewhat transitional characters of the Labua Sore specimens tend to suggest that *M. tonkeana* and *M. hecki* may intergrade morphologically in the narrow zone of contact between their adjacent ranges, but the equivocal evidence of these three juveniles does not now appear sufficiently strong to establish that *tonkeana* and *hecki* are conspecific." The records seemed to me sufficiently suggestive to be worth investigating, in spite of the rather wide gap between Labua Sore and the nearest *hecki* localities.

The road from Palu to Parigi goes 20 kilometers north along the coast, then turns east across the mountains, and on reaching the other side of the isthmus turns south again for 15 kilometers to Parigi. Halfway along the mountain road is Kebun Kopi ("Coffee garden"—which is what it was in former days), where there is a government guest-house. Due to the courtesy of Dr. B.L. Salata, Secretary to the Governor of Sulawesi Tengah, I was able to stay overnight in this guest-house after returning from a trip to Parigi and Malakosa. Monkeys seen nearby, alongside the road (Fig. 5-7), were as follows:

1. 8–10 kilometers west of Kebun Kopi, two troops about 2 km apart. In one of them, two individuals were distinctly seen to have the brown on the hind-shanks typical of *hecki*.
2. 3 kilometers east of Kebun Kopi, a troop glimpsed on two separate occasions (probably the same troop). Two individuals clearly seen to have light gray rump-patches and no contrast on the hind-shanks; another had bright orange, oval callosities; an adult male, who remained in intermittent view for 10 minutes, had light cheek-whiskers and pale rump-patch and was dark on the ventral surface; it was however im-

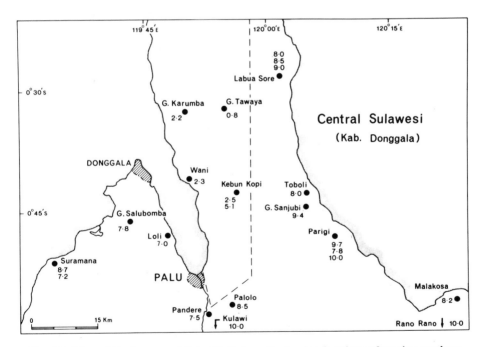

Fig. 5-7. Map of district around Palu (C. Sulawesi), showing locations of captive monkeys (according to stated origin) and scores on the 0 (= *hecki*) to 1 (= *tonkeana*) scale. The dotted line is intended to suggest the halfway boundary between the two subspecies.

possible to see his ischial callosities and difficult to determine whether his hind-shanks had any degree of lightening or not.

3. 9–10 kilometers east of Kebun Kopi: two troops, 1 kilometer apart, showing all the characteristics of *tonkeana*: 4 animals in one troop, 10 in the other, were seen clearly to have gray rump-patches and light cheeks, most were noted to have orange callosities, and some to have a dark undersurface. The second of these two troops, especially, seemed much less shy than the others seen in the area, and remained by the roadside for 15 min. permitting photography (Fig. 5-8).

It appeared on the surface, therefore, that *tonkeana* and *hecki* had a north-south boundary on at least that part of the isthmus, and that they met sharply without intergradation. Examination of village "pets," however, put a rather different complexion on the matter: almost none of the 16 captive monkeys were 100 percent one or the other.

Fig. 5-8. *M. t. tonkeana* alongside the Palu-Parigi road, about 9 km east of Kebun Kopi. The bushy pale rump patch is very conspicuous.

In order to facilitate analysis of the different degrees of intermediacy which the captives demonstrated, a scoring system was adopted, each feature being scored from 0 (*hecki*-like) to 1 (*tonkeana*-like). Most characteristics showed either one (.5) or two (.3 and .7) intermediate states; two (presence or absence of crest defining rump patch; shape of ischial callosities) were either one thing or the other. The features were scored thus:

1. Color of cheeks: dark - 0, light-tipped - .3; medium grey-brown - .7; pale - 1.
2. Color of ventral surface: distinctly paler than dorsal - 0; slightly paler - .5; dark - 1.
3. Color of forearms: darkish brown - 0; brown tones - .5; black - 1.
4. Color of external surface of hind-shanks: (gray-) brown - 0; somewhat browner/grayer than thighs - .3; brown traces - .7; black - 1.
5. Color of rump patch: dark - 0; gray or buff - .5; pale buff - 1.
6. Crest of rump patch - absent - 0; present - 1.
7. Gluteal fields - (small or) absent - 0; vaguely defined - .5; large - 1.
8. Ischial callosities, shape: reniform - 0, oval - 1.
9. Ischial callosities, color: gray-yellow - 0; pinkish tones - .5, pinky-orange - 1.
10. Tail: stub - 0, intermediate - .5; *tonkeana*-length - 1.

Table 5-3 gives the results of the breakdown. Monkeys said to have been caught on the east side of the isthmus always tended more towards *tonkeana*, those caught on the west side tended towards *hecki*, whether the village where they were examined was on the same side or not. This added to one's confidence that the reputed localities were at least approximately correct; they have at any rate been mapped in Fig. 5-7 as if they were correct.

The general picture appears to be of *tonkeana* coming up from the south and extending north along the eastern side of the isthmus, and onto the Donggala peninsula, while *hecki* comes south down the west side of the peninsula. As *hecki* reaches further and further south, *tonkeana* further and further north, each becomes more and more affected by gene-flow from the other. From Table 5-3, it might appear that the gene-flow gradient into *hecki* is more marked geographically (from Pandere north to Gunung Karumba the score falls fairly consistently, i.e., the *tonkeana* characters become fewer and fewer). The "predominantly *tonkeana*" monkeys show scores from 7 to nearly 10 wherever they come from, those from the Donggala peninsula being somewhat less "pure" than those from the isthmus region. It may be, therefore, that the true midpoint of intergradation is west-of-center on the isthmus, gradually drifting eastwards as one goes north, to judge from the descriptions of the Labua Sore skins, whose scores would presumably be 8.0, 8.5 and 9.0 as

Table 5-3. Intergradation Scores of Monkeys From The *tonkeana-hecki* Border Zone.

STATED ORIGIN	WHERE SEEN	CHEEKS	VENTER	ARMS	SHANKS	RUMP PATCH: COLOR	CREST	GLUTEAL CALLOSITIES: FIELDS	SHAPE	COLOR	TAIL	TOTAL
Malakosa	Malakosa	0.7	0.5	1.0	1.0	0.5	1.0	1.0	1.0	0.5	1.0	8.2
S. of Parigi	Palu	0.3	0.5	1.0	1.0	1.0	1.0	1.0	1.0	0.0	1.0	7.8
Parigi	Palu	1.0	1.0	1.0	0.7	1.0	1.0	1.0	1.0	1.0	1.0	9.7
G. Sanjubi	Pangi	0.7	1.0	1.0	0.7	1.0	1.0	1.0	1.0	1.0	1.0	9.4
Toboli	Tawueli	1.0	1.0	0.5	0.0	1.0	1.0	0.5	1.0	1.0	1.0	8.0
Palolo	Petobo	1.0	1.0	1.0	1.0	0.5	1.0	1.0	1.0	0.0	1.0	8.5
G. Salubomba	Watusampung	1.0	1.0	0.5	0.3	1.0	1.0	1.0	1.0	0.0	1.0	7.8
Suramana (1)	Donggala	0.0	1.0	1.0	0.7	0.5	1.0	1.0	1.0	0.0	1.0	7.2
" (2)	Donggala	1.0	1.0	1.0	0.7	0.5	1.0	0.5	1.0	1.0	1.0	8.7
Loli	Labuan	0.3	1.0	0.5	0.7	0.5	1.0	0.5	1.0	0.5	1.0	7.0
Pandere	Petobo	1.0	0.5	0.0	0.0	1.0	1.0	1.0	1.0	1.0	1.0	7.5
Kebun Kopi (1)	Tabeo	0.3	0.5	1.0	0.3	1.0	1.0	0.0	0.0	1.0	0.0	5.1
" (2)	Kebun Kopi	0.0	0.0	0.0	0.0	0.5	1.0	0.5	0.0	0.5	0.0	2.5
Wani	Wani	0.3	0.0	0.0	0.0	0.5	0.0	0.0	0.0	1.0	0.5	2.3
G. Tawaya	Palu	0.3	0.0	0.0	0.0	0.0	0.0	0.0	0.0	0.5	0.0	0.8
G. Karumba	Karumba	0.7	0.0	0.0	0.0	0.5	1.0	0.0	0.0	0.0	0.0	2.2

The animals are arranged in three groupings: first, from the east side of the isthmus and from the south; second, from the Donggala peninsula; third, from the west side of the isthmus. Descriptions by Fooden (1969) of specimens in the United States National Museum suggest scores from three skins from Labua Sore of 8.0, 8.5 and 9.0, and for a skin from Parigi of 10.0 (i.e. pure *tonkeana*).

described by Fooden (pp. 110–111). The midpoint would not reach the east coast of the peninsula for a considerable distance north of Labua Sore. Future research may be able to test this surmise.

Two particular monkeys illustrate the diversity of intergradation patterns very well. One of these, the first Kebun Kopi specimen, kept in Tabeo village, has few of the "positive" characters of either taxon. The legs are not very light, and though the rump-patch is large and pale, the gluteal fields are absent. The arms are not light, the cheeks only slightly lightened. The ischial callosities are reniform, but bright orange. By contrast, the monkey from Pandere (a lowland forest along the Palu River), kept in Petobo village near Biromaru, has the "positive" characters of both taxa: light cheeks, pale crested rump patch, big gluteal fields, oval orange-callosities, combined with striking brown hind-shanks and brown forearms and a fairly light ventral surface (Fig. 5-9). This animal, although perhaps the method of scoring underestimates this aspect, is perhaps the most remarkable mixture studied and if no other intermediate animals had been seen would by itself have constituted evidence that inter-breeding had occurred.

The fact that interbreeding does occur, and in such a manner that different degrees of interbreeding can be detected, forces us to combine the two taxa into a single species for which the prior name is *Macaca tonkeana* (Meyer, 1899); there are still, of course, two well-marked subspecies, *M.t. tonkeana* (Meyer, 1899) and *M.t. hecki* (Matschie, 1901). Nonetheless some problems are raised. From the evidence presented by Fooden, it would seem that morphologically *hecki* is part of a northern group, while *tonkeana* is closer to a southern group with *maura*—the different form of the ischium, described and figured above, supports such a morphological division. That *tonkeana* is reproductively isolated from *maura* but intergrades with *hecki*, provides a warning that speciation patterns cannot invariably be predicted from morphology alone. One may speculate that an original cline *maura-tonkeana-hecki*, "stepped" at the points of taxonomic division, was disrupted by the downfaulting of the Tempe Depression (followed by a marine incursion). By the time contact had been resumed between *maura* and *tonkeana* the differential adaptions of the two had already proceeded beyond the reproductive isolation threshhold. No such total separation is postulated on geological grounds at the *tonkeana-hecki* boundary, and it may be significant that other Celebes mammals do not show taxonomic changeover in that region (G.G. Musser, pers. comm.), so that the morphologically stronger differentiation of these two taxa—whatever the original reason for it—had no chance to develop to reproductive isolation.

A few characteristics do however unite *tonkeana* and *hecki*: bushy cheek-whiskers, short suberect crest on the crown, long but broad muzzle, and very great upper facial breadth. These may be employed to help define the composite species. Notably, Albrecht (1976) finds these two quite close in skull form.

(a)

(b)

Fig. 5-9. (a, b); *M. t. tonkeana* x *hecki* intermediate, from Pandere. Animal resembles *hecki* in color of forearms and shanks, *tonkeana* in light cheeks, tail length, rump patch and callosities, and is intermediate in ventral coloration.

THE NIGRA / NIGRESCENS BOUNDARY

The differences between *nigra* and *nigrescens*, exclusive of skull features, are as follows:

1. General body-color is dark brownish black in *nigra*, but medium to dark brown in *nigrescens*, with a black dorsal stripe.
2. In *nigra* the whole body is the same color, but in *nigrescens* the crown of the head, arms, legs and ventral surface are black, clearly contrasting with the body color.
3. Ischial callosities are orange and reniform in *nigra*, but gray in *nigrescens* with an elongated oval shape (Fig. 5-10).

Fooden suggests that the Sungei Onggak Dumoga is the boundary between these two forms; the most southwesterly locality for *nigra* is Modayag (0.45 N, 124.45 E), the two most easterly localities for *nigrescens* being the Sungei Onggak Dumoga between Doloduo and Molibagu (0.30 N, 123.54 E), and Negerilama (quoted as 0.18 N, 123.43 E, but according to J. Stickings, (pers. comm.) this is a defunct village—hence the suffix "lama" = former—on the Sungei Sangkub at about 0.45 N).

In the present study, I observed *M. nigra* in the G. Tangkoko-Batuangus Nature Reserve, Minahasa, and proceded to Imandi, on the east bank of the S. Onggak Dumoga, via Kotamobagu (see Fig. 5-1). In the short time available, only a single three-hour visit could be made to the nearby reserve (set aside to protect the Celebes incubator bird, *Megacephalon maleo*), and monkeys were not seen in the wild. Two captive specimens were examined however, in villages near Imandi.

A young male at Tonom, 3 kilometers southwest of Imandi, had the contrasting coloration of *nigrescens* but lacked a dorsal stripe; the ischial callosities were pale orange with a much enlarged supero-lateral lobe beginning to fold over in a medial direction as in *nigra*. This animal was therefore an approximate halfway mixture.

A younger (Juvenile I-II) monkey at Modamang, about the same distance northeast of Imandi, was again contrastingly colored, but the callosities were gray, not folded over but with the superior lobe still somewhat enlarged. Except for this last point, and the total lack of a dorsal stripe, the animal was very near to *nigrescens*.

Although the evidence is incomplete, therefore, it does seem that the two forms intergrade across the S. Onggak Dumoga and are to be regarded as subspecies of a single species, for which the prior name is *Macaca nigra* (Desmarest, 1822). The two subspecies are *M.n.nigra* (Desmarest, 1822) and *M.n.nigrescens* (Temminck, 1849). Both have a long, narrow muzzle with prominent paranasal ridges, and an elongated, central crest on the crown, the two most obvious distinguishing features of the amalgamated species (Fig. 5-11).

THE NIGRESCENS / HECKI BOUNDARY

Fooden places *nigrescens* at Gorontalo (0.33 N, 123.03 E) and Tulabolo (0.31 N, 123.16 E); and *hecki* at Kwandang (0.50 N, 123.53 E) and Bumbulan (on the south coast at 122.04 E, see Fig. 5-1). The differences are quite large: *nigrescens* has the baboon-like face of the Black ape, the elongated narrow crest, blackish limbs contrasting with brown body, and gray callosities, while *hecki* has a shorter, very broad face, a short, broad, somewhat upstanding crest on the

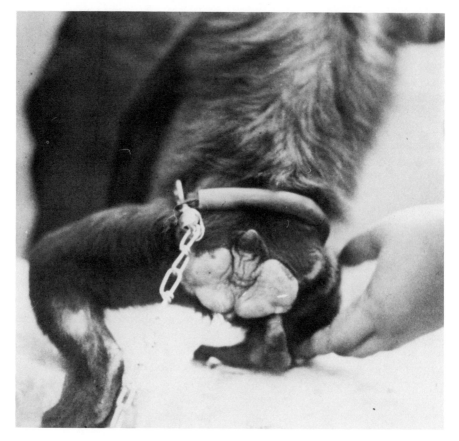

Fig. 5-10. (a). Ischial callosities of subadult female *M. n. nigra;* (b). Ischial callosities of subadult male *M. n. nigrescens.*

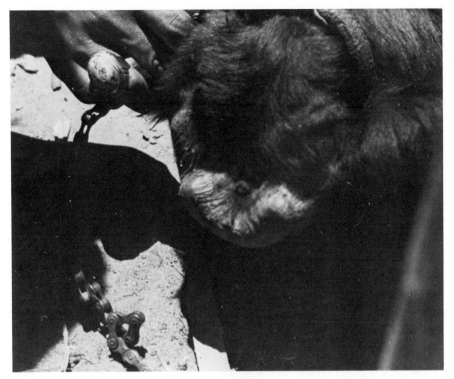

Fig. 5-10 (b).

crown, and is black with strongly contrasting brown shanks (this color extending up the back of the thighs) and weakly contrasting dark brown arms and
underparts, while the callosities are reniform but flattened and with yellow
tones, less iron-gray.

Captive monkeys from the hills to the east of Gorontalo Harbor were *hecki*,
like those from the hills to the west. Localities for *hecki* (Fig. 5-12) were:
Paguyaman valley, Tilamuta, Molosipat (west of Marisa), Isimu, Buhu,
Bumbulan, Pulubala (all these are west of L. Limboto, and so well within the
known range of *hecki*). And, to the east of Limboto: near Leato, on the coast
east of Gorontalo (about 123.05 E), two specimens; G. Modelidu, junction of
S. Paliomamuta and S. Boyade, near Longalo (about 0.45 N, 123.03 E). Both
of the Leato monkeys were held captive in houses at the foot of the hill (G.
Mantulangi) on the east side of Gorontalo Harbor, on the footpath that leads
4 kilometers to Leato. The G. Modelidu monkey was in the village of Longalo,
and its point of capture was visible from the village. Unfortunately, two visits
to the lower, eroded slopes of G. Mantulangi failed to encounter any monkeys,
but independent informants agreed that monkeys not only occurred there but
even, on occasion, were visible from the harbor.

Fig. 5-11. Adult male *Macaca nigra nigra* (canines extracted), kept by a man in Palu.

Specimens of *nigrescens* were seen in captivity as follows (Fig. 5-12): Tapadaa (0.40 N, 123.09 E); Modelomo, on the coast 3 kilometers southeast of Leato (about 123.06 E); Molingtogupo (0.35 N, 123.07 E). The latter two were held captive in villages near their point of capture. The first was in Duano village, Suwawa district, about one-quarter the way from Gorontalo to Tapadaa. Earlier on the same day that Duano was visited, wild *nigrescens*—easily identifiable through binoculars, with a strikingly disruptive coloration in the bright sunlight—were observed raiding a maize plantation at Tapadaa itself.

As all of these specimens were quite typical of their respective taxa, and the nearest locality records (Leato, cf. Modelomo and Molingtogupo) are only about 3 kilometers apart, it seems unlikely that intergradation is taking place between *nigrescens* and *hecki*. Their placement in different species by Fooden appears to be correct, although, as shown above, these species must be called *M. nigra* and *M. tonkeana* respectively.

In view of the finding that *hecki*, not *nigrescens*, is the species which occurs around Gorontalo, some examination of the basis of the "Gorontalo" record for the latter is in order. Of the collectors of examples of this species which are today labeled Gorontalo, only von Rosenberg (1878) left a record of his travels. He describes some encounters with monkeys, calling them *Cynocephalus nigrescens*, placing his account in the context of his journey from Gorontalo along the S. Bone (via Tapadaa) to Tulabolo. As even in those days Gorontalo was a town of some importance due to its harbor, it seems likely that this

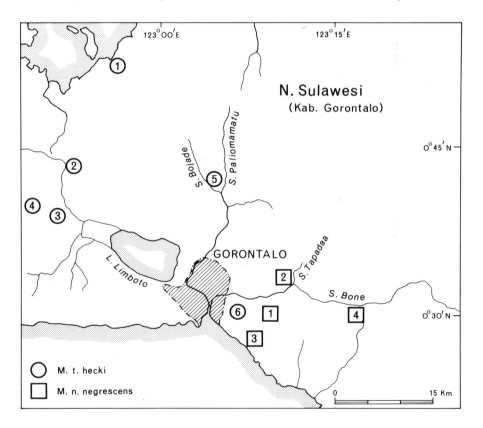

notation refers to the starting-point of his journey, and the point to which he returned and from which his specimens were exported. It is also possible that it refers to the district as a whole, which is also called Gorontalo. Very likely the same explanation applies to other specimens, collected by von Forsten who left no account of his travels, which are similarly labeled "Gorontalo" in museum collections.

If the marine inundation of the lower S. Bone, as postulated by Fooden (1969: p. 63), plus perhaps a similar inundation of the rift valley which cuts across the peninsula from Gorontalo to Kwandang, were thought responsible for the isolation of *nigrescens* from *hecki*, then clearly since the re-emergence of the inundated area *hecki* has advanced slightly at the expense of its neighbor, as, it would seem, *maura* has done in the south.

THE STATUS OF M. OCHREATA AND M. BRUNNESCENS

The differentiation of *Macaca brunnescens* relative to its nearest neighbor (*M. ochreata*) as described by Fooden is much the weakest of all the Sulawesi macaque taxa. (The differentiation of *nigrescens* from *nigra* seems the next weakest.) It therefore comes as no surprise to discover that its homeland, the islands of Buton and Muna (see Fig. 5-1), which are geologically of a piece, emerged above sea level in the Holocene only (Wiryosujono and Hainim, 1976), so that *brunnescens* can have been separate from *ochreata* for a maximum of only 10,000 years. The differences between them amount merely to the brown, rather than black, dorsal color of *brunnescens*, the short mat fur, rather than long and glossy as in *ochreata*, and its longer skull with longer face and rather larger maxillary ridges. For that matter specimens from P. Muna average slightly paler and larger than those from P. Buton (Fooden, 1969 pp. 35, 128) so that they continue the trend away from *ochreata*. In view of these facts it seems most realistic to classify *brunnescens* as a subspecies of *ochreata*.

The question of the status of *M. ochreata* is itself not clear-cut. It borders to the north only on *M. tonkeana*, as far as is known—the Matana Rift would seem to be a likely barrier—but morphologically it is much closer to *M. maura*, from which it differs only by its strange disruptive pattern, its shorter muzzle and its rather more distinct maxillo-malar fossae and maxillary ridges. Albrecht (1976) finds the two to be extremely close in skull features. M. Nicolas Herrenschmidt (pers. comm.) had to remove the adult male *ochreata* "Kiki" (Fig. 5-13) from his free-ranging troop of *M. tonkeana* at Landskron (near Bâle), as it was ill-treated by them. Moreover, "Kiki" suffered from frostbite during the winter, while the *tonkeana* all remained in good health and grew long winter coats. This seems to support the inference that the two are different species.

Provisionally, therefore, *M. ochreata* will continue to be maintained as a

Fig. 5-13. Adult male "Kiki" *Macaca ochreata ochreata,* Landskron castle, kept by M. Nicolas Herrenschmidt, University of Strasbourg.

valid species; some may prefer to class it as a race of *M. maura,* but it seems to be totally separated geographically from *M. maura.* The whole fauna of southeastern Sulawesi is poorly known. It may or may not turn out to be related to that of the southern arm, rather than to that of the center. In the present instance it is possible either that *M. tonkeana* has extended its range south to separate *maura* from its close relative *ochreata* (a theory which however conflicts with the view, supported by the existing situation at Maiwa, of the Tempe Depression as a barrier between *maura* and *tonkeana*), or else that the similarities between *ochreata* and *maura* are due simply to symplesiomorph retention.

EVOLUTION OF CELEBES MACAQUES

The resultant picture of Sulawesi macaque distribution is shown in Fig. 5-14. Fooden (1969, pp. 59-66) proposed a model of evolution whereby an ancestral type, of the *nemestrina* group, arrived by waif dispersal from Borneo during the Pleistocene, and differentiated on a number of insular or quasi-insular sections on Sulawesi into the seven "species." The *M. nemestrina* group was selected as ancestral because of similarity of bacular structure and the short tail. Subsequently, Fooden (1975) cited, in addition, external proportions, "form and proportions of the glans penis . . . and general form of estrous sexual skin swelling," and further specifies (1976) that the Celebes forms, like *nemestrina, silenus,* and *sylvanus,* have a relative breadth of glans above 59 percent.

It does not appear that tail length is in any way significant, having regard to the many macaque stocks which can be demonstrated to have shortened their tails independently (*M. sylvanus, M. arctoides, M. thibetana, M. fuscata*). While the other features cited certainly are alike in the *nemestrina* group and the Celebes macaques, it is questionable whether they might not be symplesiomorph for *Macaca* as a whole. The inclusion (Fooden, 1976, this volume) of the otherwise distinctive (and geographically isolated) species *sylvanus* in the *nemestrina* group, on just these grounds, suggests that the characters involved must be very ancient. (Lineal ancestors of *M. sylvanus* have been in Europe since the earliest Pliocene [Delson, this volume].)

In a recent paper (Groves, 1976) I suggested that the Sulawesi mammal fauna is derived from the Pliocene Siva-Malayan fauna, at a time when a land-bridge existed between Java and Sulawesi. This land-bridge then tectonically submerged, and the Sino-Malayan fauna, from which the modern Sundaland fauna derives, entered Indonesia but was unable to reach the now isolated Sulawesi. An antique origin of the Sulawesi mammalian fauna is indicated because of the high level of endemicity at both species and genus level; because of the indications that Sulawesi itself may have acted as a dispersal center for various genera (*Tarsius, Rattus*); and because many

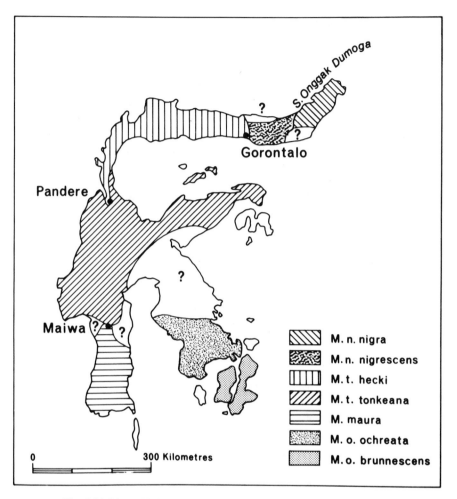

Fig. 5-14. Map of Sulawesi showing revised distributions of *Macaca* taxa.

elements of both Pleistocene and recent faunas have their closest known relatives in the Pliocene of Java (*Elephas, Stegodon*), the Middle Siwaliks (*Anoa, Celebochoerus*), or elsewhere and even earlier (*Babirusa, Tarsius*). In other words, the fauna has to be treated as a whole; unless a species is demonstrably of recent origin (such as *Cervus timorensis*), or must be derived from Borneo because there is nowhere else its congeners live (such as *Haeromys minahassae*), its origins should preferentially be sought with those of its faunistic associates. The suggestion by Simons (1970), that certain long-faced cercopithecines from the Middle and Upper Siwaliks may be involved in the origin of the Sulawesi forms, should be seriously investigated. But the finding of Albrecht (1977) that the short-faced *M. maura* resembles in its skull form the "generalized macaque" more closely than any other Sulawesi taxon, seems an equally promising point of departure.

Within the island, Fooden chooses *M. tonkeana* as the most archaic species, on grounds which seem not altogether convincing: some (callosity shape) are shared with *M. maura* and *M. ochreata*, others (baculum shape) are based on very little material and are in any case rather subjective. As with the skull (Albrecht, 1977), so with external characters, *maura* seems to me a much more convincing candidate for the species, if one must be selected, in which the ancestral characteristics have changed least. It has a plain brownish coloration without well-marked rump-patch or cheek-spots. There are no crests, cheek-tufts or bushy rump-crest. Only the presence of gluteal fields, which it shares with *tonkeana* and *ochreata*, can be definitely pointed out as a specialization.

The peculiar ischial features of *M. nigra*—prominent ischial tuberosities, enlarged callosities, excessively shortened tail—are clear specializations, but occur, in a somewhat less advanced condition, in *M. tonkeana hecki*. This seems to support Fooden's hypothesis of an original unbroken cline which later disruption (by tectonic events?) caused to evolve into a situation of reproductive isolation in some instances, and secondary intergradation in others. It is eloquent testimony to the phyletic unity of all Celebes macaques.

ECOLOGY AND BEHAVIOR: SOME NOTES

In a study such as this, where success depended on seeing as many monkeys as possible, in the wild and in captivity, it was of course not possible to record very much detail about either ecology, social organization, or ethology. A few points were noted and should be mentioned, as there is as yet absolutely no information, beyond the purely anecdotal, of the free-ranging behavior of Celebes macaques. What little descriptive detail there is, quoted verbatim or in direct translation, fills a mere 3¼ pages of Fooden's monograph (pp. 48–51).

1. Habitat

It has already been mentioned that in the Tempe Depression, as well as in the mountains to the south, the forest cover is broken by cultivated areas—both shifting and permanent—as well as by grasslands which seem to be mostly natural, resulting from the climatic regime which includes at least one month with less than 50mm of rain. There is thus more opportunity for *M. maura* to come to the ground than in the case of other species; and every evening at Maiwa, troops would emerge from the bush/forest to sit, grooming, playing and leisurely foraging in the declining sun. What were certainly three different troops were seen in approximately their previous locations on two successive evenings. At Karaenta, too, a troop was encountered in a patch of grassland at 5:30 p.m., whereas at other times of day monkeys were invariably high in the trees. (At Maiwa, *M. tonkeana* would of course come to the forest edge as readily as *M. maura*.)

Elsewhere, monkeys had less opportunity to come to the ground with any such regularity, open grasslands being absent, although the forest floor, readily utilized by *M. nemestrina* in Malaysia and western Indonesia (see sources quoted in Fooden, 1975), is still available. In the G. Tangkoko-Batuangus Reserve, three troops of *M. nigra nigra* were seen in a single morning, all high in the trees in primary forest, where the understory is nearly absent, which covers the reserve. *M. n. nigrescens* were seen on the ground at 12:30 midday and again at 2:50 p.m. raiding a maize field, and at about 1:00 p.m. one of the raiders was encountered sitting on the ground just inside the forest bordering the field. *M. tonkeana hecki* were seen both in the forest and on the road near Kebun Kopi at various times of day, but no time of day in that mountainous region appeared to be markedly more hot or sunny than any other.

Heinrich (quoted *in litt.* by Fooden, p. 51) appears to be correct in his opinion that Celebes monkeys are commonest at moderate altitudes. Rangers in G. Tangkoko-Batuangus Reserve stated that monkeys seemed generally to be at about the altitude (2–300 meters) where we met with them: not usually at sea level nor at the summit (1,000 meters). This is not, however, to be taken as absolute. Near Gorontalo, where the mountains come right down to the sea, monkeys (it was stated) come down the slopes to the vicinity of the houses—including the Forestry Office—along the harborside. Dr. G.G. Musser has seen them (pers. comm.) all the way from about 950 ft (290 meters) to the summit of G. Nokilalaki, at 7,500 feet (2,286 meters). My own impression is that on coastal slopes, at least, monkeys migrate vertically for long distances, not perhaps on a regular basis. In general these impressions accord with those of Heinrich, and with the other conclusions reached by Fooden.

2. Social Organization

In contrast to Heinrich's report, quoted by Fooden, that troops seen by him (apparently, *M. n. nigra*) were of 6 to 10 animals, troops seen in the present survey were always larger than this. I have the following records:

9.12.75, Maiwa: *M. tonkeana tonkeana.* 9 visible in a photo: incomplete.
9.14.75, Maiwa: *M. maura.* 11 visible in a photo; incomplete. (South-ernmost troop).
9.15.75, Maiwa: *M. maura.* Two incomplete counts of 21 and 19; troop seen again (not counted, but approximately same number of animals seen) 9.20.75.

9.21.75, Maiwa: *M. maura.* Estimated, about 40: a photo, taken about 10 seconds after the troop had begun seeking cover, shows 25. (Northernmost *maura* troop).

9.26.76, G.Tangkoko-Batuangus: *M. nigra nigra.* Each of three troops had 30–40 animals.

10.23.76, 9 km east of Kebun Kopi: *M. t. tonkeana.* 10 out of a troop of at least twice that number seen clearly enough for taxonomic assessment.

Troops of *M. n. nigra* were not seen at rest. However troops of *M. maura* and *M. t. tonkeana* seem not to show the clustering tendencies into groups of three or more animals reported by M.H. MacRoberts (pers. comm.) for *M. n. nigra.*

Contrary to the observations of the Sarasins (quoted by Fooden), "gray-haired" monkeys are not invariably the "leaders" of troops. Observations bearing on this are recorded in the section of Maiwa, above.

But "leadership", in the sense of "control role" (Bernstein, 1966) was quite noticeable in several instances. In a troop at Maiwa which was suddenly encountered, to the surprise of both parties, and then followed into the bush, a single large male sat 20–25 meters away from me for an hour or more, being joined from time to time by others until he finally moved off, after which no more monkeys appeared. During this time he mainly sat still, but twice dashed away, then returned, barked, and from time to time made threatening movements, dashing six feet up a small tree and shaking a branch, jumping back down noisily, and lunging. A second Maiwa troop had apparently been disturbed a few moments before contact at 4:40 p.m. by a man with a dog walking past their trees; several members barked and shook branches, but only one remained conspicuously on a tree-fork lunging first towards the man and dog and, when they went on their way, towards me and my wife. On our return to that spot one hour later the same big male was still there and began barking again; other monkeys could be seen briefly nearby, but the male was the only one continuously visible, although he shifted his position from time to time.

3. Behavior

An attempt was made to record vocalizations, but the recording equipment was inadequate to the task. All that can be said is that the alarm bark of *M. n. nigra* seemed definitely more shrill than that of *M. maura* or *M. t. tonkeana.*

Bernstein (1970) describes and figures facial gestures—apparently species-specific variants on the Silent Bared-teeth face of van Hooff (1966), a fear-threat grimace—which differ in *M. maura* and *M. n. nigra.* In the latter, where it is known as the "high grin," the upper lip is retracted exposing the upper teeth alone, the mouth being closed. In the former, the "snarl face" resembles the figure-of-eight grin of the mandrill but the corners of the mouth are not widely open, being only slightly more lifted than the center of the upper lip; both sets of teeth are visible. Bernstein also describes an exaggerated type of lipsmack, without much tongue protrusion, seen in aggressive interactions in *M. n. nigra* only.

These expressions were specially sought in the captive monkeys examined, with the following results:

M. ochreata brunnescens: an adult female at Singapore Zoo made a grin very like that of *M. maura* but with the mouth corners hardly lifted at all—that is, rather similar to the generalized fear grimace of other macaques, except that the upper gums are largely revealed.

M. tonkeana tonkeana: two young males, one at Rantepao, one at Donggala, made an apparent fear-threat grimace with the mouth closed, the upper lip considerably retracted as in *nigra*, but the lower lip slightly retracted also, and with no noticeable lifting of the mouth corners. This grimace, if its occurrence in such young animals is to be trusted, seems rather between those of *nigra* and *maura*. A young male seen in Palu made what looked like the "exaggerated lipsmack".

M. t. hecki: a juvenile male caught at Kebun Kopi made a fear grimace like those of the *tonkeana* of similar age. A subadult female at the Gorontalo offices of P. T. Tropica would make the exaggerated *nigra*-type lipsmack at her younger (male) companion, who would respond by running away.

M. n. nigrescens: the adult female from Tapadaa, owned by M.J. Mantulangi of Duaṅo, near Suwawa, makes both the "high grin" and the exaggerated lipsmack typical of *M. n. nigra*. The former would therefore seem to be a species-specific grimace for the combined species *M. nigra*.

Another monkey in Gorontalo, a six-year-old male *hecki* from Paguyaman, owned by Mr. Noldy Makalew of P.T. Tropica, twitches its stub of a tail up and down as it makes the normal type of lip-smack.

CONSERVATION

M. t. tonkeana has an extremely wide distribution in the least populated and most inaccessible parts of Sulawesi. It does not therefore seem to be in any particular danger at the moment. (The same may hold for *M. ochreata*.)

M. t. hecki and *M. n. nigrescens* live in areas which are at present fairly inaccessible as they are mountainous—extremely precipitous, in fact—and thickly forested. But the rapidly increasing population of indigenous peoples, together with the influx of transmigrants, is steadily eroding the forests. Particularly worrying today is shifting cultivation. This has been practiced since time immemorial in mountainous country in Sulawesi, and is the most reasonable type of agriculture in such a habitat, so long as populations are low. But the ever-larger populations nowadays have meant a shortening of the shifting cycle, with consequent soil erosion, meaning that the forest will take much longer to re-establish itself, or not at all. The problem is much more acute in the Gorontalo region, where hill erosion is sometimes very serious, than in the mountains near Palu, where the human population is still low, and where shifting cultivators can be seen to have employed soil conservation practices such as the crossways laying of felled tree-trunks. The Forestry Department at Gorontalo is well aware of the problem, but unless the whole human population can be enticed down to the lowlands and valleys, there is little enough that can be done.

A further problem is forestry itself. It is probable that selective logging is less detrimental to macaques than to much other wildlife, but the scandal of uncontrolled timber-cutting, which goes on even in wildlife reserves in Indonesia, is bound to have its effect even on macaques.

The habitat of *M. maura* has probably been fragmented for a long time by the inroads of a human population that has always been denser than elsewhere on the island. Although it survives well on the tops and sides of limestone blocks in areas whose valleys are fully cultivated, its status needs watching and cannot be said to be secure.

It is the situation of *M. n. nigra* which probably needs watching most closely. Its habitat within Minahasa is reduced to a few isolated pockets; not only have coconut plantations replaced most of the natural vegetation, but an additional problem exists in that the indigenous population is largely Christian, and lacking the Moslem food taboos, eats anything that moves, including monkeys. For an animal protected by law (the Black or Crested Celebes Macaque; *Cynopithecus niger* is listed as No. 42 on the List of Protected Animals and Birds in Indonesia), this cannot be tolerated, nor indeed can their capture for pets in villages.

In most villages there is at least one monkey, usually young, held as a "pet" by somebody. Usually such an animal is unmercifully teased by children, sometimes neglected or treated roughly by its owners, and in these cases it ends up a pitiful, cowed wreck of an animal, which screeches in fear every time it is approached. Abused and underfed, it does not long survive capture; on many occasions I would be taken to see such an animal only to be told, "sudah mati" (it has now died). They are caught generally be means of baited cage-traps, taking advantage of the crop-raiding habits of macaques; on one occasion, to my horror, an infant was captured in my presence by terrifying a troop until its mother dropped it. The captive animal is invariably tied to a post, underneath the houses (which among such peoples as the Bugis and Kaeli are built on stilts) or in the yard; the rope or chain is tied round the monkey's waist or, on occasion, round its neck. Most monkeys do not live long enough to grow much, but one monkey which had, against all odds, survived to maturity under such conditions, was seen in the village of Watusampung, near Donggala. It had a bicycle chain around its waist, which, apparently, had not been loosened since its capture as a youngster, and now bit deeply into its belly and spine, so that its hindlegs were paralyzed and it remained fixed in a sitting position, feebly grimacing and shuffling away on its bottom when teased, bleeding as the chain gouged further into the deep channel around its body.

Happily such spectacles were rare, and merciful death intervenes in most instances. A few cases—all too few, alas—offered a welcome contrast (see Fig. 5-11). Some monkeys were well cared-for, showed affection for their owners (who prevented children from teasing them), and had reached maturity and even considerable age because of it. One such, in Dumati, near Gorontalo, was stated by neighbors to have been in captivity for 15 years.

Even in these cases, however, the monkey is on its own, with no companionship of its own kind so that its genes are lost to the wild gene-pool. The only pet monkeys I saw which were not condemned to a solitary existence were a pair kept by a mining company in Gorontalo, and a pair housed on the common at Donggala. Whether this is any threat to the continued existence of the local wild population is doubtful, but at least in Sulawesi Utara it is illegal, as *"Cynopithecus niger"* appears on the list of protected species in Indonesia. The laws are not enforced, however, and probably are not even known to most people. It is clear that the future of wildlife conservation in Indonesia must depend on a massive campaign of public information.

It is difficult to enforce the law in remote areas, but it seems doubtful whether most people are even aware that there is such a law.

In general the following points have to be made:

1. The activities of shifting cultivators and lumber companies must be watched closely.

2. Transmigrants must be directed *away* from reserves and conservation areas.

3. Selling monkeys for food, and catching them for pets, must be stopped, as in north Sulawesi, at least, this is illegal. All species of Sulawesi macaques should be added to the list of protected animals in order to regularize the situation over the island.

4. More funds must be channeled into wildlife conservation. It is probable that the Wildlife Conservation section of the Forestry Department in Indonesia is now better organized than at any time in the past, but it is starved of money and lacks political clout. The enthusiastic, highly motivated individuals who often head the provincial sections are untrained, and not

assisted by biologically aware personnel. The existing Nature Reserves (Cagar Alam) and Game Sanctuaries (Suaka Margasatwa) are mostly too small to be effective, although there are quite a number of them, and new ones are declared each year. They have too few rangers, and there is no machinery to prosecute violators, who for example, hunt anoas in many of the reserves with impunity.

5. The activities of timber companies must be curtailed. This has been said over and over again, but it needs to be emphasized. The destructive effects on wildlife apart, huge sums of money accrue to a few individuals at little benefit to the Indonesian people as a whole.

SUMMARY

1. The history of classification of the monkeys of Sulawesi is briefly described.
2. Geological, climatic and floristic factors bearing on their taxonomy and evolution are noted.
3. Results of a field survey made over 2½ months in 1975 are reported:
   a. *Macaca maura* and *M. tonkeana* are marginally sympatric in the Maiwa region, and are valid species.
   b. *M. hecki* intergrades with *M. tonkeana*, and so is only subspecifically distinct.
   c. *M. nigrescens* appears to intergrade with *M. nigra*, so is only subspecifically distinct from it.
   d. *M. nigrescens* and *M. hecki* appear to approach each other's ranges without interbreeding, so are specifically distinct.
4. Brief remarks are made on ecology, behavior, evolution and conservation.

ACKNOWLEDGMENTS

A great many individuals and organizations assisted with this project, and grateful thanks are due to them all, as follows:

Jakarta: Mr. and Mrs. Peter Sorensen; Mr. H. Napitupulu and the staff of the Foreign Scholars' Division, LIPI (Indonesian Institute of Sciences).

Bogor: Mr. H. Prijono Hardjosentono, former Head, and Mr. I. Made Taman, former Head of Tourism, Wildlife Division; Dr. S. Kadarsan, and Dr. Sunatono, Lembaga Biologi Nasional.

Ujung Pandang: Mr. Sutarto Kadillah, Head of Wildlife Conservation; Mr. Yusri Zakaria, Head of Forestry; Mr. C.L. Bundt, Dr. W.J. Meyer, Dr. David Chessell.

Maiwa: Dr. I. Gusti Ktut Oka Ranuhs, Director, Bina Mulya Ternak; Mr. Jack Turnour, former Manager, Mr. Bahron Abas, former Assistant Manager (now manager), Mr. Mokhtar, Foreman, BMT Ranch Maiwa.

Manado: Mr. Ubus Wardyu Maskar, Head, Wildlife Conservation.

Kotamobagu: Mr. Yunus Kobandaha, Head of Forestry; Mr. and Mrs. R. Beltman; Mr. Jerry Stickings.

Imandi: Mr. Jansun Kobandaha, camat; Mr. W. Langkay, Mr. B. Wulur, Forestry Office.

Gorontalo: Mr. Bob Clynch, Administrative Head, Mr. John Dow, Head of Exploration, P.T. Tropica; Mr. Ishak Nggilu; Mr. Noldy Makalew;Mr. John Hatibi, Head of Forestry; Mr. E.J.J. Biver, Mr. Ronald Hermanus, Forestry Department.

Palu: Dr. B.L. Salata, Secretary to the Governor of Sulawesi Tengah Province; Mr. Ibrahim Madina; the staff of the Hotel Manguni; Mrs. Poppy Sapri.

Malakosa: Dr. G.G. Musser, Ms. M. Becker.

Passim: Mrs. P.R. Groves.

The study would not have been possible without a generous grant from the Australian Research Grants Commission, Australian National University, Canberra.

REFERENCES

Albrecht, Gene H. Methodological approaches to morphological variation in Primate populations: The Celebesian Macaques. *Ybk. Phys. Anthrop.,* **20**: 290–308 (1977).

Audley-Charles, M.G., Carter, D.J. and Milsom J.S. Tectonic development of Eastern Indonesia in relation to Gondwanaland dispersal. *Nature, Phys. Sci.,* **239**: 35–39 (1972).

van Bemmelen, R.W. *The Geology of Indonesia.* The Hague: Govt. Printing Office, (1949).

Bernstein, I.S. Analysis of a key role in a capuchin (*Cebus albifrons*) group. *Tulane Stud. Zool.,* **13**: 49–54 (1966).

_____. Some behavioral elements of the Cercopithecoidea. In J.R. Napier, and P.H. Napier, (eds.). *The Old World Monkeys.* New York: Academic Press, pp. 263–295 (1970).

Büttikofer, J. Die Kürzschwanz-Affen von Celebes. *Zool. Meded. Leiden,* **3**: 1–86 (1917).

Carter, T.D., Hill, J.E. and Tate, G.H.H. *Mammals of the Pacific World.* New York: Macmillan, (1946).

Fooden, J. Taxonomy and evolution of the Monkeys of Celebes (Primates: Cercopithecidae). *Bibl. Primat.,* No. 10 (1969).

_____. Taxonomy and evolution of Liontail and Pigtail Macaques (Primates: Cercopithecidae). *Fieldiana, Zool.,* **67**: 1–169 (1969).

_____. Provisional classification and key to living species of Macaques (Primates: *Macaca*). *Folia primatol.,* **25**: 225–236 (1976).

Forbes, H.O. *Monkeys.* Lloyd's Natural History, 2 vols (1894).

Groves, C.P. The origin of the Mammalian fauna of Sulawesi. *Z.f. Säugetierkunde,* **41**: 201–216 (1976).

van Hooff, J.A.R.A.M. The facial displays of the Catarrhine monkeys and apes. In D. Morris,(ed.), *Primate Ethology.* Chicago: Aldine (1967).

Katili, J.A. Geological environment of the Indonesian mineral deposits: a plate tectonic approach. *Publ. Teknik Seri Geol. Ekon.,* Direkt. Geol., No. 7, (1974).

Khajuria, H. Taxonomic studies of the Celebes ashy-black monkey—a remarkable case of convergence. *Rec. Indian Mus.,* **52:** 101–127 (1953).

Laurie, E.M.O. and J.E. Hill. *List of Land Mammals of New Guinea, Celebes and Adjacent Islands, 1758–1952.* London: Trustees, British Museum of Natural History, (1954).

Lembaga Meteorologi Dan Geofisika, Indonesia. *Mean Rainfall on the Islands Outside Java and Madura, Period 1931–1960* (1969).

_____. *Mean Rainfall and Mean Number of Rainy Days, 1961–1970* (1974).

Napier, J.R. and P.H. Napier. *Handbook of Living Primates.* New York: Academic Press (1967).

Pocock, R.I. The external characters of catarrhine monkeys and apes. *P.Z.S.Lond.,* 1925: 1479–1579 (1926).

Prawira, R. Soewanda Among, I.G.M. Tantra, Wasiat, Oetja and Momo. *Daftar nama pohon-pohonan: Sulawesi Selatan, Tenggara dan sekitarnja.* (Revision). Lembaga Penelitian Hutan, Bogor, No. 151, (1972).

_____. *Daftar nama pohon-pohonan: Menado (Sulawesi Utara).* (Revision). Lembaga Penelitian Hutan, Bogor, No. 156, (1972).

von Rosenberg, H. *Der Malayische Archipel.* Weigel, Leipzig, (1878).

Sody, H.J.V. Notes on some primates, carnivora and the babirusa from the Indo-Malayan and Indo-Australian regions (with descriptions of 10 new species and subspecies). *Treubia,* **20:** 121–190 (1949).

Stresemann, E. Die Vogel von Celebes. pt.I. *J. Ornith.,* **87:** 299–425 (1939).

_____. ibid., pt. II. *loc. cit.,* **88:** 1–135 (1940).

Thorington, R.W. and C.P. Groves An annotated classification of the Cercopithecoidea. In J.R. Napier, and P.H. Napier, *Old World Monkeys,* New York: Academic Press, pp. 629–647, (1970).

Wiryosujono, S. and Hainim, Jusril A. Cainozoic sedimentation in Buton Island: a plate tectonic interpretation. *Proc. 2nd Regional Conf. on Geol. & Mineral Res. of S.E. Asia,* Aug. 4–7, 1975 (1976).

Zakaria, Yusri; Madjo, Muh. Idris; Sila, Mappatoba; Kadillah, Sutarto; Langtang,Simon and Baga, B. *Inventarisasi dan Analisa Vegetasi Hutan Alam Karaenta Maros.* Dinas Kehutanan, Daerah Tingkat I Sulawesi Selatin (1975).

Chapter 6
Mixed Taxa Introductions, Hybrids and Macaque Systematics

Irwin S. Bernstein
and Thomas P. Gordon

Early taxonomists classified animals solely on the degree of anatomical similarity. Darwinian theory added a new tool and classifications began to consider presumed phylogenetic relationships in determining at what level taxa should be separated. Since evolution is a dynamic process, degrees of genetic differentiation during the process of speciation will fall along a continuum, and decisions will be required concerning the extent of separation which will be considered equivalent to each of the discrete categories used in taxonomy. Definitions begin at the species level, with other taxonomic levels defined in terms of populations related to each other at the species or other subsequent level of classification. A discrete definition of species allowing one to discriminate the point at which speciation processes may be said to result in two distinct species would, therefore, be highly desirable. Such a definition should rely not only on the degree of morphological similarity but also on the degree of reproductive isolation of the two populations.

Attempts to provide such a definition have emphasized a barrier to gene flow between species, and the relatively free flow of genetic material within species populations. Systematists have, however, differed with regard to the required degree of effectiveness of barriers to gene flow in order to classify two populations as distinct species. "All or none" positions seem to treat evolution as a stepwise discrete process allowing little ambiguity of definition. Sterile hybrids are accepted as evidence of recent phylogenetic separation indicating cogeneric status. Any degree of hybrid fertility would preclude division of populations at the species level, and the degree of effectiveness of any partial barrier would be used to identify taxonomic units below the species level. Alternate views, however, emphasize the speciation process as a continuum and argue that selective pressures far less than perfect would nonetheless effectively isolate populations. The existence of a barrier to gene flow, of whatever level of effectiveness, between populations living under natural conditions, is thus of significance. Attention is focused on mechanisms which may or do result in genetic separation, rather than on mechanisms which permit the flow of genetic material throughout a population.

The existence and production of hybrids, and their fertility, is thus of crucial importance to those who require an absolute barrier between species. This data is also significant to systematists who are concerned with the evolution of barriers to gene flow in the speciation process. If two species represent two populations differentially adapted for survival, then the production of hybrids will decrease the genetic fitness of individuals investing reproductive effort in them, inasmuch as these hybrids will not be able to compete successfully against either parental population. There will thus be selective pressure exerted favoring any mechanism which prevents a decrease in individual genetic fitness through participation in hybrid production. Once such a mechanism is established and proves to be 100 percent efficient, then, of course, there is no longer any selective pressure operating to produce additional barriers to gene flow.

The large number of reported primate hybrids (Chiarelli, 1973; Gray, 1971) suggests that for many primate species there is no absolute barrier to gene flow, and/or that we have incorrectly classified populations at levels of separation far greater than warranted by the degree of speciation which has actually occurred. If most of the hybrids reported occurred in the natural habitat, we should be inclined towards the latter interpretation, but inasmuch as almost all of the hybrids were produced under conditions of artificial housing, it is possible that these conditions mitigated against the operation of mechanisms serving to prevent gene flow between wild populations. In particular we might suspect that geographic isolation may account for some of the separation of gene pools in the wild and that certain types of anatomical, physiological and/or behavioral barriers might operate to separate sympatric wild populations. Separation mechanisms dependent upon geography are inoperative under the usual conditions of captive confinement, and many behavioral mechanisms may be overcome under artificial conditions.

Members of the genus *Macaca* are ordinarily quite hardy under the conditions of captivity and are well represented in many collections. Fooden (1971, 1975, 1976, this volume) has reported on the distribution and systematics of the diverse species contained in the genus, and notes several sympatric and allopatric species. The existence of a wide variety of macaque hybrids produced in the laboratory, and their fertility (Bernstein, 1974) suggests that we examine first the possible ways in which naturally occurring hybrids may be produced, and then study behavioral mechanisms which might bar such production between sympatric wild populations.

There are four possible ways in which naturally occurring hybrids may be produced. First, a single male of one species may join with a single female of another species and produce a hybrid. Although males of the genus *Macaca* are often reported to be solitary, females seldom are, and this event is therefore considered highly unlikely in natural populations. It is nonetheless exactly how hybrids are usually produced under captive conditions, i.e. a single male is caged with a single female.

The second possible way is for an aggregation of isolated animals of several species to congregate and breed. For reasons cited above this is even less likely to occur in the natural habitat, and has occurred only rarely in captivity.

A third possibility is that a single animal of one species may join a group of another species. Although not typically found in captivity, such groups are not uncommon in the wild, and there are numerous reports of a single individual of one species, usually male, associating with a group of a second species. Such associations may have some temporal stability, but the isolated male usually leaves, presumably to join a conspecific group.

The fourth possibility is that two or more intact groups of different species will form a polyspecific association in which some extraspecific mating may occur. Whereas polyspecific associations are commonly reported for New World cebids (Bernstein *et al.*, 1976; Izawa, 1975; Klein and Klein, 1973; Thorington, 1968) African cercopithecines and colobines, (Gartlan and Struhsaker, 1972; Gautier and Gautier-Hion, 1969; Hayashi, 1975; Struhsaker, 1975) and suggested for some Asian colobines (Bernstein, 1967), different species of macaque rarely form polyspecific associations in the natural habitat; sympatric macaque species do not freely intermingle (Eudey, this volume; Crockett and Wilson, this volume).

The balance of this paper will be devoted to an exploration of behavioral mechanisms which may serve to isolate macaque species, many of which have been demonstrated to be anatomically and physiologically capable of producing hybrids under captive conditions. The first possible route to hybridization, pairing of a single male with a single female, was used to determine which species would produce hybrids in the laboratory, and, in addition to gang caging, was used also to determine the fertility of the hybrids produced, to see if a delayed physiological mechanism served to prevent gene flow. Table 6-1 summarizes the list of hybrids produced and indicates their fertility (see Figs. 6-1 through 6-5). It should be noted that a sizable number of males and females of several species proved incompatible in our paired caging attempts, and the population from which our hybrids were obtained was therefore highly

Table 6-1. Macaque Hybrids Produced and Hybrid Fertility Demonstrations.

| MOTHER | | FATHER | | Sex | Birthdate | Survival | HYBRID | |
Species	Code	Species	Code				Reproductions	Survival
M. nemestrina	Oa	M. fascicularis	B	F	25 Aug. 64	Alive	18 Sept. 70, paternal backcross	Stillbirth
							23 Apr. 71 " "	Stillbirth
							Two confirmed with hybrid female	
" "	G	" "	B	M	11 Nov. 64	Alive		
" "	G	M. niger	T	F	15 Nov. 67	Stillbirth		
" "	G	" "	Y	M	26 May 68	Stillbirth		
" "	G	M. fascicularis	B	F	12 May 69	Alive		
" "	G	M. assamensis	A	?	26 Jul. 70	Stillbirth		
" "	G	M. fascicularis	B	F	22 May 71	10 weeks		
" "	G	" "	A	M	16 Aug. 72	1 year		
" "	P	M. niger	M	F	4 Dec. 68	Stillbirth		
" "	P	M. silenus	A	F	19 Sept. 69	Alive	M 11 May 74, maternal backcross	Alive
" "	P	M. fascicularis	B	?	16 Sept. 70	Stillbirth		
" "	Ja	" "	La	M	15 May 72	Alive		
" "	M	M. mulatta	700	F	11 Apr. 67	Alive	M 24 May 73, with hybrid	Alive
							M 24 Jul. 74, with hybrid	6 weeks
							M 4 Jun. 75, paternal backcross	Alive
" "	M	M. niger	P	?	21 Jun. 69	Stillbirth		
" "	M	" "	P	F	1 Feb. 70	Alive	M 30 Dec. 74	Stillbirth
							F 6 Oct. 75	Alive
M. mulatta	Fa	M. fascicularis	A	M	5 Apr. 72	Alive		
M. niger	L	M. nemestrina	Ob	F	17 Sept. 68	Alive	F 4 Jul. 73	10 weeks
							5 Dec. 74	Stillbirth
							23 Jul. 75	Stillbirth

Species	Code	Species		Sex	Date	Outcome	Second birth	
"	Na	"	?	F	20 Jun. 70	Alive	F 14 Mar. 76	Alive
"	Na	"	?	?	2 Oct. 71	Aborted	F 20 Jun. 75	Alive
"	Na	"	?	?	25 Aug. 73	Stillbirth		
"	La	M. silenus	E	F	6 Mar. 73	Alive	F 30 Mar. 76	Alive
"	I	M. tonkeanna	H	M	30 Jan. 70	Alive		
"	I	M. nemestrina	?	F	8 Jul. 71	Alive		
"	I	M. silenus	E	M	30 Nov. 73	Alive		
M. niger	?	M. nemestrina	?	?	4 Oct. 71	Stillbirth		
M. maura	C	M. fascicularis	V	F	13 Jul. 65	6 months		
M. hecki	La	M. nemestrina	?	?	6 Nov. 70	Aborted		
M. tonkeanna	J	M. maura	A	M	22 Feb. 73	2 years		
"	J	M. nemestrina	Ob	M	26 Jul. 73	1 day		
"	J	M. maura	A	F	10 Jul. 75	1 month		
"	E	M. maura	A	M	27 Apr. 65	Alive		
"	E	M. niger	Q	?	9 Jan. 68	Stillbirth		

Note: A "?" under Code means that multiple individuals of the same species were putative fathers, or, in the one case, an abandoned infant was found where several putative mothers, all of the same species were present. A "?" under Sex means the infant sex was not determined as the body was not retrieved.

129

Fig. 6-1. Adult female *Macaca nemestrina* x *M. nigra* and adult female *M. mulatta* x *M. nemestrina*, each holding their most recent infant.

Fig. 6-2. An adult female hybrid *Macaca nemestrina x M. nigra* grooms an adolescent male second generation hybrid. (Father *M. fascicularis* x *M. nemestrina.* Mother *M. mulatta* x *M. nemestrina.*)

Fig. 6-3. An adult female *Macaca silenus* x *M. nemestrina* with a second generation juvenile male hybrid (father *M. mulatta,* mother M. *mulatta* x *M. nemestrina*).

selected for toleration of extraspecific cage mates. Further, not every compatible pair produced an infant, and no listing of negative evidence is included in the table inasmuch as our failures cannot be taken as definitive.

Individual females used in producing hybrids have produced as many as seven hybrid infants, including some fathered by the same male and some by males of other species. There is therefore no suggestion of decreased fertility following a cross-species pregnancy. Further, both male and female hybrids themselves have proven to be fertile and some have produced as many as three second-generation hybrids, including infants resulting from backcrosses to both parental species. The rate of production and the viability of conceptions seems nearly identical in the hybrids and in their parents, as judged from the proportion of stillbirths and early neonatal deaths, as summarized in Table 6-2. The fertility of male hybrids was demonstrated using paternity tests conducted by the late Dr. Susan Duvall, using techniques also reported in Duvall *et al.* (1976).

Paired animal tests revealed a range of responses in males introduced to

Fig. 6-4. Second generation hybrid adolescent male (father, *Macaca fascicularis* x *M nemestrina*, mother *M. mulatta* x *M. nemestrina*) mounts adult female *M. nemestrina* x *M. nigra*.

conspecific and extraspecific female partners. One behavioral mechanism which would reduce gene flow between species was suggested by the observation that some males vigorously attacked any extraspecific female immediately upon introduction, whereas when introduced to conspecific females the same males showed little aggression, and what was expressed was restrained, such that they did not use their canine teeth. After repeated attempts to cage these particular males with extraspecific females, they were dropped from the study, but were later observed to live amiably with conspecific females. It is suggested that the differential responses of these males to conspecific and extraspecific females may reflect recognition of the differential stimulus properties of the extraspecific partners.

The second possible way to produce hybrids was tested by assembling multiple individuals of several taxa in a single group. Although an unlikely natural event, this procedure has been used repeatedly and systematically in various laboratory studies of cross-species interactions (Stynes *et al.*, 1968,

Fig. 6-5. An adult female *Macaca nemestrina* x *M. nigra* grooms her most recent offspring. Two other second generation hybrid juveniles sit nearby.

1975; Warden and Galt, 1943). These studies have indicated that play and grooming are primarily restricted to conspecifics, and that dominance and aggression may be particularly pronounced towards extraspecifics. Stynes *et al.* (1968) indicate that initial frequencies of aggressive responses may decline rapidly, but remain higher than levels directed towards conspecifics. Stynes *et al.* (1975), however, found that some sexual behavior did occur across species lines in the two macaques they used (*Macaca nemestrina* and *M. radiata*).

A series of experiments, (hereafter referred to as Series I) in which representatives of multiple taxa were caged together, was conducted to explore the nature of extraspecific interactions on both the immediate and long term levels. The number of subjects selected for each experiment and the size of the enclosures used were identical to that used in forming conspecific groups (Bernstein 1964, 1969, 1971a; Bernstein and Mason, 1963; Bernstein *et al.*, 1974a). Four mixed taxa groups were formed over a period of 12 years. The first group (M-1) included New World as well as Old World representatives. (A

Table 6-2. Comparative Success in Producing F1 and F2 Macaque Hybrids.

	MALES	FEMALES	STILLBIRTHS	TOTAL	SURVIVING	# MOTHERS	MULTIPLE PREGNANCIES
F1 Hybrids	9	12	11	32	15	14	Twice (1) Three (4) Four (1) Seven (1)
F2 Hybrids	4	5	5	14	7	7	Twice (2) Three (1) Four (1)

Table 6-3. Animals Used in Mixed Taxa Group Compositions (Never All Present at Once).

SPECIES	SEX	GROUP M1	GROUP M2	GROUP M3	GROUP M4
Macaca mulatta	M	1		3	
	F		1	1	1
M. assamensis	M	1		1	1
M. nemestrina	M	2	1	4	
	F	3	4		
M. niger	M	1	1	2	
	F	2	1	2	
M. maura	M			1	
M. tonkeanna	M			1	
	F			1	
M. arctoides	M	1	2		
	F	1	2		
M. radiata	M		1		
M. silenus	M		2	1	2
M. fascicularis	M	1	2	2	2
	F			2	
Theropithecus gelada	F				2
Ceropithecus aethiops	M			1	
C. sabaeus	M			4	
	F			3	
C. mitis	F			2	
C. cephus	F			1	
Cercocebus atys	F			1	
Lagothrix lagotricha	F	1			
Ateles paniscus	F	1			
A. fusciceps	M	1			
	F	1			
A. geoffroyii	M	1			
	F	1			
Cebus nigrivittatus	M	1			
C. albifrons	F	2			

complete list of subjects is provided in Table 6-3.) Due to medical problems resulting largely from injuries, and a change in facilities, this group was separated and reconstituted repeatedly over a period of four years. No more than two males or females of the same species were present in the group simultaneously, and the same data collection procedures as had been used with conspecific group formations were used during each introduction and reintroduction. Naturally, the repertoire of responses scored was different so as to include the species typical responses of the diverse taxa (Bernstein, 1970). Group formation used single animal additions at intervals, rather than simultaneous release of all subjects into the group. This technique was later shown in several experiments to produce the highest levels of agonistic expression in conspecific group formations (Bernstein, 1964, 1969, 1971a; Bernstein *et al.*, 1974b). Followup data were also collected during periods between introductions, and at intervals following the last reintroductions, in order to see if response levels in the group stabilized.

A second mixed taxa group (M-3) was formed with adult cercopithecines. In this case all of the males were placed under light anesthesia and released simultaneously. All of the females were added simultaneously one week later. Some subjects had served in the previous study, and many had been used in paired animal production of hybrids. Many of the animals were therefore already familiar with one another at formation. The group remained together for less than a year when the study was terminated after reviewing the accumulated record of injuries, associated illnesses and fatalities (none of which occurred during the introductory phases).

Two other groups (M-2 and M-4) were formed by placing hybrid infants together along with purebred infants of the species represented in our hybridization attempts. All subjects were weaned after six months of age, at which time they were introduced to the group. As a consequence, the groups were formed by single and occasional multiple introductions. Although animals have been removed when no hybrids of their species were produced, and others have been temporarily removed for treatment of injuries, or for paired cage breeding experiments (such as backcrossing the hybrids after maturity), the first group (M-2) has remained together for more than 10 years. The second group (M-4) was formed only after some of the adult members of M-2 began to inflict serious injuries upon newly introduced animals. This group differed from M-2 in that it included some very aged individuals as part of a nucleus to establish the group, and two *Theropithecus gelada* infants, although no gelada hybrids have been attempted. These two groups were recently merged after removal of the aged purebred animals.

The third way in which extraspecific individuals might meet to produce hybrids involves single males joining an extraspecific group. Inasmuch as this appears to be a possibility in wild populations, we undertook a series of experiments (Series II) wherein we not only introduced extraspecific adult males to established breeding groups, but we also tried single females and some non-macaque subjects in an effort to reveal any possible behavioral isolating mechanisms which preclude an adult male macaque joining an extraspecific troop and producing hybrid offspring. Inasmuch as adult macaques are sexually dimorphic and variable in size, we tried to control for the relative size of the introduced animal and host species by selecting various sized species to be introduced to the host group; and also by counterbalancing introductions so that a rhesus group served as host to a series of five pigtail males, and a pigtail group served as host to a series of five rhesus males. No

Table 6-4.   Introductions of Extraspecifics to Established Groups.

	INTRUDER		GROUP
Sex	*Species*	*Replications*	*Species*
M	Macaca nemestrina	5	Macaca mulatta
M	M. mulatta	1 (control)	” ”
M	” ”	5	M. nemestrina
M	M. nemestrina	1 (control)	M. mulatta
M	” ”	5	M. fascicularis
M	M. mulatta	1	” ”
F	M. radiata	1	” ”
M	Cercopithecus aethiops	1	” ”
F	” mitis	1	” ”
M-F	” ” and Hybrid	pair	Cercopithecus aethiops
M-F	Cebus albifrons	pair	” ”
M	Macaca niger	trio	Macaca nemestrina

male served as both host and introducee in this series, and two different rhesus groups were used as hosts for one of the pigtail males to verify the generality of the responses of the first group. An additional control was incorporated into the design by including a conspecific male in the series of extraspecific male introductions, to provide direct comparison with data on conspecific introductions available as a result of the studies previously cited which used the same design and enclosures. Each introduction was terminated after one hour, or whenever a subject was threatened with serious injury or unable to escape a concerted group attack. (Only the conspecific introductions and one extraspecific male introduction were terminated before the one-hour time limit.) A complete list of introductions is provided in Table 6-4.

Finally, possible mechanisms precluding the production of hybrids through the mingling of two or more intact troops of different species were explored by merging intact breeding groups (Series III). This is considered a possible natural event in that polyspecific associations of numerous nonmacaque primates have been reported, and many macaque species are sympatric over much of their ranges. The fact that sympatric macaques usually maintain separate existences suggests that either they are ecologically isolated (considered unlikely in omnivorous macaques) or that a behavioral mechanism precluding the formation of polyspecific macaque troops may reduce the possibility of naturally occurring hybrids in those species demonstrated to be physiologically capable of hybridizing in paired cage testing. As Moynihan (1973) puts it in his discussion of the role of evolution in behavior: "A great many questions have been left hanging in air. What do individuals of the same or different species really *do* when they come face-to-face with one another?"

In order to answer this question and to explore a possible macaque behavioral isolating mechanism, Series III consisted of the following mergers:

A. Long-term breeding groups of 40 pigtail monkeys (*M. nemestrina*) and 51 crab-eating monkeys (*M. fascicularis*), known to be sympatric in the wild (Bernstein, 1967; Crockett and Wilson, this volume) were allowed access to each other's normal living areas through an open doorway for a period of 30 days;

B. A vervet breeding unit of 11 (*Cercopithecus aethiops*) was introduced to a breeding group of 13 bonnet monkeys (*Macaca radiata*); and

C. The same vervet group was introduced to a breeding group of 40 sooty mangabeys (*Cercocebus atys*).

Experiment B was terminated after two days when serious fighting began. Experiment C followed two months after the termination of B, with a much larger host group of even greater body size, in order to separate body size and troop size from the bonnet macaque's generic status. The vervets lived with the mangabeys for more than one year before the study was terminated.

The results of the first series of experiments are summarized in Table 6-5. Comparisons with data collected during comparable conspecific group formations (Bernstein, 1964, 1969; Bernstein *et al.*, 1974a, b) reveals lower initial levels of agonistic behavior in the mixed taxa formation (M-1) with somewhat higher levels for other social interactions. In both conspecific and extraspecific formations there was a rapid decline in the initial rates of agonistic interactions, but in the extraspecific groups the rates for agonistic interactions remained absolutely higher than in established conspecific groups. In one extraspecific group (M-1) agonistic rates in followup data are almost as low as those seen in established conspecific groups (taken from Bernstein, 1971b) but in this group, containing many new world representatives, social interactions of any sort were infrequent after the initial formation period. Agonistic responses therefore remain relatively high in mixed taxa assemblages with little evidence for a reduction in agonistic behavior coupled with an increase in grooming and play, as is seen in successful conspecific groups. A rich variety of responses were recorded across taxonomic lines but the most common responses were classified as social exploration or species typical displays, rather than positive social interaction; extraspecifics sniffed one another more than they groomed or played with one another. The wound data for the M-3 group revealed a continuing record of serious wounds and injuries from formation in June of 1972 until the group was disbanded in January of 1973. Serious aggressive encounters persisted without any sign of being replaced by less damaging expressions of aggression as a function of time, and the experiment was terminated when deaths due to fighting and the indirect consequences of aggression (exposure and related illnesses) continued despite medical attention.

Table 6-5.   Social Interaction Rates in Mixed Taxa Groups.

	AGONISTIC	SEX	GROOMING	PLAY	OTHER SOCIAL
At Formation					
M I	106	11	9	3	81
Mean Conspecific					
Macaques	193	11	4	0	39
Baseline					
M I	10	2	7	1	10
M III	80	16	40	1	15
Mean Conspecific					
Macaques	7	4	22	5	7

In the hybrid groups (M-2, M-4) in contrast, there was an initial low frequency of agonistic interactions (perhaps age related) followed by a decline to a lower stable level, albeit somewhat higher than the very low levels seen in stable conspecific macaque groups. The hybrid groups have, nonetheless, remained together as coherent social units and, of all the mixed taxa assemblages, only the long-term hybrid group (M-2) has engaged in successful reproduction of macaques (21 born between 1973 and 1977, to eight female hybrids). The birth of infants into this group may also be contributing to the establishment of social mechanisms seen in conspecific macaque groups. For example, the recent merger of M-4 with M-2 revealed strong cohesive group responses and the mothers of infants in M-2 were prominent in concerted actions against the "intruder" group. This merger followed much the same course as we had seen in the mergers of conspecific groups with the defeated group members being intergrated into the social organization of the victorious group.

Perhaps the most instructive information obtained from these introductions is provided by qualitative data. Especially where different taxa use distinctively different communication signals, we wondered whether communication breakdowns would preclude the establishment of any social order. We therefore addressed ourselves to the following questions: (1) Would differences in the communication repertoires preclude effective cross species communication? (2) If extraspecifics can communicate, is it because they learn appropriate responses to the signals of extraspecifics? (3) Do individuals incorporate the signals of another species into their own repertoire?

Not all of these questions can be answered unequivocally. We did see apparent communication failures in early periods, especially those involving the most divergent taxa. An invitation to play by a pigtail male was responded to with extreme agonistic responses by a capuchin. Aggressive displays by guenons were attended to by macaques who failed to respond in kind. Despite such early communication failure, in time all group members began to respond more appropriately to the signals of extraspecific animals. The pigtail and capuchin did engage in frequent play later on, and the macaques did attack threatening vervets. It is at least an impression that those signals with immediate urgency, such as attacks, were recognized most rapidly, whereas appropriate responses to play invitations and friendly social interactions developed more slowly. The ability of monkeys to learn the signal value of extraspecific communication signals should not be surprising. After all, any animal that can learn that an arbitrary visual sign, such as a triangle, signifies food, whereas a similar sign, such as a pentagon, signals no food, should be able to recognize the significance of signal patterns already selected for high stimulus contrast in the communication repertoires of other species. Moreover, the anatomical differences of species notwithstanding, may species use basically similar signals, and the animals have proven remarkably tolerant to signal variability, e.g. they respond appropriately to even the most grotesque distortions of their own signals as produced by familiar humans.

Despite the apparent ability to learn the significance of extraspecific signals, no animal was observed to incorporate the signals of other taxa into its own communication repertoire. The results of experiments in series II are summarized in Table 6-6. It can be seen that whereas the general reduction of interactions with conspecifics and extraspecifics is similar, the rates of agonistic interaction within the hybrid groups are usually lower during initial group formation. The real differences between extraspecific and conspecific introductions with regard to initial agonistic responses lies, not in the frequency, but in the quality of the responses. Conspecific males are mobbed by the group and the conspecifics used as controls in this series had to be removed for their own safety after only a few minutes. Extraspecifics, on the other hand, did not face concerted group attacks, but were approached, sniffed, explored, and subjected to threats by a few group members. During the first few minutes, when conspecific males were engaged in the most vigorous fighting, extraspecific males often walked about freely, followed by the residents and challenged by only a few individual animals. With only one exception, all of the extraspecific males were removed from their host groups after one hour without having suffered or inflicted a single injury. The exception involved a pigtail male who was engaged in a fight by a single rhesus male supported by two females. The balance of the rhesus group did not become involved, even when the pigtail male was defeated and totally submissive.

We have no way to explain why this particular pigtail male, but no other in the series, was strenuously attacked, but it might be noted that the resident animals did not inhibit their attacks even when the extraspecific animal

Table 6-6. Social Interaction Rates for Extraspecific and Hybrid Introduction.

	AGONISTIC	SEX	GROOMING	PLAY	OTHER SOCIAL
On Introduction					
Extraspecific Macaque (N = 19)	201	2	+	0	40
Hybrid (N = 18)	111	14	14	1	116
Rhesus conspecific	200	3	0	0	12
Pigtail conspecific	187	19	9	0	66
Celebes conspecific	144	0	2	0	27
End of 1st Hour					
Extraspecific Macaque (N = 15)	44	2	0	0	12
Hybrid (N = 18)	80	12	11	1	73
Rhesus	57	0	0	0	3
Followup					
Hybrid Day 2-10	15	4	17	4	31
Hybrids, young	18	4	19	12	39
Hybrids, mature	10	3	26	5	32
Macaques	7	4	22	5	7

Note: A + sign indicates an hourly rate of less than 0.5 but greater than 0.

displayed the most extreme forms of passive submission. This had also been noted in the mixed taxa assemblages. The usual shift from contact aggression to noncontact aggression during the first hour of conspecific introductions was likewise absent. This suggests that whereas extraspecific intruders do not have the stimulus properties which ordinarily result in initial group mobbing of conspecific individuals, it is also true that extraspecific subjects are less effective in attenuating attacks, either by submissive signals or by virtue of relative familiarity to the group.

Thus, whereas conspecific intruders may be subjected to immediate group mobbing and the most severe forms of aggressive attack upon introduction, there is ordinarily a marked rapid attenuation of this group attack, perhaps due to the effectiveness of submissive signals. Further attacks on conspecifics seem directed towards establishing relative social relationships as in a dominance hierarchy, and often consist of token attacks, although these may be repeated in reinforcing the new relationships. Conspecifics may also benefit from the behavior of group members who may prevent the expression of aggressive responses by other group members. After the first concerted group effort to eject the intruder, some individuals may show positive responses to a conspecific intruder and directly defend him or redirect aggression directed at the submissively signaling intruder.

Extraspecifics, on the other hand, are neither subject to immediate mobbing, nor are they aided by group members. They are not effective in their use of submissive signals to reduce the intensity of whatever attacks are initiated against them. Extraspecifics may not be subjected to aggression which establishes relative social position, but this lack of social position may result in more severe attacks whenever a conflict does occur. The relative absence of later positive social interactions with extraspecific intruders, and between extraspecifics in mixed taxa assemblages, reinforces the speculation that the extraspecifics do not establish a social position within the first hour, whereas conspecifics generally do so, even if it is the lowest position in the group.

To be sure, species variability among the macaques is apparent from the comparative data presented. Some, such as the pigtails, may be more deliberate, show more exploratory responses and prove slower to attack than others, such as the rhesus. Pigtail host animals also showed more restraint than rhesus in threatening extraspecifics during introductions, but the general pattern and results were the same. The form of the curve for initial agonistic responses and the subsequent decline of aggressive responses, is the same for the two species, although significantly displaced relative to one another.

In Series III we introduced intact social units to one another. In conspecific mergers we had found that our living areas would not accommodate two independent social units, and that one unit defeated the other and then integrated the members of the defeated group into their social structure (Bernstein, 1969; Bernstein et al., 1974a). In the case of extraspecific group mergers we asked whether the differential stimulus properties would preclude group conflicts and mergers, and whether the two social units could therefore maintain separate social organizations in the same space which had proven inadequate for conspecific groups.

We began the series with the merger of a pigtail monkey group with a crabeater monkey group. No cohesive group-to-group aggression was seen in the initial period with members of both groups moving freely throughout both living areas. Some sniffing and social exploration were observed, and individual fights resulted in the formation of small coalitions, rather than organized group-to-group conflict, as had been seen in conspecific group mergers. Neither group was defeated, but over the 30-day period of the experiment, the crabeaters gradually showed more submissive behavior to the pigtails and reduced the space occupied by the group at any one time to about one-eighth of the available area. The pigtails, in contrast, continued to scatter widely through both living areas. The spatial cohesiveness of the crabeaters and the increase in submissive signaling did not mean that their group was subordinate to, or defeated by the pigtails, because active cross-species conflicts persisted and the crabeaters showed their full repertoire of aggressive responses in these fights. Fighting continued sporadically throughout the study and the wound data attests to serious contact aggression episodes in every week, with no clear peak, or indication of tapering off. Agonistic rates at the end of four weeks exceeded those seen in conspecific group formations after a comparable period.

We concluded that indefinite prolongation of the experiment would have resulted in continuing an unacceptably high rate of injuries which would eventually result in mortalities. It should be noted, however, that despite the discrepancy in size between these two macaques, one of the largest and one of the smallest in the genus, no animal died of wounds inflicted during the 30 days of the experiment. Some wounds, however, were very serious indeed, and the individuals may not have recovered without prompt and expert veterinary care. It should also be noted that many of the wounds requiring treatment resulted from conspecific fighting, which was perhaps increased due to the general stress produced by the experiment, and the repeated captures of the groups to remove injured animals for treatment. The agonistic and social interaction rates are summarized in Table 6-7.

The second experiment paired the vervets with the bonnets, and once again there was no immediate group-to-group conflict. Some aggression gradually developed during the first hour and both groups began to act cohesively against one another. These were both small groups and both acted as units. Aggressive interactions increased in intensity with time as the bonnets became increasingly more belligerent. The vervets responded to aggression with aggression, but initiated few episodes. Although the vervet group was not defeated as a unit and continued to fight back, they did attempt to avoid bonnet approaches and the experiment was terminated on the second day when a vervet suffered a significant injury.

When the vervets were introduced to the mangabeys, there was considerable sniffing and exploration, but by the end of the first hour there had been few agonistic encounters. Vervet charges were avoided by the mangabeys, despite the much larger size of the mangabeys, and almost all agonistic behavior involved noncontact aggression. After several days the two groups moved freely throughout the enclosure. Two months after the formation, an infant

Table 6-7. Extraspecific Group Mergers.

	AGONISTIC	SEX	GROOMING	PLAY	OTHER SOCIAL
Introduction					
Vervet to Mangabey	98	4	0	0	72
Vervet to Bonnet	156	0	0	0	18
Crabeater to Pigtail	117	0	0	0	9
Pigtail to Pigtail (N = 3)	53	8	2	0	20
Rhesus to Rhesus	474	0	0	0	12
End of hour					
Vervet to Mangabey	32	0	0	0	0
Vervet to Bonnet	138	0	0	0	0
Crabeater to Pigtail	93	3	0	0	3
Rhesus to Rhesus	261	0	9	0	3
Pigtail to Pigtail	58	0	0	0	41
Followup; same groups					
Vervet to Mangabey month 2	7	1	1	0	4
Vervet to Bonnet Day 2	60	0	0	0	0
Crabeater to Pigtail week 4	19	0	+	0	1
Pigtail to Pigtail	6	0	8	–	6
Rhesus to Rhesus week 3	22	+	+	+	0

Note: The vervet-bonnet groups were separate on Day 2 and the crabeater-pigtail groups were separated after 30 days due to serious injuries being produced by continued high levels of contact aggression. A + sign indicates an hourly rate of less than 0.5 but greater than 0. A – sign indicates that the data were not recorded for this category.

vervet was removed for treatment of diarrhea, and when returned to the group displayed some locomotor difficulties, possibly due to repeated intramuscular injections. The mangabeys approached the vervet infant en masse and the resident vervets acted as a unit in protecting the infant. The most active aggression between mangabeys and vervets occurred at this time and clearly demonstrated the existence of two independent social groups, i.e., the two groups had not formed a single intermingled social organization. The mangabey interest in the injured infant persisted and after two days the infant was removed with scalp lacerations and missing hair. The balance of the vervet group continued to live with the mangabey group for over a year, and although there were relatively few positive social interactions across taxa, a variety did occur, including copulations, and the level of agonistic interactions fell to that of stable macaque groups. The relative lack of grooming, play and other positive interactions, clearly indicates that the vervet and mangabeys did not form a single cohesive unit, but rather that they coexisted in the same area with minimal conflict. (Gene flow between vervets and bonnets or vervets and mangabeys is presumably blocked by differential chromosomal numbers.)

CONCLUSION

Hybrids reported in this study, and in the literature, demonstrate that many species within the genus *Macaca* are capable of fertile cross-matings and that the offspring of such mating are no less fertile than the parental generation. The absence of macaque hybrids under most undisturbed natural conditions suggests that the mechanisms preventing gene flow among the species of the genus *Macaca* are largely geographic, and in the case of sympatric species, behavioral. In support of the latter conclusion are the reports of naturally occurring hybrids under conditions which have been suggested to be badly disturbed through habitat destruction, or such vigorous hunting that one or more local populations was approaching extinction (Bernstein, 1966). The series of introductions of single extraspecific intruders suggests that the occasional association of a single wild male with an extraspecific group, is not an example of the socialization of an extraspecific male into a wild group. Extraspecific adult animals are not as vigorously repelled as are conspecific intruders, but neither are they actively incorporated into the social organization. Some fighting with extraspecifics does occur and it is likely to include damaging forms of aggression, not readily controlled by the normal signals, role relationships or social mechanisms effective among conspecifics. Positive social interactions with extraspecific adults are less common than between conspecifics, and the association of a solitary male with an extraspecific group is interpreted to reflect passive association rather than active social bonding. The fact that solitary males join conspecific groups despite the greater vigor of resistance to conspecific intruders underscores the loose association of solitary males with extraspecific groups. Our data from groups M-2 and M-4 do suggest that extraspecific infants could possibly be intergrated into a wild group, but it is considered highly unlikely that an infant would become separated from its natal group and survive both while on its own and after joining an extraspecific group. The occasional penetration of a group by extraspecifics under special circumstances is, however, evidenced by some field reports (Bernstein, 1968).

The data resulting from Series III suggests that the behavioral barriers to polyspecific group formations may be related to macaque behavioral attributes and do not reflect more general primate characteristics. The free association of other primate taxa with one another and their ability to not only coexist in space but also to interact positively (Abordo *et al.*, 1975; Maple and Westlund, 1975) may indicate that alternative barriers to gene flow may already be operational among sympatric New World monkeys or Old World cercopithecines such as guenons and mangabeys, which do form polyspecific associations.

The ready production of macaque hybrids (and even second-generation hybrids) in captivity, and under severely disturbed natural conditions, is thus accepted as evidence of geographical and behavioral barriers to gene flow operating within the genus under natural conditions. The alternative, accepting all the diverse populations of macaques which can hybridize as a single species, is rejected as an unparsimonious use of available information and an overemphasis on one extreme aspect of a single definitional criterion.

ACKNOWLEDGMENTS

The research programs referred to have been supported by grants from the National Institute of Mental Health, MH 13864, from the National Science Foundation NSF GB 3008, GB 1167, and in part by PHS grant RR00165 from NIH. In conducting this research, the investigators adhered to the "Guide for Laboratory Animal Facilities and Care" prepared by the Committee on the Guide for Animal Resources, National Academy of Sciences, National Research Council.

REFERENCES

Abordo, E. J., Mittermeier, R.A., Lee, J. and Mason, P. Social grooming between squirrel monkeys and uakaris in a seminatural environment. *Primates,* **16:** 217–222 (1975).
Bernstein, I.S. The integration of rhesus monkeys introduced to a group. *Folia primatol.,* **2:** 50–63 (1964).
_____. Naturally occurring primate hybrid. *Science,* **154:** 1559–1560 (1966).
_____. Intertaxa interactions in a primate community. *Folia primat.,* **7:** 198–207 (1967).
_____. Social status of two hybrids in a wild troop of *Macaca irus. Folia primat.,* **8:** 121–131 (1968).
_____. Introductory techniques in the formation of pigtail monkey troops. *Folia primat.,* **10:** 1–19 (1969).
_____. Some Behavioral Elements and the Cercopithecoidea. In J. A. Napier and P. Napier, (eds.), *Old World Monkeys.* New York: Academic Press, pp. 265–295 (1970).
_____. The influence of introductory techniques of the formation of captive mangabey groups. *Primates,* **12:** 33–44 (1971a).
_____. Activity profiles of primate groups. Chaper 2 in A. M. Schrier and F. Stollnitz, (eds.), *Behavior of Nonhuman Primates.* New York: Academic Press, Vol. 3, pp. 69–104, (1971b).
_____. Birth of two second generation hybrid macaques. *J. hum. Evol.,* **3:** 205–206 (1974).

Bernstein, I.S., Balcaen, P., Dresdale, L., Gouzoules, H., Kavanagh, M., Patterson, T. and Warner, P. Differential effects of forest degradation on primate populations. *Primates*, **17**: 401-411 (1976).

Bernstein, I.S., Gordon, T.P. and Rose, R.M. Aggression and social controls in rhesus monkey (*Macaca mulatta*) groups revealed in group formation studies. *Folia primatol.*, **21**: 81-107, (1974a).

_____. Factors influencing the expression of aggression during introductions to rhesus monkey groups. In R. L. Holloway, (ed.), *Primate Aggression, Territoriality, and Xenophobia.* New York: Academic Press, pp. 211-240 (1974b).

Bernstein, I.S. and Mason, W.A. Group formation by rhesus monkeys. *Animal Behaviour*, **11**: (1) 28-31 (1963).

Chiarelli, B. Check-list of catarrhine primate hybrids. *J. hum. Evol.*, **4**: 301-305 (1973).

Duvall, S. W., Bernstein, I.S. and Gordon, T.P. Paternity and status in a rhesus monkey group. *J. Reprod. Fertil.*, **47**: 25-31 (1976).

Fooden, J. Report on primates: Collected in western Thailand, January-April, 1967. *Fieldiana Zool.*, **59**: 1-62 (1971).

_____. Taxonomy and evolution of liontail and pigtail macaques (*Primates: Cercopithecidae.*) *Fieldiana Zool.*, **67**: 1-169 (1975).

_____. Provisional classification and key to living species of macaques (*Primates: Macaca*). *Folia primatol.*, **25**: 225-236 (1976).

Gartlan, J. S. and Struhsaker, T.T. Polyspecific associations and niche separation of rain-forest anthropoids in Cameroon, West Africa. *J. Zool.*, **168**: 221-265 (1972).

Gautier, J. P. and Gautier-Hion, A. Les associations polyspécifiques chez les Cercopithecidae du Gabon. Extrait de *La Terre et la Vie*, No. 2, 164-201 (1969).

Gray, A. P. Mammalian hybrids. A checklist with bibliography, 2nd ed., Slough, England: Commonwealth Agricultural Bureaux, pp. x, 262 (1971).

Hayashi, K. Interspecific interaction of the primate groups in Kibale Forest, Uganda. *Primates*, **16**: 269-282 (1975).

Izawa, K. Foods and feeding behavior of monkeys in the upper Amazon Basin. *Primates*, **16**: 295-316 (1975).

Klein, L. L. and Klein, D. J. Observations on two types of neotropical primate intertaxa associations. *Amer. J. phys. Anthrop.*, **38**: 649-653 (1973).

Maple, T. and Westlund, B. The integration of social interactions between cebus and spider monkeys in captivity. *Appl. Anim. Ethol.*, **1**: 305-308 (1975).

Moynihan, M. The evolution of behavior and the role of behavior in evolution. *Breviora*, **415**: 1-29 (1973).

Struhsaker, Thomas T. The Red Colobus Monkey. Chicago: University of Chicago Press, pp. 311 (1975).

Stynes, A. J., Rosenblum, L.A. and Kaufman, I.C. The dominant male and behavior within heterospecific monkey groups. *Folia primat.*, **9**: 123-124 (1968).

Stynes, A.J., Kaufman, I. C. and Reiser, S. M. Social behavior in heterospecific groups of young macaque monkeys (*Macaca nemestrina* and *Macaca radiata*): Preliminary observations. *Amer. Zoologist*, **15**: 822 (1975), (Abstract only).

Thorington, R. W., Jr. Observations of squirrel monkeys in a Colombian Forest. In L. A. Rosenblum and R. W. Cooper, (eds.), Chapter 3, *The Squirrel Monkey*, New York: Academic Press (1968).

Warden, C. J. and Galt, W. Study of cooperation, dominance, grooming and other social factors in monkeys. *J. genet. Psychol.*, **63**: 213-233 (1943).

Chapter 7
The Ecological Separation of Macaca nemestrina *and* M. fascicularis *in Sumatra*

Carolyn M. Crockett
and *Wendell L. Wilson*

The large Indonesian island of Sumatra supports two macaque species, the pigtailed macaque (*Macaca nemestrina nemestrina*) (Fig. 7-1) and the long-tailed or crab-eating macaque (*M. fascicularis fascicularis*) (Fig. 7-2), which frequently are sympatric with four or five other primate species, excluding prosimians (Crockett Wilson and Wilson, 1977). Although there is presumably some dietary overlap with the other primate species, these macaques' terrestrial adaptation allows exploitation of certain foods, such as in croplands, which are less frequently used by Sumatran langurs (*Presbytis* spp.) and rarely by gibbons (*Hylobates* spp. and *Symphalangus syndactylus*) or orang–utan (*Pongo pygmaeus*). Since macaque species exhibit adaptive patterns that are more similar to one another than to those of other primate genera with overlapping distributions, one would expect some competition between sympatric macaques. Current discussions of niche theory (Vandermeer, 1972) and resource partitioning (Schoener, 1974a) suggest that sympatric congenerics potentially capable of considerable competition should show ecological separation along one or more niche dimensions. The identification of relevant dimensions and behavioral indicators of resource partitioning is an essential though sometimes difficult task. From a theoretical viewpoint, the ideal situation is for the niche dimensions ("ordered environmental variables", Hutchinson, 1957) to be few and independent. Schoener (1974a) identifies habitat, food type, and temporal separation as the major ways by which species can partition resources and thereby minimize competition.

Fig. 7-1. An adult male *Macaca nemestrina* at Ketambe, Gunung Loeser Reserve, Aceh Province, northern Sumatra. Rijksen (1978) reports that this animal was a solitary individual observed on many occasions. (Photo courtesy H. D. Rijksen.)

This chapter is a preliminary description of the behavior of pigtailed macaques (hereafter referred to as "nemestrina") and longtailed macaques (hereafter, "fascicularis") with special emphasis on the ecological separation that we observed between these species in Sumatra.

METHODS AND SURVEY AREAS

The results presented here are based on a 14-month census/survey of the island of Sumatra, Indonesian Archipelago. Our helpful companion during most of the study was C.L. Darsono of the Indonesian Zoo Association. Our major goal was to estimate the distribution and abundance of *M. fascicularis* and *M. nemestrina* because these species were being imported for biomedical research by the Regional Primate Research Center at the University of Washington. However, we expanded the focus of the survey to include all diurnal primate species (Crockett Wilson and Wilson, 1977; Crockett and Wilson, in prep.).

Between November 1971 and January 1973 we surveyed 654 square kilometers: 32 square kilometers on foot, 79 square kilometers by boat, and 543

Fig. 7-2a. Adult male and Fig. 7-2b. adult female *Macaca fascicularis* in mangrove habitat, Belawan, near Medan, North Sumatra (photos courtesy C. L. Darsono).

square kilometers by jeep. This represents about 0.2 percent of Sumatra's area, including road travel spanning its length and breadth. Census data were collected at 18 survey areas and several brief-survey locations (Fig. 7-3). We re-surveyed an additional 10 square kilometers on foot, 49 square kilometers by boat, and 454 square kilometers by jeep during repeat travel. Time spent at the survey areas varied from two days to three weeks. We spent over 550 hours on foot and in boats searching for and observing primate groups, and hundreds more surveying roadside habitats by jeep.

We surveyed 25 different habitat types (listed in Table 7-4) representing almost all types occurring in Sumatra except those above 1,500 meters. Except for road counts, we spent little time in areas of extensive deforestation, agriculture, and human habitation. We distinguished habitat types on the basis

Fig. 7-2b

of elevation; whether they were climax forest ("primary") or disturbed/successional habitats ("secondary"); by presence of standing water ("swamp"); by proximity to rivers ("riverbank"); and in some cases by predominant flora (e.g., *Rhizophora* mangrove). Comparing the topography (Fig. 7-4) and vegetation (Fig. 7-5) provides an approximation of habitat distribution in Sumatra.

We used a transect or strip census technique modified from that employed by Southwick and his associates (Southwick *et al.*, 1961; Southwick and Siddiqi, 1966; Southwick and Cadigan, 1972; [Crockett] Wilson and Wilson, 1975a,b). We traveled through the forest and on rivers, calculating the distance by estimated speed of travel and time spent in each habitat type or occasionally by direct measurement; the jeep odometer was used to determine distance of road survey. Distance to the left and right of travel to which we could reliably detect primate groups provided an estimate of strip width. Recent surveys have

Fig. 7-3. The island of Sumatra: Provinces and areas visited during this survey. Other study areas noted include West Malaysia (Chivers, 1973), Ujung Kulon (Angst, 1973), and Umbulan Durian (W. Wilson, summer 1975).

provided a more precise measure of strip width by recording the estimated distance to the primate group detected (Muckenhirn *et al.*, 1975; Cant, 1978). Area surveyed is strip length multiplied by strip width, and density is a straightforward calculation of counts per area.

For each primate group contacted, we recorded species, number of individuals and age/sex composition (plus some indication of confidence and completeness of these figures), time of contact, habitat type, behavior, height from ground, distance from cultivated areas, elevation, terrain, and whether this was the first or a subsequent encounter with the troop. We have used the term "troop" to refer to a complete social unit of macaques and "group" to refer to monkeys seen together which may represent only a subset of the entire troop. Most of the data are based on sightings of groups rather than troops; density data are based on troops, i.e., whenever groups were considered to be subunits of the same troop they were counted as one. Because we spent little time in each survey area, our judgment regarding which subunits belonged to

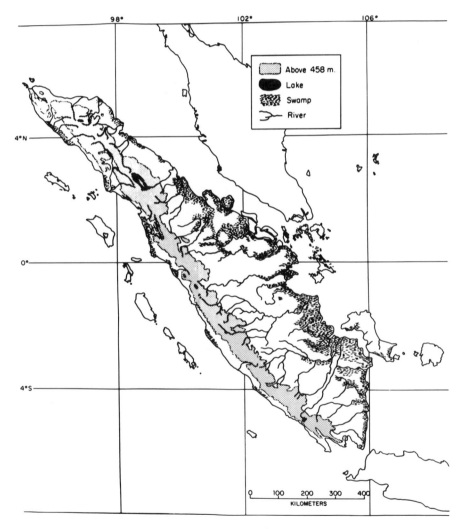

Fig. 7-4. Topography of Sumatra.

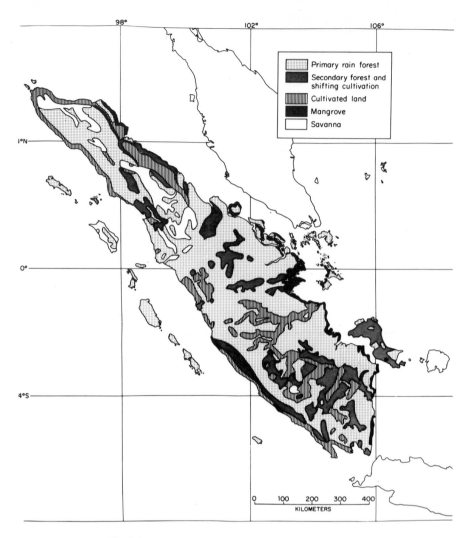

Fig. 7-5. Approximate distribution of vegetation in Sumatra.

which troops was based on times of contact of groups whose locations were plotted on maps drawn of each area. Estimates of home range size and ranging patterns are derived from these maps and our population density estimates, and by reports of local people whose information usually proved to be very reliable. *Ad lib* notes were taken on social interactions and other behaviors, especially when the same group was under observation for more than a few minutes. We recorded type of food eaten by the macaques whenever it could be determined, but dietary information provided by this study is limited.

Most of the data were transcribed from notebooks into computer card format. Each encounter with a primate group comprised one card or "case."

For data tabulation and analyses we used the Statistical Package for the Social Sciences (Nie *et al.*, 1975) and a pocket calculator.

The total observations were divided into three sets of data: original, repeat, and nondensity. Partitioning was based on whether the data were collected during the first ("original") or repeat ("repeat") travel of a survey area or stretch of road, and contributed to density calculations, or during observations made under conditions not contributing to density calculations ("nondensity"). For some analyses, only the first encounter with an identifiable troop or individual was used (e.g., in calculating mean elevation). Figures in the results represent total observations unless otherwise indicated. Additional observations, not contributing to the quantitative data, were made by W. Wilson at Umbulan Durian, Lampung Province, during July 1974.

Rijksen (1978) observed nemestrina and fascicularis in his organg–utan study area which was predominantly lowland primary forest (Ketambe, Aceh Province, Northern Sumatra: see Fig. 7-3). His qualitative description of these macaques' behavior is generally consistent with our impressions, and we will present his data in comparison with our own wherever possible.

RESULTS

The quantitative data are based on a total of 389 observations of *M. fascicularis* groups and solitary individuals (309 original, 66 repeat, and 14 nondensity, including 380 first encounters) and 36 observations of *M. nemestrina* groups and solitary individuals (16 original, 10 repeat, and 10 nondensity, including 21 first encounters).

Distribution of Macaques

The subspecies of longtailed macaques (*M. f. fascicularis*) and pigtailed macaques (*M. n. nemestrina*) found on Sumatra also occur on the Malay Peninsula and the island of Borneo. Other subspecies occur elsewhere in Southeast Asia with *M. fascicularis* having a much wider distribution (Napier and Napier, 1967).

Figure 7-6 indicates locations where we observed fascicularis and nemestrina during this study and locations reliably reported by other researchers and local residents. Both species are widely distributed throughout Sumatra, but in some locations, one or the other or both species are absent (e.g., see Table 7-6 below).

Group Size and Composition

Most contacts with an identifiable troop were for a brief period of time, and the combination of complete counts and accurate age/sex determination was

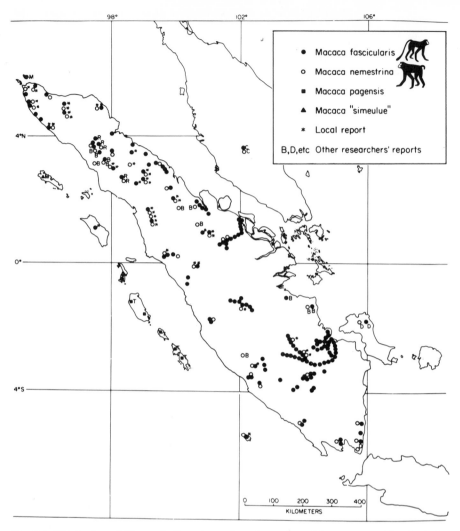

Fig. 7-6. Distribution of *Macaca* in Sumatra and nearby regions. Other sources of data include Chivers (1973; C), and personal communications from M. Borner (B), C. L. Darsono (D), J. T. Marshall (M), H. D. Rijksen (R), and R. R. Tenaza (T). See Crockett Wilson and Wilson (1977) for brief discussion of *M. pagensis* and *M. "simeulue."*

never achieved. The mean group size, all encounters combined, was about five individuals for both species (Rijksen, 1978: *M.f.* = 6.4, *M.n.* = 4.04); single individuals comprised over one-third of all recorded observations (Table 7-1). Troops in a sleeping tree, at the forest fringe, crossing the road in front of our jeep, or walking along a riverbank, could be counted much more reliably than those encountered briefly in thick vegetation. Under these more-or-less "ideal" census conditions, counts were considered "complete" or "nearly

Table 7-1. Distributions of Group Sizes Encountered, All Encounters Combined, Complete and Incomplete Counts[1].

MACACA FASCICULARIS			MACACA NEMESTRINA		
N OF INDIVID-UALS IN GROUP	N OF GROUPS	%	N OF INDIVID-UALS IN GROUP	N OF GROUPS	%
1	134	34	1	16	44
2	54	14	2	2	6
3	46	12	3	3	8
4	31	8	4	3	8
5	23	6	5	1	3
6	17	4	6	2	6
7	17	4	7	1	3
8	12	3	—		
9	8	2	9	2	6
10	6	2	—		
11	9	2	—		
12	6	2	—		
13	5	1	13	2	6
14	2	1	—		
15	2	1	—		
16	3	1	16	1	3
17	3	1	—		
18	5	1	18	2	6
19	2	1	—		
20	3	1	—		
21	1	0[1]	21	1	3
22	1	0[1]			
27	3	1			
28	2	1			
30	1	0[1]			
43	1	0[1]			
46	1	0[1]			
Total n of groups	398	100%		36	100%
$\bar{X}$ individuals encountered		4.89		5.06	
S.D.		5.97		5.78	
Median		3.50		2.73	

[1] Percents adjusted to nearest whole %.

complete" depending on the duration of our contact. Mean troop size based on these counts was similar for the two species, 18.6 ± 8.5 S.D. (n = 34; range 9-46) for fascicularis and 18.3 ± 2.5 S.D. (n = 3; range 16-21) for nemestrina (Rijksen, 1978: *M.f.* = 19.2, n = 4; *M.n.* = 23, n = 1).

Troop sizes for fascicularis seemed larger in secondary habitats. Because there are few complete counts, the variation in group size as a function of habitat was analyzed in the following manner. All counts of individually

Table 7-2. Composition of Sumatran Macaque Troops.

SPECIES AND HABITAT	ADULT M	F	?	SUBADULT M	F	?	JUVENILE	INFANT YEARLING	BLACK	UNIDENTIFIED	TROOP SIZE
M. f. fascicularis (5 groups)											
Rhizophora mangrove	4	2	1	1		2	3			0	13
	1	1	2			1		1	1	5	12+
	1	2						1	2	6	12+
Mixed mangrove	2	6ᵃ					5		2	3	18+
Secondary lowland forest	1	8ᵇ					15	8	4	10	46+
M. n. nemestrina (2 groups)											
Primary lowland forest	2	6ᶜ		1		2	6	1		0	18
Primary hill forest	2	1	4	1			4	1		4	16

ᵃ 1 pregnant, 2 with infant
ᵇ 4 pregnant, 4 with infant
ᶜ 1 pregnant, 1 in estrus

identifiable groups greater than eight monkeys contributed to mean group size for three major habitat categories. Therefore, some troops were underestimated because incomplete counts were included, and a few troops were omitted because their actual sizes were eight or smaller. Nevertheless, the differences based on this truncated sample support our impression: mean troop size of fascicularis was 12.33 ± 3.06 S.D. (n = 12) in mangrove habitats, 14.56 ± 5.39 S.D. (n = 16) in primary forest habitats, and 18.28 ± 9.73 S.D. (n = 29) in secondary habitats. However, because none of the comparisons are statistically significant, we used the mean troop size based on complete and nearly complete counts, all habitats combined, to estimate population density (see Table 7-4 below).

Our troop composition data are scanty. Table 7-2 presents the best troop composition data obtained for the macaques during the survey. These data, and data from groups for which less complete counts were made, indicate that both species are typically characterized by multimale troops in which more than one fully adult male can be present. Three adult and one subadult nemestrina males were confirmed to be solitary (observed sleeping alone) and others were suspected to be at least semisolitary. Single fascicularis males were observed, but none were confirmed to be traveling independently from a troop. Rijksen (1978) observed solitary males of both species (Fig. 7-1) and the troops in his study area were multimale.

Behavior

Our quantitative data on behavior are limited to what the majority of the group was doing when we first contacted it (Table 7-3). When first detected, fascicularis were most likely to be stationary and possibly feeding, while nemestrina were more likely to be traveling or fleeing.

Table 7-3. Behavior of Macaque Groups at Initial Contact, All Encounters Combined.

		NOT RECORDED[1]	SIT	FEED	TRAVEL	FLEE	HEARD ONLY	N
M. fascicularis								
	n	175	77	75	44	25	2	398
	%	44.0	19.3	18.8	11.1	6.3	0.5	
M. nemestrina								
	n	6	4	2	14	9	1	36
	%	16.7	11.1	5.6	38.9	25.0	2.8	

[1] In most cases, "not recorded" implies stationary behavior (usually sitting and sometimes feeding).

Table 7-4. Habitats Occupied and Density Estimates for Sumatran M. Fascicularis (MF) & M. Nemestrina (MN).[1]

HABITAT TYPES	KM² SURV	(A) MF	(B) MF	(C) MF	(D) MF	(A) MN	(B) MN	(C) MN	(D) MN
Within village									
Swamp:									
Rhizophora mangrove	2.76	0.22	8	4.35	80.9				
Mixed mangrove	4.41	4.41	100	7.71	143.4				
Primary freshwater	11.14	11.02	99	6.47	120.3				
riverbank	0.49	0.49	100	2.04	37.9				
selective logging	10.10	9.79	97	3.29	61.2				
Secondary forest	1.24	1.24	100	1.61	29.9				
riverbank	0.08			P	P				
scrub, grassland	23.31	23.31	100	1.37	25.5				
riverbank	0.16								
Lowland:									
(0–458 m)									
Primary forest	7.61	7.61	100	1.31	24.4	1.21	29	2.48	36.7
riverbank	4.19	2.01	48	2.48	46.1				
selective logging	12.38	12.38	100	2.02	37.6			P	P
Secondary forest	2.20			P	P				
riverbank	0.75	0.75	100	4.00	74.4	5.20	80	P	P
rubber grove	6.50	6.50	100	1.54	28.6			0.38	5.6
scrub, grassland	1.31	0.78	60	3.90	72.5				
riverbank	5.81	5.05	87	1.78	33.1	1.33	23	1.52	22.5
Hill:									
(458–915 m)									
Primary forest	3.32	3.32	100	6.02	112.0				
Secondary forest	4.55	0.59	13	5.26	97.8	1.31	29	2.24	33.2
rubber grove	3.68	0.95	26	7.29	135.6	2.50	68	1.61	23.8
scrub, grassland	0.73	0.73	100	2.74	51.0				
riverbank	2.25	1.68	75	0.60	11.2			P	P

Submontane: (915–1525 m)	Primary forest	2.38					
	Secondary forest	0.10			P		P
Total Km²; Mean % and Density	111.45	92.83	83%	2.98	P	55.4	11.55	10%	1.21	17.9

[1] Density estimates are based on counts made during original walk and boat survey (Km² Surv). Observations made during repeat, nondensity, and road surveys are indicated by a P ("present"). (a) Km² Where Observed (Km² Obs); (b) percent of habitat type where species were observed (Km² Obs ÷ Km² Surv); (c) estimated number of troops per Km² Obs; (d) estimated number of individuals per Km² Obs. Although habitat types where roadside groups were encountered and total area surveyed by jeep were recorded, the area of each roadside habitat type was not calculated. Therefore, estimated species density per habitat type was calculated for walk and boat survey only. In some areas we found that the macaques were absent from a habitat type although they had been encountered in a similar habitat type at another location. Because averaging in "zero" values for the former locations would underestimate densities in occupied areas, densities were calculated by dividing number of groups encountered by the area per habitat type within locations where the species was observed (Km² Obs). (Dividing Km² Obs by Km² Surv yields the percent of the total area surveyed of a particular habitat type where the species in question was actually observed.) Group density equals number of encounters during original travel divided by Km² Obs. Because partial groups were frequently encountered, calculating a realistic individual density required some extrapolation. For *M. fascicularis*, group density was multiplied by the mean troop size (18.6). Because 44 percent of total encounters of *M. nemestrina* involved single animals, over half of which were determined to be truly solitary, a weighted mean troop size was used to avoid overestimates of population density. Based on the number of known solitaries, we assumed that 20 percent of all encounters were of truly solitary animals, while some unknown proportion of the other single sightings was of individuals belonging to a troop whose mean size was 18.3. Therefore, to calculate estimated individual density for *M. nemestrina*, we multiplied group density by 14.8 [= 18.3 (0.80) + 1.0 (0.20)].

Habitat Separation

Table 7-4 presents troop and individual density for habitats surveyed on foot and by boat (see footnote of table for description of density calculation). Fascicularis were observed in 22 of the 25 habitat types surveyed, while nemestrina were observed in only eight. Nemestrina occur in at least some of the other habitat types, but because of their lower densitites and inconspicuous habits they are less likely to be detected. Fascicularis were observed in 83 percent of the total area surveyed on foot and by boat at an estimated density of 54.5 individuals per km^2 in areas where they occurred; in contrast, nemestrina were encountered in only 10 percent of the area surveyed, where they had an estimated density of 17.9 individuals per km^2. Density of fascicularis in Rijksen's study area was $48/km^2$; for nemestrina, $19/km^2$ or less.

Based on highest densities, fascicularis prefer disturbed habitats, especially riparian habitats (which we have defined as riverbank, lakeshore, or along the seacoast) and secondary forest near cultivated areas. Their highest densities occur in mangrove swamp in which they are sometimes the only primate species, or share the habitat with only one other primate species, the folivorous silvered langur (*Presbytis cristata*). (If troop sizes in mangrove swamp prove to be significantly smaller than 18, then mangrove density may be overestimated in Table 7-4.)

Nemestrina occur at highest densities in lowland and hill primary rainforests. They apparently are not regular residents of swamp forest, but farmers in one area reported that nemestrina did enter freshwater swamp during the months of lowest rainfall (the flowering/fruiting period for many Southeast Asian tree species; Whitmore, 1975). Although they occasionally range into secondary forest and scrub far from primary forest, their more common pattern is to sleep in primary forest and enter secondary habitats to raid croplands. Three of four observations of bedded nemestrina found them in broad-crowned emergents at the fringe of primary forest adjacent to cultivated land; the fourth was in a similar, remnant primary tree several hundred meters from primary forest. (Fascicularis, the langurs, and the hylobatids all were observed to sleep in broad-crowned emergents at one time or another.)

Figure 7-7 shows that over 70 percent of encounters with fascicularis found them in riparian habitats. Nemestrina were about equally likely to be observed within the forest, in riparian habitats, or in forest fringe; over 20 percent of encounters occurred when they were raiding farms. However, because they can be so silent, nemestrina are observed more readily in open areas than within the forest where they presumably spend more than 25 percent of their time.

Fascicularis are more likely to be found at a lower elevation and on flatter terrain than are nemestrina (Table 7-5). Elevation and terrain are somewhat correlated, but in Sumatra there are extensive high elevation plateaus in the

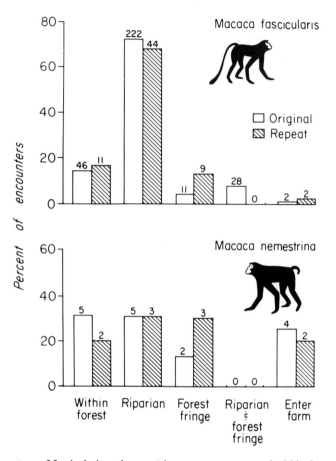

Fig. 7-7. Percentage of fascicularis and nemestrina groups encountered within forest versus in fringe or agricultural habitats. *Within forest* includes primary and secondary forest habitats farther than 100 meters from cleared area or "water." *Riparian* includes all habitats within 100 meters of river, lake or sea ("water"). *Forest fringe* includes primary and secondary forest habitats within 100 meters of cleared area (savanna or cultivated). *Enter farm* includes all terrestrial travel in cultivated areas excluding rubber groves which are considered secondary forest; a group was recorded as "entering farm" even if only part of the group entered and whether or not eating of crop foods occurred. Number of encounters during *original* travel compared with encounters during *repeat* travel appear above percentage bars (for *original*, $X^2 = 59.8$, p < 0.0001; for *repeat*, $X^2 = 9.6$, p < 0.025).

western mountain chain (Bukit Barisan) which would be classified as flat terrain (elevation, soil, and topography can all influence the structure, physiognomy and floristic composition of the habitat; Whitmore, 1975).

This quantitative separation of the macaques with respect to habitat, elevation, and terrain was apparent qualitatively at survey locations where a

Table 7-5.   Species Comparison of Elevation and Terrain Based on First Encounter With Macaque Groups.

	M. FASCICULARIS	M. NEMESTRINA	TEST	P
Elevation				
mean meters	70.2 m	364.5 m	F = 4.04	< .001
S.D.	± 138.4	± 278.1		
range	0-946 m	0-793 m		
n	380	21		
Terrain				
% flat	96.1%	47.6%	X^2 = 69.2	< .0001
% hilly or				
mountainous	3.9%	52.4%		
n	380	21		

variety of topography and vegetation types was present. Separation was particularly obvious where a heavily cultivated, level-terrain lowland was adjacent to a relatively undisturbed hilly or mountainous area with scattered farms: nemestrina raided hillside farms while fascicularis feasted on the crops planted in flat areas (e.g., Moro Batin, Suka Banjar, and Batang Garut). Nevertheless, there is a great deal of overlap between these two species which commonly occur together, sometimes simultaneously, in lowland primary and secondary forests. In such locations, another difference between the species is most apparent: our observations confirm those of others that when startled within the forest, fascicularis flee noisily through the trees while nemestrina (of the subspecies *M. n. nemestrina*) flee quietly on the ground (see Fooden, 1975). Within the forest, fascicularis generally travel in the trees while nemestrina travel on the ground. When encountered during the survey, nemestrina were more likely to be terrestrial than were fascicularis (Fig. 7-8). Most observations of nemestrina on the ground occurred while they were raiding farms. Although fascicularis also raid farms, the majority of terrestrial observations were of groups walking along riverbanks or along mangrove mudflats where they capture crabs during low tide.

Rijksen's (1978) data are similar to ours. Fascicularis troops were less than 10 meters from the ground (including terrestrial) during about 20 percent of contacts while nemestrina were observed in the lower strata nearly 50 percent of the time. He rarely observed fascicularis more than 300 meters from the river; nemestrina "preferred the dry ground at the foot of hills and slopes" (p. 132).

Observations by investigators in other parts of these subspecies' distributions are also compatible with our data on habitat separation. Rodman (1973)

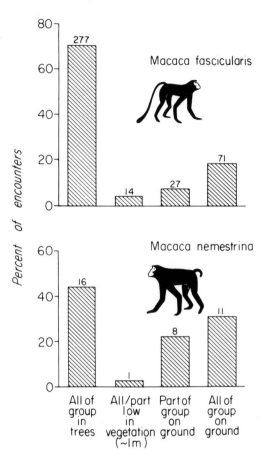

Fig. 7-8. Arboreal versus terrestrial encounters of fascicularis and nemestrina groups. Number of groups encountered appears above percentage bars ($X^2 = 15.6$, p < 0.002).

found a significant negative correlation between the presence of nemestrina and fascicularis in the 4-hectare quadrants of his 3-km^2 study area in East Kalimantan (Indonesian Borneo). The presence of fascicularis was negatively correlated with distance from the river and elevation of quadrant while the presence of nemestrina was positively correlated with these two factors. Rodman also found nemestrina traveling or foraging on the ground during 67 percent of contacts and observed fascicularis coming to the ground along riverbanks. In Wheatley's (1976) nearby study area, fascicularis were in hectares with streams or rivers 75 percent of the time.

Medway (1972) and Chivers (1973) present evidence for altitudinal zonation of the two macaques in West Malaysia. Surveying from the lowlands (213 meters), to the peak of Gunong Benom (2,109 meters), Medway (1972) found fascicularis at 213 meters and nemestrina at 732 meters and 1,098 meters. Chivers (1972), during a 5-month survey throughout the Malay Peninsula, observed fascicularis at a mean elevation of 203 meters and nemestrina at a mean elevation of 366 meters. These elevations are compatible with those we obtained for the macaques in Sumatra.

Foods Eaten

When sharing the same habitat, it is possible that the two macaques eat many of the same food items, particularly forest fruits. However, the extent of habitat separation we observed suggests that complete dietary inventories would include many items eaten by one species and not by the other. The two species seem to prefer different crop items even when they are raiding in the same area. Fascicularis eat rubber fruits *(Hevea brasiliensis)*, rice shoots *(Oryza sativa)*, and new growth on young corn plants *(Zea mays)* and some mature ears. Nemestrina raid oil palm (*Elaeis guineensis*) plantations, eat only the mature corn ears, as well as the thick and spiny-skinned durian fruit (*Durio* spp.), and papaya (*Carica papaya*) which are quickly picked and carried to the forest fringe for undisturbed consumption. Nemestrina also dig up tapioca roots (cassava, *Manihot esculenta*); at Umbulan Durian an adult male nemestrina dug up a large tapioca root measuring 35 x 15 centimeters. Crop-raiding fascicularis eat more parts of the same food item, e.g., the young leaves and ears of corn, while nemestrina select the large fruits, ears, or tubers while ignoring the edible leaves of papaya, corn, and tapioca.

We suspect that nemestrina exploit food resources that their relatively greater size and strength aid them in obtaining, such as large, nutritious fruits. Farmers in West Sumatra (and other areas of Southeast Asia; Bernstein, 1967; Bertrand, 1967; Corner, 1955) trap juvenile nemestrina and train them to pick coconuts (*Cocos nucifera*) and fruits from other cultivated trees (Fig. 7-9). (We saw one fascicularis "in training," but its owner reported that it was much less "clever" than the preferred nemestrina.) Whether their strength aids nemestrina in obtaining noncultivated dietary items is not known. We speculate that they extract grubs and insects from fallen tree trunks, and pulling apart a rotten trunk can require considerable strength. Once, several nemestrina were observed to ingest items obtained under the bark of a fallen tree; we suspected that the items were grubs or insects. Bernstein (1967) observed Malayan nemestrina eating swarming termites, grasshoppers, and other insects and spiders in addition to fruits, seeds, young leaves, leaf stems and fungus. Nemestrina at Ketambe consumed a variety of food items including many figs (*Ficus* spp.) also eaten by the apes and fascicularis (Rijksen, 1978).

Fig. 7-9. *Macaca nemestrina* picking coconut near Padang, West Sumatra (photo by W. Wilson).

Macaca fascicularis is also a frugivore-omnivore. In the mangrove swamp, we saw fascicularis eating fruits of pedada (*Sonneratia* spp.) and nipa palm *(Nypa fruticans)*, aerial roots of *Avicennia* spp. and other mangrove trees, and crabs (e.g., *Uca* spp.). When foraging for crabs, fascicularis sits near a hole until a crab emerges, then seizes it and quickly tears off its pinchers with its hands. We heard several shrieks when monkeys apparently were pinched.

Wheatley (1976) found that fascicularis in his secondary forest riverine study area consumed 86 percent fruit, supplementing their diet with grass, leaves, flowers, insects, fungus, vinestems, and clay; in north Sumatra, Rijksen (1978) observed them to spend considerable time searching for insects in addition to consuming a great deal of fruit. Their smaller size enables them to exploit food items on small branches which probably are inaccessible to the larger nemestrina; their long tails seem to provide balance and support during small-branch foraging.

Ranging Patterns and Crop-Raiding

The two macaques differ in their home-range sizes and ranging patterns. The survey data do not allow for precise calculation of home-range size, but estimates can be made using troop density data and assuming some overlap of ranges. Home ranges of fascicularis generally appear to be 50–100 hectares and may be as small as 25 hectares in mangrove swamp, although they may actually be considerably larger there than revealed by our view from the river (i.e., deep but with limited waterfront footage). The extent of overlap may be small (Fittinghoff, 1975) or extensive (Angst, 1973, 1975). Fascicularis use their home range in a fairly regular pattern, e.g., foraging for crabs when the tide is out or raiding the adjacent farm every day at a particular time. The fascicularis in Rijksen's (1978) study area came to the river every afternoon at 4:00.

Nemestrina, on the other hand, appear to have very large ranges of at least 100–300 hectares which they use on a longer and probably more variable time scale. They raid farms in circuit fashion—every day, once a week, once a month, or once a year, depending on the location and availability of crop foods. If a favored food source such as corn is ripe, the field may be raided daily until completely destroyed; then, if no other farms in the area are producing desirable crops, the nemestrina troop(s) may move out of the area for months. From our camp at Batang Garut, we observed members of a nemestrina troop on three occasions in one week; there were a few fruit-laden eggplants (*Solanum* spp.) among the hundreds of chile-pepper plants (*Capsicum fru-tescens*) surrounding our tent. A month later we returned to the same camp for a three-week stay and saw thousands of ripe chiles, no eggplant fruits, and a solitary male nemestrina.

McClure's (1964) observations support our own suspicions that nemestrina are widely ranging; during 625 hours of observation spanning 33 months, a nemestrina troop was seen on only two occasions separated by a nine-month interval. McClure never saw fascicularis in his study area, which was within primary rainforest in West Malaysia at an elevation of 610 meters. Rijksen's (1978) study area supported three fascicularis troops but only one nemestrina troop.

Nemestrina tend to raid with more stealth than do fascicularis, although both may raid at any time of day. They frequently survey a field for some time before entering it, usually one animal at a time and rarely as a large group. A lookout, often a subadult male, may keep watch from a tree at the forest fringe, vocalizing a warning bark when a frantic farmer comes running. Nemestrina have a reputation of raiding during rainstorms while farmers have sought shelter in their huts, and we saw them doing so on several occasions.

Although fascicularis occasionally will enter cropland during rainy weather, their basic strategy for raiding differs. They usually enter a field as a group (not necessarily the entire troop), a few animals leading with the rest following if no alarm is sounded. There seems to be a general alertness among the group rather than a specific lookout. However, after detecting us, it was usually an adult male that gave the "kera" warning vocalization which could be repeated for many minutes. Although hard data are lacking, the size of the raiding groups gave the impression that all or most of the fascicularis troop was participating and that even though the troop might be dispersed over 100 or more meters, each member had visual contact with at least one other individual.

Nemestrina troops, on the other hand, split up into smaller foraging parties on several occasions. For example, troop B at Moro Batin (hill primary forest and adjacent field) was composed of 23 individuals during cohesive troop travel along a gully. While 10 or more individuals of that troop were observed raiding a cornfield 100 meters down the mountainside, one of us (C.C.) recorded the following while observing a forested ravine from a cleared ridgetop:

9:10 a.m.: On the ground amongst tall trees and undergrowth in the ravine I see a small group of nemestrina (members of troop B). When one sees me, it moves slowly and silently out of sight. Others are about 15 meters uphill in the cornfield. The group is incredibly quiet. Only the juveniles make some rustling in the bushes.

10:20 a.m.: A farmer yells at the monkeys he has just discovered in his cornfield. They run from the cornfield, very quietly, along the bottom of the ravine allowing me to count more accurately. There are two adult females, one adult male, two juveniles, and three other individuals. The adult male runs last, then stops, climbs about 6 meters up the trunk of a tree and looks uphill in the direction of the man. The monkey is a large, healthy looking animal with much white fur on his rump and cheeks.

10:32 a.m.: The man apparently has left. The big male descends from the tree and goes back up the hill toward the cornfield. Immediately after, there is some rustling near that tree as the rest of the group comes out from hiding or relaxes. The monkeys had run less than 30 meters but had completely disappeared.

While fascicularis probably raid as a group partly because of the potential protection offered by it, an adult nemestrina, especially a male, is of formidable size and can easily defend itself against an unarmed human. To successfully consume the types of crop foods that seem to be their specialty, nemestrina need an uninterrupted period to reap the harvest. The corn stalk must be broken down to reach the ripe ears which then must be husked. Tapioca roots require some time to be unearthed. Stealth is more successfully accomplished if fewer, more widely dispersed animals enter the crop area.

Maples *et al.* (1976) describe the crop-raiding strategies of baboons (*Papio cynocephalus cynocephalus*) who sometimes divide into subunits during raids.

Rijksen's (1978) observations suggest that the two macaque species differ in foraging strategies within the forest as well: fascicularis troops "gave the impression of moving through the home range over a wide front with dispersed individuals who foraged for many different items along the way" (p.112). The nemestrina in his area generally foraged in small, widely dispersed subgroups that traveled on the ground and seldom stayed in the same place for very long; although they searched for food on the ground, more often they fed in the trees. "Pigtailed macaques kept in contact during their dispersed foraging by uttering a soft, low sounding moaning vocalization, which could be heard at distances varying from 30–80 metres" (p.113). Once we heard a bout of similar vocalizations at dusk when members of a nemestrina troop were approaching trees where they apparently spent the night.

Fittinghoff (1975) reported that fascicularis troops in East Kalimantan not only travel as a unit, but frequently fragment into separately foraging subgroups which might not rejoin the main troop for several days. Fascicularis occupy a wide variety of habitat types and appear to have flexible foraging strategies.

Reproductive Strategies

Fascicularis and nemestrina neonates are black, making them conspicuous against their mothers' paler pelage (Fig. 7-10). Within the first few months of life, the infant's pelage changes to transitional brown and then to a tan (nemestrina) or grayish (fascicularis) agouti similar to that of adults. We recorded all observations of black infants, pregnant females, copulations, and estrous swellings (the latter noticed only for nemestrina). Table 7-6 presents these data arranged by month and latitude. We traveled Sumatra from south to north during approximately one year, seriously limiting any conclusions that may be drawn from these data.

Many Southeast Asian tree species fruit and flower during the drier months (Whitmore, 1975), and primates might be expected to evolve reproductive cycles that capitalize upon the cyclicity of their food resources, with gestation

Fig. 7-10. Black infant *Macaca fascicularis* at mother's breast; mangrove habitat, Belawan, North Sumatra (photo courtesy C. L. Darsono).

Table 7-6. Reproduction Data Recorded for Sumatran Macaques.

LOCATION	MONTH	YEAR	RAINY SEASON	N OF GROUPS ENCOUNTERED[1]		N OF GROUPS WITH:			
				MN	MF	BLACK INFANTS	PREGNANT FEMALES	PREGNANT F & BLK INF	ESTRUS AND/OR COPULATION
Sungsan	Nov	71	—	0	5				
Kiambang	Nov	71	—	2	10				
Moro Batin	Dec	71	beg	2	4				
Suka Banjar	Jan	72	mid	4	12	1			(1)[2]
Road count, South	Jan	72	mid	1	4	(1[3])	1	1	
Musi upstream	Feb	72	mid	0	102	5	(1)	1	1
Musi downstream	Feb	72	mid	0	101	5	1	1	
Jeleket	Mar	72	end	1	2				
Kelabung	Mar	72	end	1	0				
Road, south/Bengk.	Mar	72	end	1	4				1 (1)
Road, south/Bengk.	Apr	72	end	0	3				
Road, Jambi	Apr	72	end	0	11	1			
Teluk Kayu Putih	Apr/May	72	end	0	3	2 + 1[3]			1
Landai	May	72	—	0	0				
Pekan Selasa	May	72	—	1	2				
Road, West Sumatra	May–Aug	72	—	1	5				
Pinagar	July	72	—	0	3				
Bangko	July	72	—	0	1				
Libo	July	72	—	0	0				
Siak River	July	72	—	2	80	2			

Road, Riau	July	72	—	0	4			
Rimba Panti	Aug	72	—	2	0			
Road, North Sumatra	Sept	72	beg	0	1			
Road, North Sumatra	Oct	72	beg	1	4	2		
Batang Garut	Oct	72	beg	1	1		(1)	
Bukit Lawang	Oct	72	beg	1	2			
Road, Aceh	Nov	72	beg	0	7			
Batang Garut	Dec	72	mid	1	0			
Ketambe	Jan	73	mid	0	2	1		(1)

[1] First encounter: includes single individuals.
[2] Data for *M. nemestrina* in parentheses.
[3] Brown infant.

and/or lactation coinciding with an optimum period (Lancaster and Lee, 1965). Accordingly we have indicated whether our survey took place during the rainy season ("beginning," "middle," or "end") or during the period of less rain ("-"). Most of Sumatra lacks a distinct dry season (Whitmore, 1975). Annual rainfall in the western mountains and coast is over 300 centimeters compared with 200–300 centimeters in the eastern lowlands and 100–200 centimeters in the northeastern tip of Sumatra (The Atlas Editorial Committee—Army Topographical Directorate of the Republic of Indonesia, 1963). The driest months throughout Sumatra are June and July, when 5–25 centimeters fall per month compared with the wettest months when 20–48 centimeters are recorded, depending on location (Wernstedt, 1972). According to local lore, it rains most in the months whose names include the letter "r," i.e., September through April. Rainfall records (Wernstedt, 1972) show that this is generally the case. However, in southern Sumatra, the rains begin in earnest in November or December and are fairly consistent through March or April. In northern Sumatra, the rains tend to begin in October, lull in February, and increase again in March and April, sometimes continuing until May.

The reproduction data do not provide a clear picture. The apparent peak in births for *M. fascicularis* during the rainy season may be an artifact since over half the groups counted were observed during that period. Copulations were noted for both species during the rainy months. Two black infant fascicularis were seen in July along the Siak River where we counted many fascicularis groups, further supporting the hypothesis that the probability of seeing a black infant is a function of number of groups encountered rather than rainfall.

Although evidence for birth seasonality *per se* is equivocal, in six fascicularis troops more than one female was pregnant and/or had similar-aged young infants, suggesting birth synchrony. In contrast, one nemestrina troop contained a pregnant and an estrous female, and another included a pregnant female and a mother with a brown infant.

Wheatley (1975) suggested that fascicularis show characteristics of "r-selected" species (MacArthur and Wilson, 1967; Pianka, 1970), displaying higher fecundity and more rapid development than some other primate species. He related this reproductive strategy to the ecological instability of the fascicularis' preferred habitats. The abundance of fascicularis in Sumatra attests to the fact that many monkeys are regularly reproducing, and suggests that many females give birth every year. Adapted to a more stable primary forest habitat, nemestrina might be expected to show more *K*-selected characteristics (MacArthur and Wilson, 1967; Pianka, 1970), with their rate of reproduction producing a population at equilibrium. Bernstein (1967) estimated that only half of the adult females in the nemestrina troop he studied gave birth in a year, suggesting an interbirth interval of two years. He pointed out that most female nemestrina in captivity become pregnant every year. However, a female's pregnancy may depend on the availability of food which in captivity would not be subject to seasonal or annual fluctuations. The age at which females first reproduce also influences the rate of population growth. If fascicularis macaques are more "*r*-selected," females should bear their first offspring at a younger age than do nemestrina.

These hypothesized differences in reproductive rates coupled with decreasing areas of primary forest could easily account for the discrepancy in population sizes for the two species in Sumatra which we estimate to be at least several million for *M. fascicularis* and no more than several 100,000 for *M. nemestrina* (Crockett and Wilson, in preparation).

DISCUSSION

Resource Partitioning

In spite of the fact that, in Sumatra, fascicularis and nemestrina were sometimes found in the same forests, the comparative data collected during this survey strongly suggest that, overall, ecological separation is marked. This separation is best discussed within the context of interspecific competition and resource partitioning.

The two species show considerable habitat separation based on relative density and occupancy of habitat types. Fascicularis are abundant in riparian, forest fringe, and secondary forest habitats at lower elevations, and nemestrina are found more often in primary forests, especially in the mountains.

The macaques might also reduce competition by partitioning food types or by using the same resources at different times. Our data on these aspects of ecological separation are limited. We do know that nemestrina and fascicularis raiding the same farms tend to select different food items, but this may not be characteristic of their foraging in the forest. We also observed juvenile nemestrina accompanying fascicularis groups who were raiding crops. This sort of association seems unlikely if severe competition for food were present, especially since the juveniles are closer to the size of the adult fascicularis. However, fascicularis densities are very high in mangrove habitats where nemestrina and the apes are absent, perhaps reflecting reduced competition.

Temporal separation on a diurnal basis is not apparent, since both species may raid the farm simultaneously. (Temporal partitioning is rarely a major contributor to ecological separation among mammals; Schoener, 1974a, b.) Temporal separation on an annual basis may be achieved through different ranging patterns whereby nemestrina and fascicularis troops with overlapping ranges do not forage in the same place at certain times during the year. Because nemestrina have large ranges, parts of which may be visited infrequently, a troop may travel for extended periods of time without contacting a fascicularis troop. However, it would be difficult to determine whether the species were avoiding one another during times of potentially high competition, or whether nemestrina were attracted to another area with a better food resource during these periods. Annual temporal separation is unlikely to be a major aspect of resource partitioning for these two species.

We speculate that the major factor contributing to different ranging patterns is that nemestrina are more selective in their diets, specializing to some extent on food that comes in larger, high-energy packages. The widely ranging habits of nemestrina can be viewed as closely tied to their foraging strategy. Troops exploit a preferred food supply until it is depleted and then move to another area. Presumably, traditional ranging patterns have developed in areas of ecological stability whereby the older troop members can lead the troop to locations where food is predictably available at a certain time of year. Separately foraging subunits that maintain periodic vocal contact may increase the probability of detecting dispersed food resources. Terrestrial travel allows great mobility over varied terrain, and this tendency probably evolved in conjunction with the behavior of ranging widely. Locating fruiting trees may be facilitated by finding fallen fruits on the ground. The generally quiet behavior of nemestrina may have evolved because silence reduced the probability of being detected by predators while on the ground, where they are especially vulnerable. The absence of nemestrina in swampy areas may be a result of the short supply of favored fruits during the wetter season and the difficulty of terrestrial travel through swamps when ground water is substantial.

Wheatley's (1976) study of fascicularis in East Kalimantan suggests that their ecological strategy favors smaller ranges. He speculates that fascicularis, by feeding on secondary forest fruits characterized by abundant, small seeds, are acting as dispersal agents and help to maintain their own food supply (and that of their descendents) by feeding in the same areas throughout the year. The idea that natural selection could favor foraging strategies which increase one's food supply is supported by observations of *Cebus capucinus* who apparently increase the branching and foliage of *Gustavia superba* trees by picking the terminal leaf buds (Oppenheimer and Lang, 1969).

Wheatley's (1976) transect surveys revealed that fruit was significantly more abundant and more continuously available in streamside areas (within 100 m) than in nonstreamside areas, and he found no fruit on more than half of his ridge trail surveys. The fascicularis spent only 25 percent of the time in nonstream areas. If Wheatley's data on riverine versus ridge fruit abundance is characteristic of Sumatra as well, the macaques may be partitioning the available habitats along lines that are highly correlated with fruiting patterns.

An inventory of plant species found in both secondary and primary habitats probably represents a small percentage of the total diversity of each habitat type. Thus, when nemestrina are foraging on ridges, they are usually eating different fruits than fascicularis eat along rivers.

We do not know the extent to which nemestrina and fascicularis consume different foods when they are foraging in the same locations (food-type partitioning), and to what extent ranging patterns (annual temporal partitioning) are related to minimizing competition for the same food sources or are simply an artifact of spatial and temporal availability of foods. While resource partitioning seems to take place along at least two of the three dimensions summarized by Schoener (1974a), these dimensions do not seem to be wholly independent.

Theoretical Considerations

Although this study does not provide quantitative data for the other niche dimensions, the primary separation between the Sumatran macaques is clearly by habitat type. This finding is in accord with the "compression hypothesis" (MacArthur and Wilson, 1967; Schoener, 1974b) that the initial, nonevolutionary response in competing species upon becoming sympatric is to partition habitat types rather than food types. Considerable differences in these macaques' morphology suggest that initial habitat partitioning was followed by selection favoring phenotypes best suited for exploitation of a different range of habitat and, perhaps, food types. For example, terrestrial travel through large home ranges in mountainous primary forests may require adaptations such as large body size, which nemestrina possess but fascicularis do not. At the other end of the macaque weight range, fascicularis may have become lean and lanky in order to penetrate the dense vegetation of edge habitats and to facilitate arboreal travel (Fittinghoff and Lindburg, this volume).

Because our habitat types are not orderable along continuous dimensions, the niche separation distance (d/w; Schoener, 1974a) between these two Sumatran macaques cannot be calculated. Nevertheless, the extent of habitat separation alone suggests less overlap and greater niche separation than might be expected for a two-species complex. Perhaps their particular pattern of ecological separation evolved before they dispersed to the Sunda Shelf, in some other location such as Thailand where five macaque species are sympatric and potentially competing (Eudey, this volume). In that location nemestrina and fascicularis might not occupy adjacent niches, and the intervening species have not dispersed to the south. Niche shift or expansion by the macaques reaching Sumatra might not have occurred yet because of insufficient evolutionary time. On the other hand, Roughgarden's (1976) theoretical analysis shows that niches may not always shift close to the point where they are separated only by the minimal distance possible for coexistence, and in fact will not where the range of resources is great, a condition probably met in Sumatra.

Futhermore, niche overlap has been shown to decrease with increasing "diffuse competition," the aggregate competition produced by related and unrelated species in an assemblage (Pianka, 1974). The Sumatran macaques' niche overlap may be minimized by diffuse competition from other primates, as well as from squirrels (e.g., *Callosciurus* spp.), civets (e.g., *Paguma larvata* and *Articus binturong*), fruit bats (e.g., *Pteropus vampyrus*), and some hornbills (*Bucerotidae*) (Rijksen, 1978).

These are not entirely alternative explanations since all involve the mechanisms of resource competition. The "diffuse competition" hypothesis would be favored over the "evolutionary lag" hypothesis if the Philippine macaque (*M. f. philippinensis*) shows greatest niche expansion (e.g., farther from rivers, deeper in mountainous primary forests) and the Javan macaque (*M. f. mordax*) shows intermediate niche expansion relative to Sumatran (or Malayan or Bornean) *M. f. fascicularis*. We predict this because there are no *M. nemestrina* on Java, and *M. fascicularis* is the only nonprosimian primate in the Philippines, where there are also fewer nonprimate competitors (Delacour, 1947; Delacour and Mayr, 1946; Medway, 1969; Ewer, 1973). Niche expansion on Java or the Philippines still does not rule out the possibility that the limited degree of overlap between nemestrina and fascicularis observed in Sumatra originated in a more competitive mainland community. *Macaca fascicularis* dispersed more recently than did *M. nemestrina* (Fooden, 1975, 1976, this volume) and failure to find niche expansion among fascicularis in the absence of nemestrina would lend some support to the "evolutionary lag" hypothesis. However, diffuse competition, resource diversity and the evolutionary compromise of habitat shift could still operate to determine the extent of future overlap.

At the moment we cannot determine whether these two species have filled the macaque niche to the exclusion of new immigrant species. The greater diversity of macaques in the Thailand region is likely elated to the fact that this is an area of floral transition, where deciduous forests are replaced by evergreen (Whitmore, 1975). Of the five macaque species there, *M. mulatta, M. assamensis,* and *M. arctoides* may be better adapted to a deciduous habitat. *Macaca nemestrina* appears to be adapted to climax evergreen forest as is its close relative, the liontailed macaque *(M. silenus)* of western India. Nemestrina, however, can exploit secondary forests and farmland while liontailed macaques never do (Green and Minkowski, 1977).

Given that *M. nemestrina* or its ancestor had already dispersed to the Malay Peninsula and the larger Sunda Islands, perhaps only a small-bodied macaque preadapted to successional evergreen forest could coexist in sympatry. The relatively rapid and widespread dispersal of fascicularis throughout Southeast Asia likely reflects this species' adaptation to successional habitats which are currently expanding, particularly due to deforestation by humans. We found ample evidence for this in Sumatra, where fascicularis are found far inland and at elevations of over 1,000 meters in areas where agriculture has replaced rainforest. Given the adaptability of these macaques as demonstrated by the variety of habitats that they occupy and their high population density, we would predict a rapid, nonevolutionary broadening of fascicularis' niche in the absence of nemestrina; the major constraint determining suitable habitat would be the distribution of food types that these smaller monkeys are able to use.

Fooden (1976, this volume) hypothesizes that subsequent waves of dispersal of species of different macaque subgenera may have contributed to the reduction and disjunction of the ranges of those dispersing earlier, presumably through some sort of ecological competition. Since much of Asia has become deforested rather recently through human activities and climatic changes, we hypothesize that the role of competition was indirect. We believe that the reduction in range of the *silenus* subgenus is due to its inability to adapt to encroaching successional habitats; the dispersal into the area of macaques whose ecological strategy capitalized on such habitats further contributed to the decline of the *silenus* group.

In Sumatra, Borneo, and the Malay Peninsula, where primary rainforests are still extensive, *M. fascicularis* and *M. nemestrina* display considerable ecological separation, especially along the habitat dimension, and coexist in sympatry. Only detailed ecological studies can determine niche breadth and niche separation for these species. Areas exist where one or the other or both species are present; the answers await an ambitious and dedicated field worker.

ACKNOWLEDGMENTS

This research was supported by grant RR00166 from the National Institutes of Health, U. S. Public Health Service, to the Regional Primate Research Center at the University of Washington. For their assistance, cooperation and hospitality, we thank the personnel of the Indonesian Institute of Sciences (Lembaga Ilmu Pengetahuan Indonesia), the Directorate of Forestry and Department of Conservation, and many other Indonesians, especially our friend C. L. Darsono. We also thank the staff of the Regional Primate Research Center, especially Dorothy Reese and Orville Smith, for support throughout the study. Richard Holm and Colene McKee helped with the computer analyses, and Phyllis Wood, AIM, and Robyn Tarbet did the illustrations. We appreciate the use of H. D. Rijksen's and C. L. Darsono's photographs. This chapter has benefitted greatly from critical comments by T. W. Schoener, D. G. Lindburg, and Kate Schmitt.

REFERENCES

Angst, W. Pilot experiments to test group tolerance to a stranger in wild *Macaca fascicularis*. *Am. J. Phys. Anthrop.*, **38**: 625–630 (1973).

_____. Basic data and concepts on the social organization of *Macaca fascicularis*. In L. A. Rosenblum, (ed.). *Primate Behavior, Developments in Field and Laboratory Research, Vol. 4*. New York: Academic Press, pp. 325-388 (1975).

Bernstein, I. A field study of the pigtail monkey (*Macaca nemestrina*). *Primates*, **8**: 217–228 (1967).

Bertrand, M. Training without reward: traditional training of pig-tailed macaques as coconut harvesters. *Science*, **155**: 484–486 (1967).

Cant, J. G. H. Population survey of the spider monkey *Ateles geoffroyi* at Tikal, Guatemala *Primates*, **19**: 525–535 (1978).

Chivers, D. J. An introduction to the socio-ecology of Malayan forest primates. In R. P. Michael and J. H. Crook, (eds.). *Comparative Ecology and Behaviour of Primates*. London: Academic Press, pp. 101–146 (1973).

Corner, E. J. H. Botanical collecting with monkeys. *Proc. Roy. Instn.*, **36**: 162 (1955).

Crockett Wilson, C. and Wilson, W. L. Behavioral and morphological variation among primate populations in Sumatra. *Yrbk. Phys. Anthrop.*, **20**: 207–233 (1977).

Delacour, J. *Birds of Malaysia*. New York: Macmillan (1947).

Delacour, J. and Mayr, E. *Birds of the Philippines*. New York: Macmillan (1946).

Ewer, R. F. *The Carnivores*. Ithaca: Cornell University Press (1973).

Fittinghoff, N. A., Jr. Riverine refuging in East Bornean *Macaca fascicularis*. *Am. J. Phys. Anthrop.*, **42**: 300–301 (1975).

Fooden, J. Taxonomy and evolution of liontail and pigtail macaques (Primates: Cercopithecidae). *Fieldiana Zoology*, **67** (1975).

_____. Provisional classification and key to living species of macaques (primates: *Macaca*). *Folia primatol.*, **25**: 225–236 (1976).

Green, S. and Minkowski, K. The lion-tailed monkey and its South Indian rain forest habitat. In Prince Rainier III and G. H. Bourne, (eds.). *Primate Conservation*. New York: Academic Press pp. 289–337 (1977).

Hutchinson, G. E. Concluding remarks. *Cold Spring Harbor Symp. Quant. Biol.*, **22**: 415–427 (1957).

Lancaster, J. B. and Lee, R. B. The annual reproductive cycle in monkeys and apes. In I. DeVore. (ed.). *Primate Behavior*. New York: Holt, Rinehart and Winston, pp. 486-513 (1965).

MacArthur, R. H. and Wilson, E. O. *The Theory of Island Biogeography*. Princeton: Princeton University Press (1967).

Maples, W. R., Maples, M. K., Greenhood, W. F., and Walek, M. L. Adaptations of crop-raiding baboons in Kenya. *Am. J. Phys. Anthrop.*, **45**: 309–316 (1976).

McClure, H. E. Some observations of primates in climax dipterocarp forest near Kuala Lumpur, Malaya. *Primates*, **5**: 39–58 (1964).

Medway, Lord. *The Wild Mammals of Malaya*. London: Oxford (1969).

_____. The Gunong Benom expedition 1967, VI. The distribution and altitudinal zonation of birds and mammals on Gunong Benom. *Bull. Brit. Mus. Natur. Hist.*, **22**: 103–151 (1972).

Muckenhirn, N. A., Mortensen, B.K., Vessey, S., Fraser, C.E. O., Singh, B. Report on a primate survey in Guyana, July–October 1975. Manuscript prepared for Pan American Health

Organization/World Health Organization by National Academy of Sciences, Washington, D.C.

Napier, J. R. and Napier, P. H. *A Handbook of Living Primates.* New York: Academic Press (1967).

Nie, N. H., Hull, C. H., Jenkins, J. G., Steinbrenner, K., and Bent, D. H. *Statistical Package for the Social Sciences, Second Edition.* New York: McGraw-Hill (1975).

Oppenheimer, J. R. and Lang, G. E. Cebus monkeys: effect on branching of Gustavia trees. *Science,* **165**: 187-188 (1969).

Pianka, E. R. On r- and K-selection. *Amer. Nat.,* **104**: 592-597 (1970).

_____.Niche overlap and diffuse competition. *Proc. Nat. Acad. Sci.,* **71**: 2141-2145 (1974).

Rijksen, H. D. *A Fieldstudy on Sumatran Orang Utans (Pongo pygmaeus abelii Lesson 1827), Ecology, Behaviour and Conservation.* Wageningen, The Netherlands: H. Veenman & Zonen B. V. (1978).

Rodman, P. S. Synecology of Bornean primates I. A test for interspecific interactions in spatial distribution of five species. *Am. J. Phys. Anthrop.,* **38**: 655-660 (1973).

Roughgarden, J. Resource partitioning among competing species—a co-evolutionary approach. *Theoret. Pop. Biol.,* **9**: 388-424 (1976).

Schoener, T. W. Resource partitioning in ecological communities. *Science,* **185**: 27-39 (1974a).

_____. The compression hypothesis and temporal resource partitioning. *Proc. Nat. Acad. Sci.,* **71**: 4169-4172 (1974b).

Southwick, C. H., Beg, M. A., and Siddiqi, M. R. A population survey of rhesus monkeys in northern India. II. Transportation routes and forest areas. *Ecology,* **42**: 698-710 (1961).

Southwick, C. H. and Cadigan, F. L., Jr. Population studies of Malaysian primates. *Primates,* **13**: 1-18 (1972).

Southwick, C. H. and Siddiqi, M. R. Population changes of rhesus monkeys (*Macaca mulatta*) in India, 1959 to 1965. *Primates,* **7**: 303-314 (1966).

Vandermeer, J. H. Niche theory. *Ann. Rev. Ecol. Syst.,* **3**: 107-132 (1972).

Wernstedt, F. L. *World Climatic Data.* Lemont, Penn.: Climatic Data Press (1972).

Wheatley, B. P. The ecological strategy of the long-tailed macaque, *Macaca fascicularis* in the Kutai Nature Reserve, Kalimantan Timur. *Frontir,* **5**: 27-32 (1976).

Whitmore, T. C. *Tropical Rain Forests of the Far East.* Oxford: Clarendon Press (1975).

Wilson, C. (Crockett) and Wilson, W. L. Methods for censusing forest-dwelling primates. In S. Kondo, M. Kawai, and A. Ehara, (eds.). *Contemporary Primatology.* Basel: S. Karger, pp. 345-350 (1975a).

_____. The influence of selective logging on primates and some other animals in East Kalimantan. *Folia Primatol.,* **23**: 245-274 (1975b).

Chapter 8
Riverine Refuging in East Bornean
Macaca fascicularis

N. A. Fittinghoff, Jr.
and D. G. Lindburg

INTRODUCTION

Macaca fascicularis is the most southerly macaque species. It occupies a large geographic range (Fig. 8-1) from continental Southeast Asia eastward to the Philippines and to Timor, though not to Celebes (Napier and Napier, 1967). Together with its geography, its morphology and behavior have evoked many species and subspecies names (Hill, 1974), which naturally create confusion. Taxonomists have commonly called it "*Macacus cynomolgus*" and "*Macaca irus*," but *M. fascicularis* is the appropriate name (Fooden, 1976, and citations therein). Current handbooks list 20 or 21 subspecies (Hill, 1974; Napier and Napier, 1967), but macaque taxonomy is subject to revision (e.g., Fooden, 1976). The English common names of the species include "long-tailed," "Java," and "crab-eating" monkey or macaque. In eastern Borneo, where we studied it in Indonesia's Kutai Nature Reserve, villagers call it *kra*, after its alarm call, and this short word, which is both singular and plural, is the vernacular name we shall use for convenience.

The Kutai Reserve

The Kutai Reserve (Figs. 8-1, 8-2) is a forest preserve and wildlife refuge located less than 35 minutes north of the equator. When established in 1936, it contained over 300,000 hectares (Wheatley, this volume). In recent years its size has decreased as its borders have been changed to accommodate Indonesian logging companies, but it still contains large tracts of only moderately disturbed forest. It also contains many rivers and streams, which, except for aircraft and loggers' roads, are the chief means of human transportation in such areas, and which are important, also, in the ecology of *kra*.

The northernmost Reserve, along the Sengata River, was the region of our study. Rodman (1973b), Kurland (1973), Pearson (1975), and Wheatley (this volume) together characterize this region, commenting on its climate, geology, topography, flora, and fauna, and Davis (1962) and Medway (1965) further describe the mammals. Here and later, therefore, we provide little information on the Kutai, except for observations incidental to our field research.

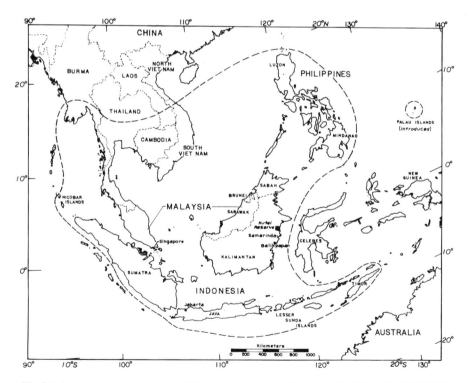

Fig. 8-1. Approximate species range of *Macaca fascicularis*. (From Fooden, 1971, Fig. 2; Hill, 1974, Map 5; Napier and Napier, 1967; Poirier and Smith, 1974; Wilson and Wilson, 1977; *National Geographic* map, "Asia.")

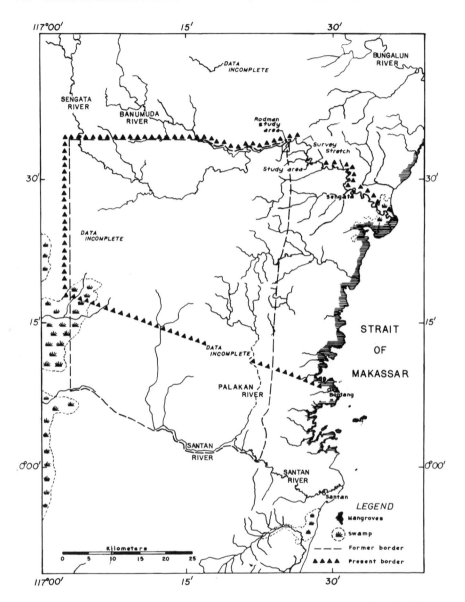

Fig. 8-2. Kutai Reserve borders and waterways. (From Rodman, 1973b; Indonesian Forestry Department sketch map; U.S. Army Map Service Series T503 (1961), Maps NA 50-15 and SA 50-3.)

Duration and Location of Research

Our stay on the river began on January 26, 1973. The initial lodgment was a tent camp on the south bank about 10 kilometers below the area where Rodman (1973a,b) studied the orang-utan and four other anthropoid species (cf. Fig. 8-2). On March 2 we relocated about five kilometers downstream, moving to a study area that offered more *kra* and a more accessible location for a permanent research facility (the Kutai Research Station, which we established with matching funds from the University of Washington School of Medicine and the Indonesian Departments of Forestry and Conservation). We then studied *kra* jointly until July 15, 1973, when Lindburg returned to the United States. Thereafter Fittinghoff continued the study until January 31, 1974. During the study year we accumulated 247 man-hours of fairly close observation of *kra* in or near their sleeping-trees and 71 man-hours of less close observation in the forest. In addition, Fittinghoff logged about three hours of visual contact with *kra* during 130 hours of boat survey along a measured stretch of river.

River-Survey Stretch and Study Area

The research station (Figs. 8-3, 8-4) was placed two kilometers from the downstream end of the survey stretch, 10 kilometers of river front that was surveyed on 48 afternoons for the presence of *kra*. To establish the stretch, Fittinghoff and two assistants chained it off 25 meters at a time and planted a tall post, striped with paint so as to be numbered, every 100 meters along the reserve side. At the time of the study, no one lived within the stretch above the station, but Indonesian farmers and a logging company occupied areas below the station on the opposite side.

The study area itself (Fig. 8-4) occupied about 125 hectares, and it contained areas of forest which we identified as "primary" or "secondary," depending on whether tall trees or second growth predominated. These areas were first identified subjectively; then we measured trees to place the identifications on a more objective basis. In our "primary" forest, working along 1,000 meters of transect 2 meters wide, we found 30 tree trunks at least 30 centimeters in diameter encroaching on the transect. The comparable figure for tall "secondary" forest was eight trees. In addition, the mean diameter of the 30 trees was 50.4 centimeters; of the eight, 36.1 centimeters. Finally, the mean height of 15 trees that formed part of the main canopy was 30.8 meters in the "primary" forest, but only 17.8 meters in the "secondary."

The edge of the primary forest was mapped approximately (Fig. 8-4). In the Holdridge Life Zone system (Holdridge *et al.*, 1971), this forest would be identified with the Tropical Moist Forest Life Zone, and in general, it could be

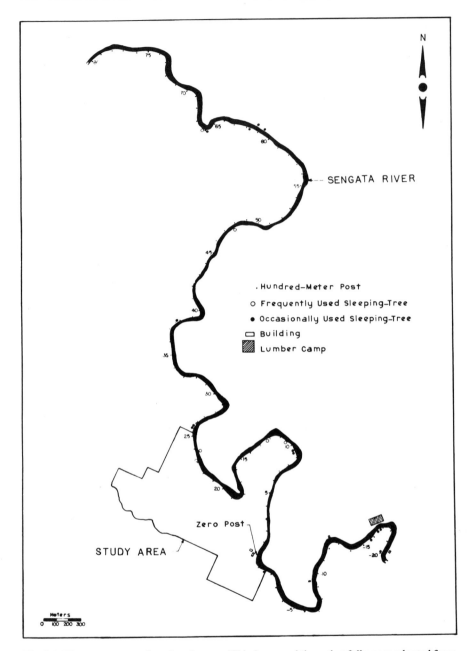

Fig. 8-3. River-survey stretch and study area. (This figure and those that follow are adapted from Fittinghoff, 1978.)

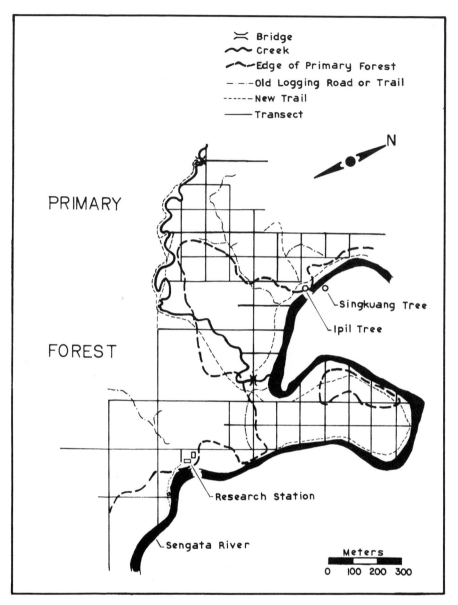

Fig. 8-4. The study area.

classed as mixed lowland rainforest. It was highly variable, containing many tree species of different sizes and shapes. At one extreme, it encompassed all stands of trees, including some individuals with trunks more than 1.5 meters in diameter, that we could regard as climax forest; at the other, small stands of second growth.

In contrast, the riverine forest included all large stands of second growth.

Some of these probably reflected repeated flooding, but others probably arose after the river margins were abandoned by slash-and-burn farmers. The former presence of agriculture was suggested by the remains of a house, the presence of pepper bushes whose fruit is relished as a spice, and the presence of large, emergent trees in areas of low, trashy, second growth. These emergents would have been hard for peasant farmers to remove; so they were presumably left standing when the forest was cleared.

No contour map was made of the study area, but the edge of the primary forest marks a change in profile. Except near the creek that bisects the area (Fig. 8-4), an observer who heads into the deep forest has to climb almost at once, because the ground rises in intersecting series of ridges. Some of the ridges are broken by steep escarpments, but none is high. The river level at the station is about 20 meters above sea level, and the highest ground of the study area may be no more than 50 to 60 meters above the river.

The largest trees were found on the slopes of the ridges, and several slopes showed signs of hand logging. Although none of the logging appeared recent, we assessed the degree of disturbance by censusing 16 primary-forest hectares for cut trees. Fifty-six cut trees were found, of which 55 had been cut for wood and one for honey. The mean number of trees cut per hectare was therefore 3.5; however, there was no sign that this degree of disturbance had harmed the resident fauna. Overall, the study area was rich in reptiles, birds, small mammals, deer, wild pig, and primates.

Anthropoid Species

In particular, the area proved to contain parts of the home ranges of three large groups of *kra*. However, four other anthropoid species were also seen. These were the orang-utan (*Pongo pygmaeus*), a gibbon (*Hylobates muelleri*), the Sunda Island leaf monkey (*Presbytis aygula*), and the pig-tailed macaque (*Macaca nemestrina*). Of these four, the apes and the leaf monkey were seen frequently, but the macaque only occasionally.

The proboscis monkey (*Nasalis larvatus*) is common along the Sengata estuary, but was not seen in the study area. Five individuals did appear briefly along the survey stretch 500 meters below the research station.

KRA OF THE NORTHERN KUTAI

Physical Appearance

Kra are small, gracile, long-tailed monkeys, clearly one of the lightest macaque species (Napier and Napier, 1967). In their equatorial range their size and proportions seem highly adaptive; for example, they are much more abundant

(Southwick and Cadigan, 1972; Wilson and Wilson, 1977), and probably represent more biomass, than pig-tailed macaques, which conform less well to Bergmann's and Allen's rules. Moreover, feral *kra* spend most of each day in trees, and small size and long tails are common adaptations among arboreal cercopithecines, such as *Cercopithecus* species. Sengata River *kra* belong to the Borneo-Sumatra-Malay-Peninsula subspecies, *M. fascicularis fascicularis*, which one might expect to be larger than the subspecies from the smaller islands of Indonesia. Our impression (based on comparisons with captive *M. mulatta* males) was that very big Sengata River males might weigh 10 kg. Adult females usually ranged from half to three-fourths the size of adult males (cf. Napier and Napier, 1967; Crook, 1972).

Pelage colors vary on the Sengata. Infants are black at birth, but lighten within a few months to gray or gray-brown, except for the pate, which may remain dark for many months. Maturing monkeys lighten further, but some remain gray dorsally, while others become a tawny agouti. In mature animals the ventrum and the medial surfaces of the legs become very light buff (Kurland, 1973) or, much more often, a shade of gray so light as to be effectively white. The pate often turns rich brown or chestnut, though this colorful "cap" apparently fades to a grizzled golden brown in aged monkeys. The cap varies in width: some individuals have bordering stripes, which are usually blond in immatures and gray or white in adults. Kurland (1973) describes the face as black, but we recall no dark-faced adults: typically the faces of adults are gray or tawny, and so are those of many immatures. The eyelids and the skin above them are light, in contrast to the face, and these areas function in communication (Shirek-Ellefson, 1967; personal observations). Some individuals have a dark malar stripe below each eye; perhaps, in a riverine species, this reduces glare from the open sky and the water.

Mature females often have a ruff of long hair at the sides of the face and beneath the mandible; the effect is that of "mutton-chop" whiskers or a beard. Kurland (1973) reports that adult males may have cheek whiskers and moustaches, but these were inconspicuous or absent in the males we observed.

Sexual swellings were either absent or so inconspicuous as to be of little help in identifying estrous females. A. Eudey (pers. comm.) has photographed red-faced *kra* in Thailand, but we saw no red faces along the Sengata. The scrotal color is pink, and color changes are slight, if they occur at all.

Response to Observation

Sengata River *kra* pay little heed to passing boats, but respond to persistent observers with various escape tactics, suiting the tactic to the occasion (Fittinghoff, 1978). Six months or more of pursuit and intermittent contact may be required to habituate a group to close observation (Wheatley, pers. comm.).

RIVERINE REFUGING

Borneo naturalists from Wallace (1869), Hornaday (1885), and Beccari (1904) to the present, describe *kra* as common river monkeys. A reminiscence of Hornaday's (1885, p. 358) from Sarawak might almost stand as a report of our own first encounter with *kra* on the Sengata:

> Just before sunset we passed the last Dyak village and clearing, and came to where the large trees and dense undergrowth clothed the banks to the water's edge and even beyond. Then we began to see monkeys by the score, and as evening approached their numbers seemed to increase as they began to perch in the branches that overhung the river, and settle themselves for the night. Sometimes as many as five or six would be seen sociably huddled together on a single bough, and often one small tree-top contained fifteen to twenty of the little animals. They were all of one species, *Macaca cynomolgus*, the commonest in Borneo, and by the natives it is called the *krah*. They are about the color of a gray squirrel, and three times as large . . . . I counted them as we paddled along until in a few minutes I ran the number up into the eighties, and was obliged to give up the attempt.

At sunset of our first day on the river, as we cruised upstream from the coast, we repeated Hornaday's experience. *Kra* were roosting or coming to roost in riverine trees, and it soon became clear that they are a refuging species.

"Refuging" has been defined by Hamilton and Watt (1970) as "the rhythmical dispersal of groups of animals from and their return to a fixed point in space." This applies to *kra*, which leave familiar trees to forage during the day and return to them to sleep. Hamilton and Watt go on to distinguish "radial-pack" refuging systems, like that of savanna baboons, in which a large troop travels as a unit, from what might be called "pure" refuging systems, in which even larger groups of adult animals disperse and return as individuals. With respect to these categories, *kra* have an intermediate system: a large group (20 or more monkeys) often travels as a unit, but also often fragments into separately foraging subgroups of various sizes, compositions, and durations (personal observations; cf. Kurland, 1973). Often the scattered subgroups remain loosely associated, and usually they come together before nightfall, but sometimes a subgroup forages separately for several days before rejoining the main group, and sometimes peripheral or semisolitary individuals join or leave the group temporarily. As a result, the size of the sleeping-group may vary from night to night at any communal roost. The animals that forage and sleep together display social organization, but the looseness of the group somewhat recalls *Pan, Ateles,* or *Cercopithecus mitis*; it seems atypical of *Macaca*. We want to suggest that this flexible socioeconomic system is a response to patchy food distribution, food competition, and low predation pressure, but before invoking functions and causes, we shall need to present *kra* refuging in more detail.

Refuging in the Study Area

In the study area we focused our attentions chiefly on two large groups of *kra*, which we called *L* group and *R* group, and which slept, respectively, in an *ipil* tree (*Intsia palembanica*) and a *singkuang* tree (*Dracontomelon mangiferum*) that stood opposite each other across the river (Figs. 8-4, 8-5). We observed these groups at nightfall or at dawn, or at both times; so that for 167 nights we knew whether each sleeping-tree was used. Members of *L* group used the *ipil* tree on 90 percent of those nights, and members of *R* group used the *singkuang* tree on 68 percent. On the nights that the groups were absent, they presumably used one or more auxiliary trees in remote parts of their ranges: we were able to verify the use of auxiliary trees occasionally.

During our observations at the sleeping-trees we obtained 51 good counts of *L* group and 19 good counts of *R* group. The distributions of these counts are

Fig. 8-5. The *ipil* tree (left) and *singkuang* tree (right), opposed across the Sengata River. Inset: An adult female *kra* peers at Fittinghoff's passing boat from a tangle at the water's edge.

given in Fig. 8-6. The variation in the counts reflects the sleeping of subgroups apart from the main group. As suggested above, most separations appear to have been temporary, but the highest departures from the modal group counts suggest the presence of monkeys that slept with the main group only rarely.

Refuging Along the River

We watched L group and R group most often from the forest edge, but these and other groups were also observed during the 48 boat surveys. Each survey consisted of a 20-kilometer round trip along the survey stretch, starting and ending at the research station. Odd-numbered runs traveled upstream first, and even-numbered runs traveled downstream first, so as to avoid, or at least reduce, spatial and temporal bias in observing *kra*. The following variables were recorded for each observation: side of river, 100-meter segment number, vertical location (whether arboreal, quasi-terrestrial, or terrestrial), time of day, and number of animals seen. From these records we have obtained the

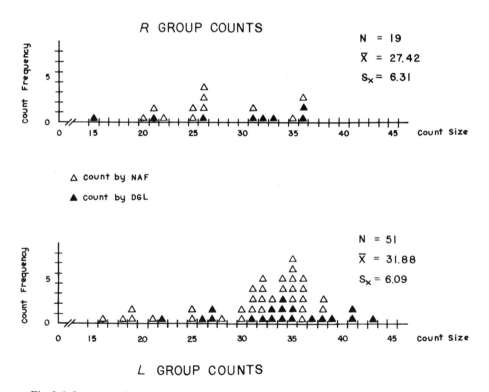

Fig. 8-6. *L*-group and *R*-group counts, taken as the monkeys returned to their sleeping-trees. N, sample size; $\overline{X}$ mean; S_x, standard deviation.

Table 8-1. Observed Frequency Distribution of Sighting Sizes Compared with a Zero-Truncated Negative Binomial Distribution.

SIZE	OBSERVED	EXPECTED
1	116	130.77
2	122	107.97
3	83	86.47
4	83	68.16
5	64	53.14
6	31	41.12
7	31	31.71
8	29	24.36
9	18	18.63
10	10	14.20
11	9	10.84
12	4	8.22
13	7	6.29
14	5	4.74
15	3	3.61
16	1	2.74
17	0	2.06
18	0	1.56
19	1	1.18
$\geq$ 20*	6	5.23
	623	623.00

Df. = 17**	X^2 = 21.28***	.30 > p > .20

* Specifically, 21,24,26,27,27,30
** Two degrees of freedom were sacrificed because two parameters used to calculate expected frequencies were estimated from the data.
*** Often repeated observations of the same subjects vitiate a X^2 test. By inflating the sample, they artificially increase the test's power to reject the hypothesis of no difference between observed and expected values. Here, despite an unknown number of such repetitions, the test failed to reject at any usual level of significance.

frequency distribution of initial-sighting sizes and the frequency distributions of *kra* sightings in time and space.

A sighting is defined here as one or more *kra* seen within a given 100-meter segment on either side of the river, and a sighting size as the number seen. The distribution of initial-sighting sizes (Table 8-1) supports the suggestion that *kra* groups are loosely organized. Small counts preponderate, and the mean of the distribution is 4.34, much less than one would expect if *kra* invariably traveled, foraged, and roosted in cohesive groups of, say, 15 or more monkeys.

This sample mean must underestimate the true mean slightly. Some inaccuracy is inevitable when riverine monkeys are counted from a moving boat, and most errors are omissions. Some members of a group or subgroup may be present at the river but concealed by foliage; some may still be approaching through the forest; a large group may be hard to count if alarmed; rarely, a cohesive group may occupy two adjacent segments or both banks of the river, etc. Yet 73.5 percent of initial sightings involved monkeys in trees, and almost all such monkeys were roosting quietly when seen. Roosting monkeys were relatively easy to count, for many groups favored *bajur* (*Pterospermum* spp.) or similar tall, spindly, lightly foliated tree species that grew in clumps along the river bank. Each segment was observed twice each run, and counts were made whenever monkeys were sighted. Often a group or subgroup was counted both times a segment was observed. If one includes only the larger of two such counts in the sample, then the mean of 356 "arboreal" counts made after 1730 hours, when most *kra* were in their sleeping-trees, was 4.97; the median, four; and the mode, two. These figures, again, are far smaller than one would expect if all members of each *kra* group invariably slept together in a single tree or two or three adjacent trees.

Our own confidence in this finding is enhanced by the form of the initial-sighting size distribution. Such a form (approximately negative binomial) has been repeatedly associated with equilibrium size distributions of subgroups within nonhuman primate species (e.g., Cohen, 1969a, b; 1972). Thus, if replication studies were to show that *kra* subgroups usually have such a size distribution, one might infer that the sampling procedure did not seriously falsify the actual distribution of subgroup sizes.

Fig. 8-7 shows how initial sightings were distributed with respect to time. Each "A" represents the first arboreal sighting of an evening within a given 100-meter segment; duplicate sightings obtained on the return trip are excluded. Similarly, each "T" represents the first terrestrial or quasi-terrestrial sighting. Nevertheless, these data are biased, because it proved impossible to observe each five-minute interval an equal number of times during 48 runs. But by using a regression equation (Fittinghoff, 1978), one can remove the bias, as shown in Fig. 8-8. The figure suggests that a slight peaking of the "T" distribution occurs after 1600 and that the "A" distribution starts to climb near 1700. The "T" distribution simply reflects terrestrial or quasi-terrestrial activity along the river, but the "A" distribution reflects both the final climb into the sleeping-trees and late arrivals from the forest canopy.

Fig. 8-9 presents the spatial distribution of 100-meter segments occupied by *kra*; again, duplicate sightings are excluded. For either side of the river, it is easy to show that the distribution of occupied segments is nonrandom. If it were random, the frequencies of segments with 0,1,2,3, ... occupancies would have a Poisson distribution, with mean and variance equal, or nearly so. Actually the ratio of variance to mean is greater than 10 for the Kutai Reserve side and greater than five for the opposite side. These ratios indicate distributions more clumped than random distributions would be, and if one goes on to examine the distributions of terrestrial, quasi-terrestrial, and arboreal

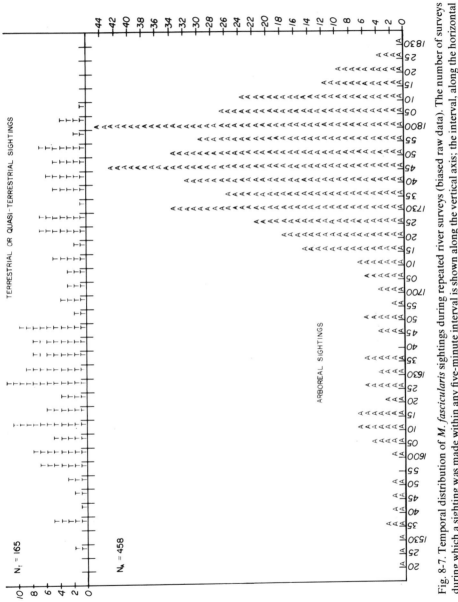

Fig. 8-7. Temporal distribution of *M. fascicularis* sightings during repeated river surveys (biased raw data). The number of surveys during which a sighting was made within any five-minute interval is shown along the vertical axis; the interval, along the horizontal axis.

TERRESTRIAL OR QUASI-TERRESTRIAL SIGHTINGS

$N_T = 165$

ARBOREAL SIGHTINGS

$N_A = 458$

195

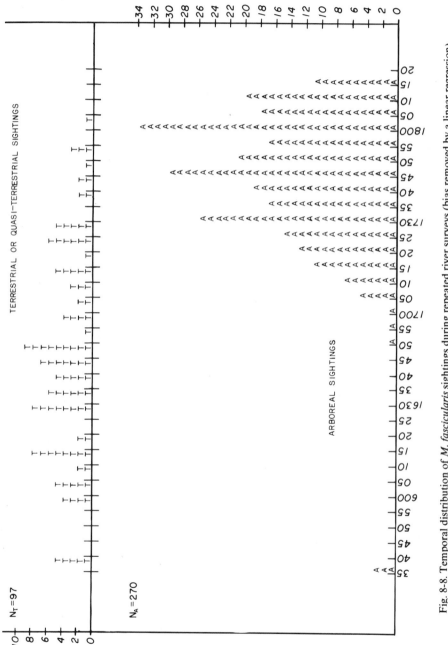

Fig. 8-8. Temporal distribution of *M. fascicularis* sightings during repeated river surveys (bias removed by a linear regression).

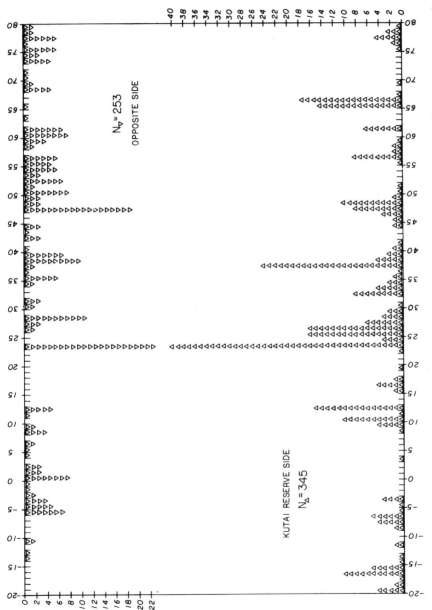
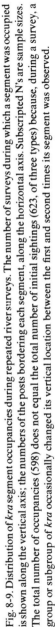

Fig. 8-9. Distribution of *kra* segment occupancies during repeated river surveys. The number of surveys during which a segment was occupied is shown along the vertical axis; the numbers of the posts bordering each segment, along the horizontal axis. Subscripted N's are sample sizes. The total number of occupancies (598) does not equal the total number of initial sightings (623, of three types) because, during a survey, a group or subgroup of *kra* occasionally changed its vertical location between the first and second times its segment was observed.

197

sightings in space, it becomes clear that the clumping is due to the monkeys' converging on their sleeping-trees. In Fig. 8-10, for example, the 41 arboreal sightings of segment 24 are sightings of *L*-group monkeys in their *ipil* tree, and in Figs. 8-10 and 8-11 each segment with six or more arboreal occupancies reflects the presence of a sleeping-site.

SUPPLEMENTARY OBSERVATIONS AND DISCUSSION

Given this description of *kra* refuging, it still remains to account for the form it takes. Here the right question to ask appears to be, Why do the monkeys come to the river to sleep? Why should they roost just there? *Kra* forage arboreally in tall forest, whether primary or secondary, and even there, as we shall try to show, their food is patchily distributed. This being so, why don't they sleep in or near their food trees, moving from one food source to the next as supply dictates? Lindburg (1977) describes a foraging pattern more like this alternative for rhesus monkeys in a North Indian forest. Even though rhesus monkeys forage lower and range farther each day than *kra*, such a pattern makes more overt sense, in bioenergetic terms, than returning each night to the river. One's sense of anomaly is sharpened by the discovery that *kra* day-journeys are often dependent. In the early summer of 1973 *L* group made repeated morning visits to an area along the creek where a number of bottom-land trees were fruiting heavily. Some seldom-used sleeping-trees were located nearby, within the forest and beside the same creek, but they were occupied infrequently, if at all, by *L*-group monkeys.

Kra typically forage both up- and downstream from their main sleeping-site, as well as inland. Thus the location of the sleeping-site is somewhat central. In the long run, because food sources are temporally and spatially variable, or because *kra* may require a varied diet, returning to a central place may save energy (Horn, 1968). But, again, why should that place be riverine, rather than central to the entire home range?

Predation

One possible explanation, the first that occurred to us, is that the familiar riverine tree offers safety from predators. Potential predators exist in the form of a python species, large monitor lizards, small cats, a small leopard, a small bear, and raptorial birds. Yet in 15 months at his study site Rodman (pers. comm.) found no predation on primates, and we found none at ours. Wheatley's later, 20-month study at our site had the same result (Wheatley, pers. comm.). Moreover, our chief assistant, who was also a game warden, suggested that nothing eats *kra*. This seems doubtful, but *kra* themselves, though cautious in some situations where they might be surprised, often seem

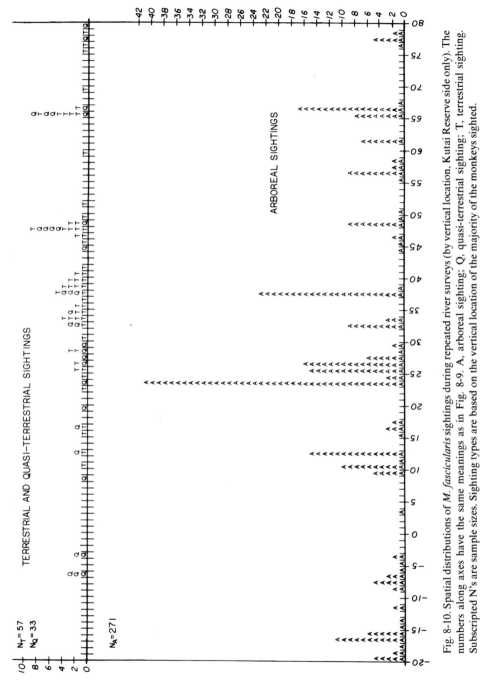

Fig. 8-10. Spatial distributions of *M. fascicularis* sightings during repeated river surveys (by vertical location, Kutai Reserve side only). The numbers along axes have the same meanings as in Fig. 8-9. A, arboreal sighting; Q, quasi-terrestrial sighting; T, terrestrial sighting. Subscripted N's are sample sizes. Sighting types are based on the vertical location of the majority of the monkeys sighted.

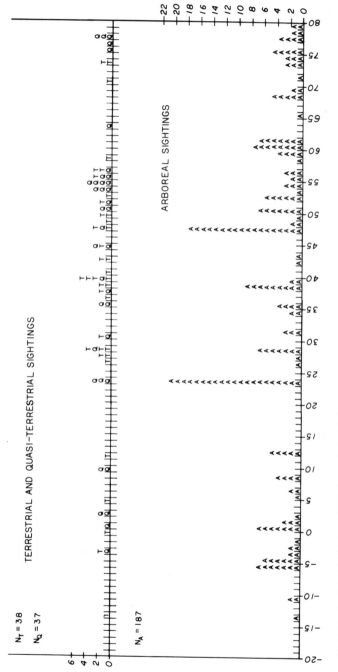

Fig. 8-11. Spatial distributions of *M. fascicularis* sightings during repeated river surveys (by vertical location, side opposite Kutai Reserve only). To be interpreted like Fig. 8-10.

remarkably careless of predators. The first author, for example, has seen pairs of monkeys or lone individuals conspicuously asleep in small riverine trees, seen a pair of *L*-group yearlings come by themselves to the *ipil* tree just at full dark, and seen *R* group proceed by moonlight along the river bank well after dark. Finally, more reptiles occur near the river than elsewhere in the forest. On 10 different days one or both of us found one to four small *Python reticulatus*, easily large enough to kill and eat young *kra*, resting in a shrub below the *ipil* tree, and on seven of those days *L* group went to sleep above them, even though the crown of the tree was accessible from the shrub. For reasons such as these, even though some predation almost certainly occurs, we think predators have only minor influence on the refuging pattern.

Competition

A clue to a more important influence comes from Rodman's study (1973a). He found the same five anthropoid species in his study area, and he showed, by pairwise partial correlations, that seven of the 10 possible pairs of species avoided each other more than they should have done by chance. In each case he concluded that the mutual avoidance reflected niche overlap, and that the species concerned were competing for the use of the forest. In particular, *kra* competed for food with orang-utans, gibbons, and leaf monkeys, though not with pig-tailed macaques: Rodman found almost no overlap in the areas ranged by the macaque species.

Although our observations of primates other than *kra* were unsystematic, they support Rodman's; so *kra* presumably competed with orang-utans, gibbons, and leaf monkeys, as well as other frugivores like hornbills and squirrels, in our study area also. Direct competition was observed between orang-utans and *kra*. Both species eat *ipil* pods, and on two occasions orang-utans invaded the *ipil* tree in search of pods and remained there with *L* group overnight.

Rodman's *kra* observations concerned one group of 16 monkeys, and he estimated that their diet was 98 percent fruit (Rodman, 1973b). Similarly, a high proportion of fruit probably characterized the diet of *L* group, which foraged arboreally most of the time. Knowing that Wheatley planned a detailed study of *kra* diets, we studied them only briefly, mainly with the idea of identifying reliable food sources that might account for the presence of the monkeys. After we determined some of *L* group's favorite morning foraging routes (Fig. 8-12), our assistant was asked to census 14.3 primary-forest hectares along the main route for mature food trees. This assistant, Sahar, was familiar with *kra* foods, for he had previously assisted Rodman in his study. Also, after 54 years on the Sengata, he knew the local names of many plants, especially those of medicinal, commercial, or nutritional value. He accom-

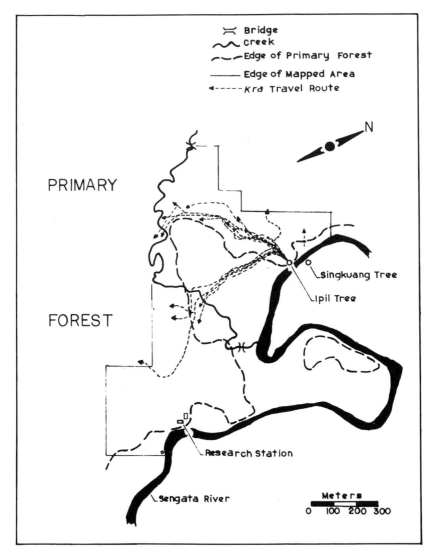

Fig. 8-12. Routes traveled by *L* group during morning foraging. The routes end where the monkeys scattered for prolonged feeding, where they eluded the observer, or where the observer voluntarily broke off observation. Southwesterly routes predominated during the second and fourth quarters of 1973; southeasterly routes, during late June and early July.

plished the census by walking the perimeter and diagonals of each hectare quadrat with the aid of our map.

The census produced a mean of 4.62 large food trees per hectare. In focusing on large trees, it overlooked many smaller specimens of food-bearing species (cf. Wheatley, this volume), but it still makes clear that *L* group exploited an area rich in potential food sources. Even in this area large food trees were unevenly distributed, but the real patchiness of food distribution is revealed by 10 hectares of low secondary forest (Fig. 8-13). These were not censused, but a census would probably have found very few mature food trees. These

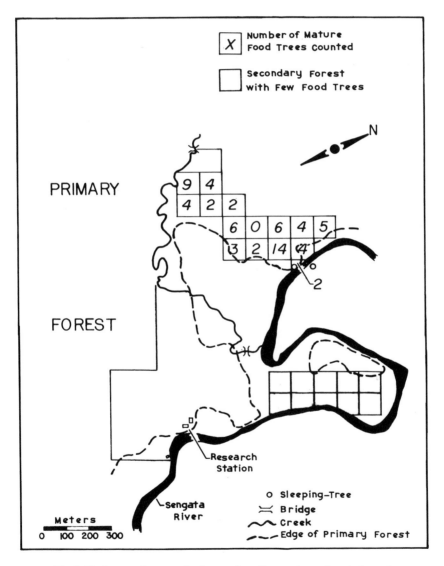

Fig. 8-13. Census of mature food trees along *L*-group's southwesterly route.

quadrats are noteworthy, also, because we never encountered *kra* in them, even though *kra* often used the adjacent taller forest and sleeping-trees along the river (Figs. 8-3, 8-4). The apparent disuse was probably only relative, for the 10 quadrats were not canvassed regularly. Yet in laying out their bordering transects, mapping them, and searching them repeatedly for monkeys, we encountered none. Moreover, Fittinghoff boated past the four riverine quadrats more than 120 times without ever sighting a primate.

Besides *L* group and *R* group, we occasionally observed *P* group, a third large group that ranged near the research station. In contrast to *L* group, *P* and *R* both foraged extensively along the river (Fig. 8-14). *R*-group monkeys, for example, often emerged from the forest around 1600 at points up to 300 meters from their *singkuang* tree; then foraged their way home for as much as two hours before dark. Across the river, *L* group usually came home downridge through the primary-forest canopy. *L* group may have been inhibited from terrestrial foraging by our presence on their side, not far from the *ipil* tree, but we think the difference was chiefly occasioned by a greater proportion of immature forest in the ranges of the other groups and by more attractive riverine vegetation.

Fig. 8-15 shows the structure of the riverine vegetation along the survey stretch. There are three fairly distinct zones: one of water grass or reeds; one of ground vines, bushes, and shrubs; and one of forest, either primary or secondary. The significance of the two rivermost zones is that among the five anthropoid species present, only *kra* exploit them. At a minimum, exploiting these zones effectively requires small size, which allows an animal to penetrate vine clusters, bushes, and shrubs without becoming entangled and without breaking them down; quadrupedalism, which permits rapid terrestrial and quasi-terrestrial locomotion; and omnivory. Although forest fruit is the staple food, *kra*, like the other macaque species that have been studied, are opportunists. Along the survey stretch, they ate not only forest flowers and fruits, but also the flowers and fruits of vines, bushes, and shrubs; the new growth of water grass, reeds, and vines; the immature seed heads of water grass; bark; dried sap or gum; insects; birds' eggs or nestlings; and man's crops and garden products.

Among the primates that might compete with *kra* at the river, the orang-utan is too large, too slow and clumsy on the ground, and too demanding of food to be effective. The gibbon would have to give up the great locomotor advantage it enjoys in tall forest. Although larger than the *kra*, the pig-tailed macaque might still compete, but it hardly needs to, being spatially almost separate. (In the Kutai region pig-tails range farther from rivers, often traveling on the ground, and Rodman (1973c) has suggested that ground travel is an adaptation to feeding in widely separated fruit trees.) The leaf monkey, however, often ranges near the river, and it is most like *kra* in size and weight. It is likewise quadrupedal—but its folivorousness makes descent to the river unnecessary,

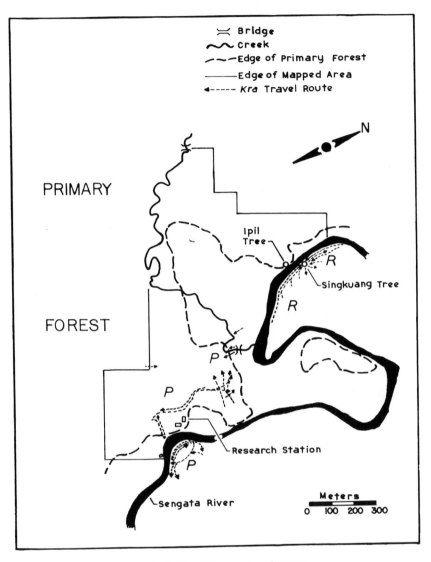

Fig. 8-14. *P* and *R* group travel routes.

not to say disadvantageous. In short, *kra* enjoy a locomotor-feeding advantage at the river that goes far towards accounting for the refuging pattern.

Any given group of *kra*, however, must also compete with neighboring groups of its own species. Although the evidence is circumstantial, it appears that roosting at the river helps in this competition too, particularly with regard to river crossings. As already noted, Sengata River *kra* are highly arboreal, but

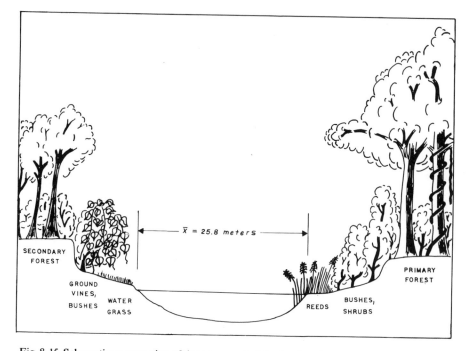

Fig. 8-15. Schematic cross-section of river-survey stretch, showing characteristic vegetation zones and types. Mean width based on 164 range-finder measurements.

they are also excellent swimmers. On separate occasions each of us saw them dive and swim under water to avoid men in boats, and each occasion involved a mother crossing with an infant on her back. Small juveniles were also seen to swim the Sengata unaided. In other words, the river is no barrier: its current is slow, and its variable width makes it passable throughout most of its length. A group of *kra* is free, if unopposed, to range across the river from its sleeping-site, and a group that hopes to maintain an exclusive home range must contend with invaders from across the river as well as from up- and downstream.

A logical strategy for would-be invaders, therefore, is to monitor *kra* activity across the river, in the hope of finding none; and a logical strategy for would-be defenders is to be conspicuously present, at least part of the time. *Kra* are not classically territorial; we saw no agonistic encounters between groups, although, like Kurland (1973), we often saw or heard encounters between group members during foraging or travel. Our groups lived in moderately overlapping home ranges (Fig. 8-16), and they secured adequate resources chiefly by discouraging the encroachments of other groups and by avoiding them, just as they avoided other primate species.

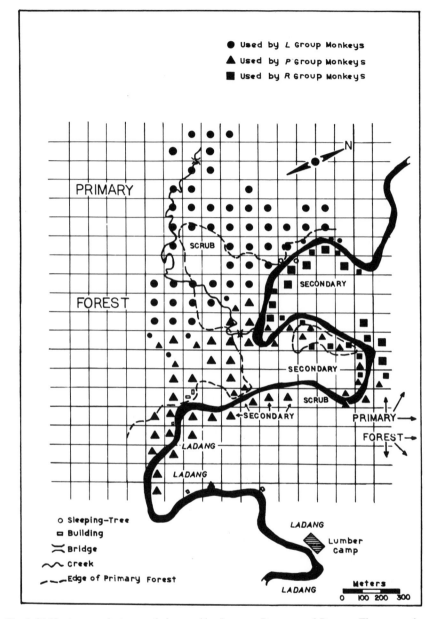

Fig. 8-16. Hectare quadrats recorded as used by *L* group, *P* group, and *R* group. The ranges shown are incomplete; also, the areas of overlap are approximate and "conservative," because subgroups and individuals of unknown provenance were assumed to belong to the nearest resident group in the absence of other clues to group membership. *Ladang:* a field cleared for swidden agriculture.

To avoid one another and still forage efficiently, however, groups of *kra* need to know one another's whereabouts. Leaf monkeys, gibbons, and male orangutans can exchange location information by means of calls that carry long distances through the forest, but *kra* must obtain it visually for the most part, because they lack such calls. The loudest sounds that they make are their alarm calls and the shrieks, screeches, snarls, and roars that occur when group members fight among themselves, as they often do in their sleeping-trees. These sounds carry better over water than through the forest, but their significance as intergroup signals is unclear, unless they serve to remind other groups that the fighting monkeys are quarrelsome. The need for visual information, incidentally, may be one reason why *kra* form subgroups when foraging. Such subgroups not only find more food than a single large group would; they also see and are seen throughout more of the group range.

In place of distance-maintaining calls, *kra* rely on branch-shaking to discourage approach. Branch shakes were seen chiefly at the river, where they are most easily seen and presumably most effective. Depending on the stimulus that evokes it, a branch shake may be prolonged, repetitive, and violent or somewhat gentler and brief. A passing observer who stares at unhabituated *kra* in their sleeping-tree may receive the first kind, usually from a big male. However, neither of us saw a violent shake exchanged between *L* group and *R* group, which knew each other well. Instead, we repeatedly saw a brief shake incorporated into a ritual. Big males would enter the sleeping-trees at the base of the crown, move slowly up main branches to favorite perches, and give the branches quick shakes before sitting down. These shakes were often given before the opposing group arrived; so they may be primarily an intra-group signal. However, they were often visible to both groups, especially after the sleeping-trees (which were deciduous) shed their leaves. Presumably the shakes conveyed assertions of ownership.

Over time, therefore, these displays might discourage river crossings. But before accepting that they help account for riverine refuging, one ought to ask whether conspicuousness at the river was effective and whether it was general.

With regard to effectiveness, the answer is that it seems to have been. Fig. 8-17 shows the river crossings that were observed or inferred near the study area. No doubt this sample is far from exhaustive. Yet, even though most of our observation time was given to *L* and *R* groups, we never saw either group invade the other's range. This observation held even though both the *ipil* tree and the *singkuang* tree provided food for *kra*, even though their fruit ripened in different months, and even though on occasion the absence of one group would have permitted a raid by the other. By contrast, the northeast sector of the study area, which we called The Loop, was invaded from several directions. *Kra* sometimes slept at the closed end of The Loop, but no group occupied it at all times, probably because the small area of tall forest that attracted the

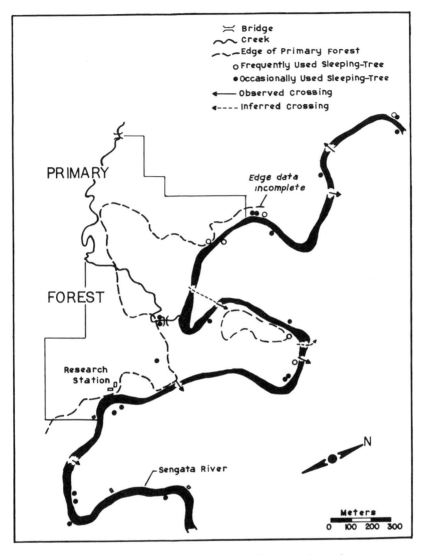

Fig. 8-17. River crossings by one or more *kra* in or near the study area.

monkeys could not support a permanent population. From the directions in which monkeys entered and fled The Loop, we judge that both *R* group and *P* group used it on separate occasions.

Although conspicuousness seems to forestall large-scale invasions, it may be less effective against individuals or subgroups. Wheatley (this volume) later saw adult males twice invade *L* group, wrest group control from resident males, and mate with adult females. He reports that the successful invaders came from *R* group. It is hard to estimate the frequency of such takeovers, for a flood toppled the *singkuang* tree early in Wheatley's study, and the takeovers occurred within the next 14 months. Loss of the sleeping-tree may have altered the ranging patterns of *R* group. If so, it may have produced some group disintegration, which, in turn, may have fostered the takeovers.

With regard to generality, both branch-shaking and the use of leafless sleeping-trees were part of a pattern of conspicuousness that obtained all along the survey stretch. The pattern also included episodes of playing and resting on beaches or piles of driftwood, processions along the river bank, and in the case of one group, the choice of a bare snag as a sleeping-tree. Ridley (1906, p. 142) comments on choices of sleeping-tree, but interprets them differently.

> . . . The wild monkeys always sleep in particular trees, those with bare branches and very lofty, and towards evening they may be seen slowly moving along, stopping here and there to eat, till they reach the sleeping place about sundown, they then settle down for the night, sitting usually in pairs or singly on the bare boughs. The same tree is occupied every evening for weeks at a time, and wherever they are in the evening they make for the same spot. They never sleep in a bushy tree, probably for fear of being surprised at night by snakes.

Having seen *kra* roost in large riverine bushes beside farmers' fields, move from bare branches to foliated terminal forks for sleeping, and apparently ignore nocturnally active pythons, we prefer another interpretation. Choices of sleeping-tree figured prominently in the pattern of conspicuousness, two elements being particularly clear. One was that *kra* chose a tree with a view. Sleeping-trees stood close to the river, often overhanging it so that a *kra* above the water could see, and be seen from, both sides of the river. Second, *kra* suspiciously often chose trees within easy view of sleeping-trees on the opposite bank. Figure 8-9 displays these choices with fair accuracy, since its frequently used segments contained sleeping-trees.

Not all frequently used segments in Fig. 8-9 are opposed by frequently used segments; nor is it crucial that they should be; yet the exceptions can be partly explained away. For example, on the opposite side no frequently used segments oppose segments -17, 11 and 13, or 66–67, which are frequently used on the Kutai Reserve side. But, in the first case, Indonesians had cleared the forest on the opposite side; in the next two (located at the end of The Loop), such opposed sightings as occur probably represent the same group; and in the last, a small group or subgroup was opposed by sparse forest lacking continuous canopy. That opposed sightings can represent the same group is troublesome; however, among the prominent oppositions, segments 23–27 represent *L* and

R groups; segments 47–48 (within which Fittinghoff camped throughout February, 1973) also represent opposed groups; and segments 37–38 probably represent opposed groups (sizable groups were seen roosting simultaneously). Of course, even buttressed by these explanations, Figure 8-9 cannot prove the present argument concerning refuging, but it is highly suggestive, and it inspired it.

Other Possible Explanations of Refuging

The competitive advantages offered by riverine refuging may best account for it, but they need not exclude other advantages. A possible additional advantage involves health, for persistent group use of the same sleeping-tree brings with it the problem of accumulating feces. In the forest dung-beetles might bury this hazard, but rainforest trees have shallow roots, and persistent churning of the soil under a sleeping-tree might also be hazardous to *kra* in a region where trees fall during any prolonged rain. Riverine trees may likewise fall during floods, but using one helps prevent disease, for at dawn, where they can, the monkeys usually defecate into the river.

Not all the advantages so far adduced may be *necessary* for the occurrence of riverine refuging, either on the Sengata or elsewhere. *Kra* appear to have been refuging to Borneo rivers for a very long time. Yet, in the past, not all Borneo rivers have been as safe or easy for them to cross as the present-day Sengata. According to Sahar, a child was killed by a crocodile near Sengata village some 20 years before our visit. To judge from the reports of Beccari (1904) and other early naturalists, before the advent of rifles most rivers of any size harbored freshwater and large estuarine crocodiles, and Beccari notes reports of sharks in the Rejang and Sarawak Rivers. The presence of predators may not always deter crossings: Beccari says that individual wild pigs (*Sus barbatus*) fall prey to crocodiles when herds of pigs cross rivers to reach fallen fruit, and Shelford (1916) reports the same for *kra* caught foraging for estuarine *Sesarma* crabs at low tide. Still, it seems reasonable to suppose that *kra* crossings become less frequent as such predators multiply. Additionally, some navigable rivers, such as the Mahakam of East Kalimantan, are so wide that opposed conspecifics, even if they could be seen, might deter *kra* less than the swim. It is a worthwhile question, then, whether *kra* refuge to rivers that they do not cross.

We imagine they do refuge, but apart from anecdotes, directly relevant data are lacking. If they refuge, advantages other than opposing conspecifics *across* the river must be locally significant. Rodman (1973b) and Wheatley (this volume) find that riparian forest may be remarkably productive, since floods *r*-select the fruiting plants to bear early, abundantly, and often. During our study, for example, some specimens of *jambu air* (*Syzygium jambos*), a riverine shrub very attractive to *kra*, bore twice in less than a year. Much the same productivity obtains along the small tributaries of a river, as Fig. 8-12 suggests. The morning foraging routes of *L* group took the monkeys to the creek, which flooded whenever the river rose. Thus, at a river like the Mahakam, the advantages of foraging in riparian forest, of foraging along the river banks, and of seeing and being seen by conspecifics *along* the river may be crucial.

INTERSPECIFIC COMPARISONS AND CONCLUSIONS

Refuging and communally nesting or roosting species are common among primates. *Microcebus murinus* (Martin, 1973), *Macaca radiata* (Simonds, 1965), *Macaca mulatta* (Southwick, Beg, and Siddiqi, 1965), *Papio cynocephalus* (Altmann and Altmann, 1970), and *Papio hamadryas* (Kummer, 1968) are familiar examples, and others could be listed. Riverine-refuging species have been reported less often. Some of those that we have noticed, besides *kra*, are *Miopithecus talapoin* (Gautier-Hion, 1970), *Papio anubis* and *Papio hamadryas* (Nagel, 1973), and *Nasalis larvatus* (Kern, 1964; personal observations). Among talapoins in Gabon, elements of the same group sleep along both sides of narrow streams, and no contacts between groups were observed. Gautier-Hion interprets talapoin sleeping-trees as true refugia, i.e., as means of avoiding predators, rather than as means of resource competition. Nagel reports from Ethiopia that both anubis and hamadryas baboons sleep, when possible, directly opposite conspecifics across the Awash River, which had a minimum width of about 10 meters in his study area. He suggests a positive attraction between opposed groups, which exchange choruses of loud grunts almost nightly. By contrast, mutual avoidance characterizes relations between adjacent groups on the same side of the river. The contrast is not explained, but it is noteworthy that the opposed groups may be prevented from competing by a local abundance of crocodiles. Apparently no river crossings occur; Nagel did not hesitate to treat the populations on opposite sides as independent.

These examples suggest that the causes of riverine refuging are likely to be ecospecific wherever refuging occurs. With regard to Sengata River *kra*, a riverine refuging pattern has developed both above and below Sengata village. Upstream, in areas of moist lowland forest, it appears to have evolved and to persist chiefly because (1) it helps *kra* exploit a locomotor-feeding advantage over competing anthropoid species in riverine vegetation; and (2) it helps groups of *kra* maintain relatively exclusive home ranges against conspecific groups. Downstream from Sengata village, moist lowland forest gives way to a nipa-mangrove association; yet both *kra* and proboscis monkeys roost in riverine trees there. Presumably the resources, competitors, and potential predators differ considerably from those upstream; yet the pattern of conspicuousness persists and appears in a second species. Moreover, the second species also swims very well (Napier and Napier, 1967). In the light of these findings, it is tempting to regard the strategy of conspicuousness as the predominant adaptation, but more work, particularly dietary studies, will be needed before the functions and causes of riverine refuging are fully understood.

ACKNOWLEDGMENTS

The field research was supported by National Science Foundation grants GB-35384 and GB-40156 to Lindburg and by a University of California Regents Fellowship and a Graduate Education Allowance awarded to Fittinghoff. In Indonesia, invaluable help was provided by the Indonesian government (L.I.P.I., the Institute of Sciences, and the Departments of Forestry and Conservation), as well as by two oil companies, P. T. Kalimantan Shell and P. T. Pertamina. In the United States we owe particular thanks to Dr. Orville A. Smith, who obtained funds for the Research Station, and to Dr. Peter S. Rodman, whose field experience has been generously shared and whose publications have provided key insights.

REFERENCES

Altmann, S. A. and Altmann, J. *Baboon Ecology.* Chicago and London: University of Chicago Press (1970).

Beccari, O. *Wanderings in the Great Forests of Borneo.* London: Archibald Constable (1904).

Cohen, J. E. Grouping in a vervet monkey troop. In *Proceedings, 2nd International Congress of Primatology* (Atlanta, 1968), Basel: S. Karger, pp. 274–278 (1969a).

_____.Natural primate troops and a stochastic population model. *Amer. Naturalist,* **103**: 455–477 (1969b).

_____.Aping monkeys with mathematics. In R. Tuttle (ed.), *The Functional and Evolutionary Biology of Primates,* Chicago and New York: Aldine/Atherton, pp. 415–436 (1972).

Crook, J. H. Sexual selection, dimorphism, and social organization in the primates. In B. Campbell (ed.), *Sexual Selection and the Descent of Man: 1871–1971,* Chicago: Aldine, pp. 231–281 (1972).

Davis, D. D. Mammals of the lowland rain-forest of North Borneo. *Bull. Singapore Natnl. Mus.,* **31**: 1–129 (1962).

Fittinghoff, N. A., Jr. *Macaca fascicularis of Eastern Borneo: Ecology, Demography, Social Behavior, and Social Organization in Relation to a Refuging Habitus.* Unpublished Ph. D. Dissertation, University of California, Davis (1978).

Fooden, J. Report on primates collected in western Thailand, January–April, 1967. *Fieldiana Zool.,* **59**: 1–62 (1971).

_____.Provisional classification and key to living species of macaques (Primates: *Macaca*). *Folia primatol.,* **25**: 225–236 (1976).

Gautier-Hion, A. L'organisation sociale d'une bande de talapoins (*Miopithecus talapoin*) dans le nord-est du Gabon. *Folia primatol.,* **12**: 116–141 (1970).

Hamilton, W. J. III and Watt, K. E. F. Refuging. *Ann. Rev. Ecol. Systemat.,* **1**: 263–286 (1970).

Hill, W. C. O. *Primates. Comparative Anatomy and Taxonomy,* **VII**. *Cynopithecinae.* New York: John Wiley (1974).

Holdridge, L. R., Grenke, W. C., Hathaway, W. H., Liang, T. and Tosi, J. A., Jr. *Forest Environments in Tropical Life Zones: A Pilot Study.* New York: Pergamon Press (1971).

Horn, H. S. The adaptive significance of colonial nesting in the Brewer's blackbird (*Euphagus cyanocephalus*). *Ecology,* **49**(4): 682–694 (1968).

Hornaday, W. T. *Two Years in the Jungle.* New York: Scribner's (1885).

Kern, J. A. Observations on the habits of the proboscis monkey, *Nasalis larvatus* (Wurmb), made in the Brunei Bay area, Borneo. *Zoologica,* **49**(11): 183–191 (1964).

Kummer, H. *Social Organization of Hamadryas Baboons.* Chicago and London: University of Chicago Press, (1968).

Kurland, J. A. A natural history of kra macaques (*Macaca fascicularis* Raffles, 1821) at the Kutai Reserve, Kalimantan Timur, Indonesia. *Primates,* **14**: 245–262 (1973).

Lindburg, D. G. Feeding behaviour and diet of rhesus monkeys (*Macaca mulatta*) in a Siwalik forest in North India. In T. H. Clutton-Brock (ed.), *Primate Ecology: Studies of Feeding and Ranging Behaviour in Lemurs, Monkeys, and Apes,* London: Academic Press, pp. 223–249 (1977).

Martin, R. D. A review of the behaviour and ecology of the lesser mouse lemur (*Microcebus murinus* J. F. Miller 1777). In R. P. Michael and J. H. Crook (eds.), *Comparative Ecology and Behaviour of Primates,* New York: Academic Press, pp. 1–68, (1973).

Medway, Lord. Mammals of Borneo: Field keys and an annotated checklist. *J. Malay. Br. Roy. Asiatic Soc.,* **36** (3): 1–193 (1965).

Nagel, U. A comparison of anubis baboons, hamadryas baboons, and their hybrids at a species border in Ethiopia. *Folia primatol.,* **19**: 104–165 (1973).

Napier, J. R. and Napier, P. H. *A Handbook of Living Primates.* New York: Academic Press, (1967).

Pearson, D. L. A preliminary survey of the birds of the Kutai Reserve, Kalimantan Timur, Indonesia. *Treubia,* **28**(4): 157–162 (1975).

Poirier, F. E. and Smith, E. O. The crab-eating macaques (*M. fascicularis*) of Angaur Island, Palau, Micronesia. *Folia primatol.,* **22**: 258–306 (1974).

Ridley, H. N. 1906. The menagerie at the botanic gardens. *J. Straits Br. Roy. Asiatic Soc.,* **46**: 133–194 (1906).

Rodman, P. S. Synecology of Bornean primates. I. A test for interspecific interactions in spatial distribution of five species. *Amer. J. Phys. Anthrop.,* **38** (2): 665–660 (1973a).

_____. *Synecology of Bornean Primates, with Special Reference to the Behavior and Ecology of Orangutans.* Unpublished Ph. D. Dissertation, Harvard University, Cambridge, Mass, (1973b).

_____. Population composition and adaptive organization among orangutans of the Kutai Nature Reserve. In R. P. Michael and J. H. Crook (eds.), *Comparative Ecology and Behaviour of Primates,* London and New York: Academic Press, pp. 171–209, (1973c).

Shelford, R. W. C. *A Naturalist in Borneo.* London: T. Fisher Unwin (1916).

Shirek-Ellefson, J. *Visual Communication in Macaca irus.* Unpublished Ph. D. Dissertation, University of California, Berkeley, (1967).

Simonds, P. E. The bonnet macaque in South India. In I. DeVore (ed.), *Primate Behavior: Field Studies of Monkeys and Apes,* New York: Holt, Rinehart and Winston, pp. 175–196 (1965).

Southwick, C. H., Beg, M. A. and Siddiqi, M. R. Rhesus monkeys in north India. In I. DeVore (ed.), *Primate Behavior: Field Studies of Monkeys and Apes,* New York: Holt, Rinehart and Winston, pp. 111–159, (1965).

Wallace, A. R. *The Malay Archipelago.* London: MacMillan, (1869).

Wilson, C. C. and Wilson, W. L. Behavioral and morphological variation among primate populations in Sumatra. *Yearbook of Phys. Anthrop.,* **20**: 207–233 (1977).

Chapter 9
Feeding and Ranging of East Bornean Macaca fascicularis

Bruce P. Wheatley

INTRODUCTION

There is a recent realization that models on the evolution of primate social behavior and ecology are very incomplete. There are several reasons for this deficiency. One is that there are as yet too few studies of free-ranging primates in their natural habitat. Another is that a disproportionate share of the studies that have been done are biased in favor of species and habitats where observational conditions are optimal. These biases are unfortunate since most primates are arboreal, rainforest animals. This study attempts to counter such biases and, combined with other studies of arboreal primates, such as those of Clutton-Brock (1974), Struhsaker (1969, 1975) and Waser (1977), challenges existing hypotheses on the nature of arboreal primate societies. For example, not long ago it was thought that large multimale troops in an arboreal forest monkey were atypical (Eisenberg *et al.*, 1972). This view now appears to be in need of modification, because large, multimale troops are not just specialized adaptations to terrestrial foraging.

The crab-eating macaque, *Macaca fascicularis*, is one of the most successful of nonhuman primates. It is distributed throughout all of Southeast Asia between 20° N and 10° S latitude and between 92° and 128° E longitude (See distribution map, Chapter 8). Although this monkey is one of the most numerous and widely distributed primates of Southeast Asia, it is very inadequately known. The most useful information on its social behavior and ecology is from the reports of Angst (1973, 1975), Chiang (1968), Fittinghoff (1978), Fooden (1971), Furuya (1962, 1965), Kurland (1973), Medway (1970, Rijksen (1978), Rodman (1973, 1978a), Shirek-Ellefson (1967, 1972), Southwick and Cadigan (1972), and Wilson and Wilson (1975). Most of these reports, however, are either short-term studies, or are on animals living in botanical gardens or near villages or other habitats disturbed by man. This chapter presents some of the first detailed observations of the feeding and ranging behavior of *M. fascicularis* in the wild.

METHODS

Kutai Nature Reserve

Fieldwork was conducted near the Hilmi Oesman Memorial Research Station in the Kutai Nature Reserve from October 1974 to June 1976. For a description of the reserve and the study site refer to Fittinghoff and Lindburg (this volume).

Rowell (1967) said that a description of social organization of an animal is most useful if it is accompanied by a description of the environment in which it occurs. One reason for this is the diversity of intra- and interspecific variation in social organization. Primates existing in the same general habitat can be widely diverse in their adaptations. For this reason, the physiognomy, composition, and structure of the habitat of the crab-eating macaque was intensively investigated. The results of this investigation are reported elsewhere (Wheatley, 1978a, 1978b). Briefly, the features of a sampled hectare of forest used by the study troop were congruent with those characteristics typical of secondary forests as defined by Richards (1966). One of the important general features of some of the pioneer trees that dominated this hectare of forest is that they fruit year around. Systematic, biweekly surveys for one year showed that fruit was more abundant and predictable in streamside areas than in non-streamside areas (Wheatley, 1978c). The rainfall patterns in this part of the Kutai Nature Reserve are indicative of the wettest climate type for Indonesia, and the monthly distribution of rainfall is relatively constant throughout the year.

Macaca fascicularis is common along the riverbanks of the Sengata, especially in the early morning and late afternoon. They can be sighted among the mangroves from the mouth of the Sengata up to the Mentoko, about 35 km upstream. These monkeys can be pests in the *ladang*, or slash and burn gardens, since they are fond of eating beans, corn, and other crops. It is probably for this reason that they are considered "harmful vermin" throughout parts of Indonesia according to the Animal Protection Ordinance and Regulations of

Fig. 9-1. Several members of the study troop sit in the early morning sun in their sleeping tree. The photograph was taken from a platform blind in the top of a nearby tree.

1931, No. 134 and 266 (Hoogerwerf, 1970). Several *M. fascicularis* troops occur in the vicinity of the research station. The study troop, also known as the "L", or Ipil troop, is the same troop of animals investigated by Lindburg and Fittinghoff (this volume).

Troop Size and Composition. Research on small, arboreal, rainforest animals is often very frustrating. Data collection on the study troop was no exception. The troop had minimal contact with occasional local villagers and was extremely wary. Their typical reaction to humans was a mass retreat followed by hiding in silence 50 meters or more in the treetops. Fortunately, however, this troop returned quite often to the same sleeping tree along the river's edge. Nevertheless, six to seven months of hard work at habituation was required before reliable troop composition and other behavioral data could even begin to be taken. Between October 1974 and February 1975, I gradually exposed

Fig. 9-2. An adult female poses prior to eating a *Dracontomelon mangiferum* fruit. Note the large, bushy beard on her face.

myself to their view while they were in their sleeping tree. The animals gradually stopped fleeing from the tree when they saw me and eventually took little notice of my activities from about 75 to 100 meters away.

The most accurate method of counting the number of individuals in the troop was to count them when they climbed into their sleeping tree each afternoon. Even this method had drawbacks, however, since the animals could climb into the tree via different entrances, or they could climb into the tree when it was too dark to observe them, or unrecognized individuals could climb into the tree, leave, and then re-enter the tree. Troop counts were taken whenever possible and whenever compatible with other types of data collection, and only 56 troop counts were reasonably accurate.

Reliable estimates of troop composition were obtained after the month of March, 1975, when the animals were habituated. Forty-one reliable samples on troop composition were obtained between April, 1975, and May, 1976. Leitz Trinovid, 10 by 40 binoculars were utilized throughout the duration of the study. In July, 1975, observations continued at close range from a platform blind. The platform was erected about 20 meters up in a riverine tree about 30 meters from the animals' sleeping tree (Fig. 9-1).

The animals were aged and sexed according to the criteria of Shirek-Ellefson (1967). Young infants, for example, are between zero and six months of age, while old infants are between six months and one year of age. Young juveniles are between one and two years of age, while old juveniles are between two and four years of age. Subadult males, adult males, and adult females are easily identified by their unique facial morphology and coloration. Adult females are easily distinguished from adult males by the presence of large, bushy beards in the females (Fig. 9-2).

Feeding, Foraging, and Ranging. Food items were classified as different kinds of fruits, insects, flowers, grasses, clay, and fungi. The more direct method of measuring feeding by observing one monkey for the entire day and continuously recording the numbers of food items ingested (Hladik, 1977) was not possible under the difficult observation conditions in the rainforest. This study utilized two frequency methods in the collection of dietary information. The first method, similar to that used by Struhsaker (1974) and Oates (1977), scored one individual feeding on one food item in a tree or food patch as one feeding observation. If another individual fed on the same item, this was also scored as a feeding observation. Another feeding observation was also scored if the same individual switched to a different food item in the same tree. The second frequency method of estimating the relative proportion of dietary items was a count of the frequency of a food item eaten by an individual on a per hectare basis. If another individual ate a different food item in the same hectare, then this was scored, but if it ate the same food item in the same hectare, no record was made. The latter method severely reduced the sample size, but it also removed some of the bias of the first method. Data on actual feeding times on food items were also obtained from activity profile information.

Approximately three weeks were spent at the end of the study in the pursuit of activity profiles and other information by observing a focal male from 6 a.m. to 6 p.m. A male was followed because he was more visible than other animals, and could be readily identified on the basis of morphology and sounds. Continuous focal animal activity profiles were extremely difficult to obtain, and only three days of data were reasonably accurate. These three days were also used in the determination of average number of feeding bouts per day, median distance between successive major food sources and average travel speeds. By using a stopwatch the data were recorded as transition times of behavior. Consequently, the activities could be expressed as activity durations (Altmann, J., 1974). The activities were categorized as follows: feeding, traveling, and resting. Feeding activity consisted of harvest and process time, thus including gathering and masticating if done in the same tree. The animal usually stayed in the same tree to masticate its food, but occasionally harvested from one tree and processed food from check pouches in another tree. Feeding terminated when the animal either ceased feeding on an item or left the tree and this was scored as one feeding bout. Travel was movement between food sources, usually between trees. Resting was neither feeding nor traveling, and sometimes included behaviors such as grooming.

The 1 1/2 km² study area was dissected by a grid system of trails every 100 meters. This facilitated recording of the study troop's location and the duration of time spent in a hectare. The distance of each hectare from a stream was determined by measuring from the midpoint of each. The study troop was followed from 6 a.m. to 1 p.m. and from 4 p.m. to 6 p.m. whenever possible. For consistency, the ranging data used in this report were taken from 6 a.m. to 1 p.m. for 40 days spread out over a 14 month period. As the animals entered each hectare, the following information was recorded: (1) the location of the majority of the troop; (2) the time of day most of the animals entered and left the hectare; and (3) the feeding duration of the troop or an individual. Additional information on the numbers of animals feeding in a tree, interanimal distances, and group diameter was collected whenever possible. The number of minutes spent in terrestrial locomotion was also noted.

TROOP SIZE AND COMPOSITION

The sleeping tree, a large *Intsia palembanica* or Ipil tree, sometimes contained as many as 42 animals and as few as one individual. The average troop count was 30 individuals for 56 good counts, with a standard deviation of 3.89. The largest count for another *M. fascicularis* troop that ranged on the other side of

Table 9-1.   Troop Size and Composition.

AGE-SEX CLASS	MEAN	STANDARD DEVIATION	SAMPLE SIZE
Adult males	3.16	0.62	40
Adult females	9.87	1.39	39
Subadult males	2.50	1.01	22
Old juveniles	7.6	2.15	39
Young juveniles	5.34	2.33	37
Old infants	1.00	0.00	18
Young infants	1.29	0.67	35
Troop count	29.14	2.31	42

the Sengata River was between 28 to 30 individuals, including four adult males and two subadult males. (See Fittinghoff and Lindburg, this volume, for additional counts of these two troops.) A troop that ranged near the research station had 27 individuals, counted when they swam the Sengata River. Another nearby troop had 22 individuals.

The average study troop composition is shown in Table 9-1. This table shows the average troop count and the average count of individuals in different age-sex classes on habituated animals. If animals were not sighted in the appropriate age-sex class, then they were not included in the sample. This troop had an average socionomic sex ratio of one adult male for 3.3 adult females. The breeding sex ratio, however, was one adult male to two adult females if one includes the subadult males who were seen to copulate.

Troop counts varied from week to week. These counts probably overemphasized the variation in numbers and troop composition for a variety of reasons. The largest variation in troop counts occurred after an aggressive invasion of stange males, when 14 animals left the sleeping tree and ranged farther upriver throughout the remainder of the study. Nevertheless, a small part of the variation in counts was real because individuals on occasion slept apart from the rest of the troop.

FEEDING BEHAVIOR

Activity Patterns and Feeding Techniques

The time budgets of the male focal animals showed a strong emphasis on travel. The relative amount of time spent in the activities of traveling, resting, and feeding was 45 percent, 42 percent, and 13 percent respectively. Throughout the day, the study troop passed through approximately 20 hectares on average, and rarely spent longer than an hour in a single tree. Their average day length was 11 hours and 22 minutes. They left their sleeping tree at an average time of 6:25 a.m. on 84 mornings and settled down for the night at a mean time of 5:47 p.m. on 55 nights.

Table 9-2. Feeding Bout Lengths.

FRUIT TREE		BOUT LENGTHS IN TOTAL MINS.	FREQ.	BOUT LENGTHS AVERAGE (MIN.)
LOCAL NAME	SCIENTIFIC NAME			
Beringin	*Ficus sp.*	90	2	45
Bungkal	*Callicarpa farinosa*	34	7	4.86
Ipil	*Intsia*	15	1	15
Kedut		25	1	25
Kuranya	*Leea sp.*	66	14	4.71
Marsesat		68	4	17
Medang klema	*Phoebe spp.*	20	2	10
Medang lelan	*Phoebe spp.*	5	1	5
Murop		12	3	4
Pisang punan (liana)		30	7	4.29
Ponten	*Cratoxylon sp.*	5	1	5
Pumadam larak		20	1	20
Seragam		2.5	1	2.5
Singkuang	*Dracontomelon mangiferum*	120	6	20
Tebu hitam	*Koordersiodendron pinnatum*	205	8	25.63
Tincau hari		55	2	27.5
Ulin	*Eusideroxylon zwageri*	22	2	11
TOTAL KNOWN FRUIT		794.5	63	$\overline{X} = 12.6$
TOTAL UNKNOWN FRUIT		473.0	29	$\overline{X} = 16.3$
TOTAL FRUIT		1267.5	92	$\overline{X} = 13.78$

NON-FRUIT ITEMS	BOUT LENGTHS IN TOTAL MINS.	FREQ.	BOUT LENGTHS AVERAGE (MIN.)
Clays	.25	2	.125
Flowers (Kundikara = *Baringtonia sp.*)	8.0	3	2.67
Grasses	9.47	25	.38
TOTAL CLAYS, FLOWERS, GRASSES	17.72	30	$\overline{X} = .59$
GRAND TOTAL FOR ALL FOOD ITEMS	1285.2	122	$G\overline{X} = 10.53$

These macaques are primarily terminal branch feeders. Food at the ends of sturdy branches was harvested by sitting in place and picking the food with their hands or directly with the teeth. In order to reach food at the ends of less sturdy terminals, the monkeys often distributed their weight by grabbing several branches with their hands and feet, using their tails as a strut in order to reach their food. When feeding in the more precarious positions, individuals quickly stuffed their cheek pouches and retired to a more stable support to masticate the food, or ate while they moved between food items. When foraging in small trees where they were more visually exposed, it appeared that the animals retired to larger nearby trees to masticate their food.

There were several interesting observations on food preparation. On one occasion, an adult female dipped a fruit into the river and then ate it. It is possible that there had been sand on it. The animals also briskly rubbed small objects that appeared to be insects between the palms of their hands. Several instances of what was similar to "leaf-washing" (Chiang, 1967) were also seen, but the rubbing of objects in leaves was not seen.

The average time of a feeding bout over all food items in 122 bouts was 10.53 minutes (Table 9-2). The data for this table were collected whenever possible on a per tree basis except for feeding bouts on clay or grass which were determined on a per clump basis.

The average feeding bout length on all fruits for which names were known or unknown was 13.78 min. The average feeding bout length on clays, flowers, and grasses was .59 min. or 35 sec. Feeding bout lengths in some cases were underestimated since they only reflected harvest time and not necessarily process time. When feeding on the small, berrylike fruits of *Callicarpa farinosa*, for example,the animals filled their cheek pouches and retired to a larger tree to process the fruit. This bias in estimating the actual length of feeding bouts was probably cancelled by another bias which overestimated feeding bout lengths since the animals were very active during feeding bouts, and spent approximately one–third of their feeding time walking along the tops of the branches. The average number of feeding bouts per day was 18.3. This figure was determined by averaging the number of bouts by the focal male on May 14, 23, and 27, 1976. The number of feeding bouts on these days was 16, 18, and 21 respectively. This figure is probably an underestimate because short bouts could easily have been missed. Crude as it is, it still contrasts markedly with the average figure of about seven bouts per day for the orang-utan, another arboreal sympatric frugivore (Rodman, 1977).

DIET

The diet of the study troop was diverse. The animals ate a wide variety of fruits, flowers, insects, leaves, fungi, several kinds of grasses, and clay. The

Table 9-3. Composition of the Shannon-Wiener Information Measure.

	KUTAI FRUIT NAME	FREQ.	$P_i \, LOG_e \, P_i$
1.	Ara	7	.078
2.	Belau	7	.078
3.	Basa	4	.051
4.	Bayur	9	.094
5.	Beringin	20	.163
6.	Bungkal	60	.302
7.	Emporan	22	.173
8.	Emus	1	.017
9.	Gelongan	2	.029
10.	Ipil	9	.094
11.	Jambu air	8	.086
12.	Jelentikan	1	.017
13.	Kacang kerau	1	.017
14.	Katan	1	.017
15.	Kedut	1	.017
16.	Kumping	1	.017
17.	Kundikara	3	.041
18.	Kuraktabu	1	.017
19.	Kuranya	18	.152
20.	Leban	4	.051
21.	Letop	1	.017
22.	Mahang	2	.029
23.	Marsesat	4	.051
24.	Matahari	3	.041
25.	Medang	7	.078
26.	Medang klema	5	.061
27.	Medang lelan	2	.029
28.	Mintow hari	1	.017
29.	Mocong burung	15	.135
30.	Murahkladi	5	.061
31.	Murahpadi	2	.029
32.	Murop	12	.115
33.	Mujung	1	.017
34.	Nyatoh	3	.041
35.	Nyerumit	5	.061
36.	Pisang punan	17	.147
37.	Ponten	7	.078
38.	Pumadam larak	3	.041
39.	Putat	3	.041
40.	Seragam	3	.041
41.	Singkuang	13	.122
42.	Tebu hitam	42	.254
43.	Terupuk	1	.017
44.	Tincau hari	4	.051
45.	Ulin	10	.101
		N = 351	H' = 3.186
46.	Unknown fruits	23	
		N = 374	

frugivorous component of their diet was most diverse. The Shannon-Wiener diversity measure of the 45 different kinds of fruits for which local names were known is 3.19 (Table 9-3). This figure is probably about 3.42 when considering other fruits for which local names were not known.

Insect foods were seldom identifiable, but they included termites and caterpillars. On a number of occasions it appeared that the animals searched for trees typified by a lack of fruit and a high density of dead foliage, through which the animals searched for insects. Only two species of grass were eaten; most of this feeding was on *Calamagrostis* sp.

Small amounts of clay were eaten. X-ray diffraction analysis of the entire ground material showed an unusually high crystalline quartz fraction, as well as large peaks of kaolin. Other minor components of the sample included feldspar and illites. The animals prefer to eat the conspicuously white, kaolin-rich material. Analysis of the soil sample for available potassium and total potassium was 109 ppm and 1.15 percent respectively. Neither of these figures are particularly high.

Based on sampling of a focal male on three days, 96 percent of his feeding time was on fruit. He spent 92, 93, and 99 percent of his feeding time on fruit for a total of 249 minutes out of 260 minutes feeding on these three days. Data from two additional methods of diet estimation also indicate a heavy dependence on fruit (Table 9-4). This table shows the frequency of dietary items using the two different methods for determining frequency. Both methods of estimating the diet indicate a frugivorous percentage of between 87 to 88 percent. The difference in dietary percentage based on these two

Table 9-4.   Diet.

ITEM		METHOD ONE	(N)	METHOD TWO	(N)
1.	Fruit	88.31	748	86.98	374
2.	Insect	2.24	19	3.95	17
3.	Flower	2.72	23	3.26	14
4.	Grass	3.42	29	2.09	9
5.	Leaf	1.42	12	1.63	7
6.	Clay	1.54	13	1.39	6
7.	Fungus	0.35	3	0.70	3
	N =	100%	847	100%	430

measures of feeding is small. Nevertheless, it is interesting to speculate on the different biases that may account for some of the differences between the two methods. For example, feeding on fruit and grass may have been more observable than the other dietary items. Falling fruit is easily detected, and might result in a higher probability of being scored. Similarly, in grass-feeding the animals were exposed on sandy beaches along water courses. Overemphasizing grass and fruit feeding would tend to decrease the percentage of more difficult-to-observe food items such as insects, flowers, and fungi.

Monthly variation in proportions of food types also revealed heavy utilization of fruit in the diet of this troop (Table 9-5). The data for this table are from the more stringent of the two frequency methods of estimating the diet, Method II. The monthly median percentage of frugivory for 18 months of data is 91.5 percent.

Half of the feeding observations came from nine fruit tree species (Table 9-3). The density of these nine species, using figures obtained from one sample hectare, was 186/hectare, despite the fact that four did not occur at all in the sample. The most utilized fruit source was from *Callicarpa farinosa* trees. Sixteen percent of all feeding observations occurred in these trees.

Limited observations suggested that these animals selected fruit on the basis of ripeness. For example, they chose the red fruits of *Callicarpa farinosa*, whereas the orang-utan ignored color differences. These macaques were selective in other ways as well. They did not, for example, ingest seeds of *Koordersiodendron pinnatum*, again, in contrast to the orang-utan.

The percentage of crude protein in fruit items was determined by nitrogen analysis, using the Kjeldahl method on oven-dried fruit. The nitrogen percentage was converted to crude protein by multiplying by the standard factor of 6.25. The percentage of crude protein of 17 selected fruits and seeds ranged from 4.46 to 30.18 with a mean of 8.99 (Table 9-6).

FORAGING BEHAVIOR

The study troop generally foraged as a unit throughout its daily range. The troop members kept in contact during foraging by visual and auditory cues. The most useful auditory call for locating other troop members was the contact or "coo" call, very similar to that described by Lindburg (1977) for *M. mulatta*. Although the troop was generally cohesive, it was on numerous occasions widely dispersed. A troop spread of several hundred meters was common under certain conditions such as when foraging in stands of small trees. Sometimes this dispersion could be as great as 500 meters or more. Small subgroups were known on occasion to break off from the troop during the day, and these subgroups slept by themselves for short periods of time.

Table 9-5. Monthly Variance in Proportions of Food Types.

MONTH:	1974			1975												1976					TOTALS
	10	11	12	1	2	3	4	5	6	7	8	9	10	11	12	1	2	3	4	5	
Fruit	0	100	0	91.0	76.9	100	100	95.7	94.4	94.7	86.8	0	100	75.0	67.6	58.2	58.8	77.3	100	92.0	374: 87.0
Insects									1.8	1.8	9.4			25.0	10.8		5.9	9.0		2.3	17: 3.9
Flowers				9.0	7.7			4.3							10.8	25.0				4.6	14: 3.3
Grass					7.7				1.8			100			5.4		17.6	4.5			9: 2.1
Leaves					7.7				1.8	1.7	1.9				2.7		5.9	4.5			7: 1.6
Clay											1.9				2.7	16.7	11.7	4.5			6: 1.4
Fungus										1.7	1.9									1.1	3: .7
N:	0	3	0	11	13	20	7	23	54	57	53	1	2	4	37	12	17	22	7	87	430: 100%

Table 9-6. Percentage of Crude Protein of Fruits Eaten By *M. Fascicularis.*

1.	*Callicarpa farinosa*	11.35
2.	*Dracontomelum mangiferum*	4.46
3.	*Eugenia aquatica*	6.50
4.	*Ficus sp.*	6.30
5.	*Intsia palembanica* (small sample, seeds only)	30.18
6.	*Koordosiodendron pinnatum*	6.07
7.	*Leea sp.*	8.95
8.	*Macaranga hypoleuca*	8.32
9.	Marsesat	9.74
10.	Mocong burung	5.21
11.	*Nephelium sp.*	5.81
12.	Pisang punan	9.76
13.	*Phoebe sp.*	7.90
14.	Putat spp.	8.21
15.	Putat spp.	4.91
16.	*Pterospermum sp.*	8.75
17.	*Xylopia sp.*	10.48
		$\overline{X} = 8.99$

Foraging Patterns and Fruit Availability in Streamside Secondary Forest

Many streamside tree species were continuously distributed in space and provided an abundant and rapidly renewing food supply. One of these, *Callicarpa farinosa*, had a high density in streamside areas. One hundred and forty individuals of this most frequently utilized food source were counted in the sample hactare. Its fruit was available at all times of the year. The fruits are small with a diameter of half a centimeter, and were borne in clusters weighing about 5 grams.

Fruit from *C. farinosa* trees were also continually available in time. Two frequently visited hectares (ranked numbers two and three in Fig. 9-3) almost always had ripe *C. farinosa* fruits on the trail over a one-year period. At least one fruit cluster was found on the trail in 33 out of 34 surveys in both of these two hectares. To test the abundance and temporal availability of fruit per tree, one needs to continually estimate each one's total fruit production. Estimates were obtained by using the "probabilities proportional to area" method (Jesson, 1955; Kelly, 1959). This method is based on the reciprocal of the probability with which a sample branch is selected and the number of fruits that are counted on that branch. To undertake this analysis, two sample trees in the hectares ranked two and three (Fig. 9-3) were randomly selected. The probablility of selecting the appropriate sample limb on tree number one was

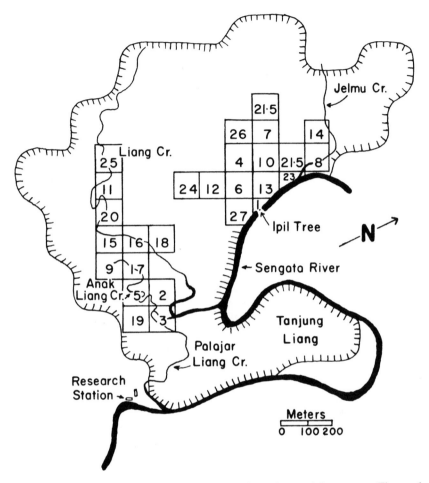

Fig. 9-3. Map of the study area showing the extent of the study troop's home range. The number within each hectare refers to the rank order of those hectares most utilized by the study troop for a sample of 35 days between the hours of 6 a.m. and 1 p.m. This figure was redrawn and modified from Figure 1-7 of Fittinghoff (1978).

.422 and its reciprocal was 2.37. The reciprocal was multiplied by each sample limb fruit count. Five systematic fruit counts were conducted between August, 1975, and January, 1976. Fruit clusters were always present on each sampling. The estimated number of fruit clusters on the entire tree varied from 14.1 to 20, with a mean of 15.8. Unfortunately, fruit estimates could not be continued beyond January because the sample limb suddenly died. The second tree was much larger and fruit estimate counts were not taken because of the risk in climbing this tree. Since the risk of harm to both tree and tree climber is often great, a better method should be devised for estimating fruit production, e.g., counting fruits from a photograph of the sample tree.

One other *Callicarpa* tree was chosen for examination of the temporal production and availability of fruit. When observation conditions were good, it was noticed that the macaques selectively chose ripe fruits. To estimate the proportion of ripe fruit, I sat in my tree platform near the Ipil tree and looked down onto the top of a nearby *Callicarpa* tree. I counted the first visible fruit clusters and noted whether they were red or green. The four scan samples began in March and lasted until the end of the study in June, 1976. On each scan fruit was always available, but on one scan none was ripe. The proportion of ripe fruits per visible fruits averaged 14 percent (N = 338). Interestingly, the tree flowered again while it was fruiting and fruited again in June.

The characteristics of these and similar streamside trees greatly affected the animals' ranging and foraging patterns. If streamside areas contain abundant, continually available, and rapidly renewing food supplies, then one might expect these areas to be frequently exploited. This relationship is examined below.

The relatively few ripe bunches of fruit on each tree probably caused the small size of foraging groups and the relatively short feeding bouts per tree. In small streamside trees such as *Callicarpa, Leea,* and *Macaranga,* over 70 percent of feeding time was by solitary foragers. Groups of three or more were never seen to forage in the same tree. The average time spent foraging in two of these trees (*C. farinosa* and *Leea* sp.) was just under five minutes (Table 9-2), in contrast to the average feeding bout length in larger trees more typical of the primary forest, such as *Koordersiodendron pinnatum.* The limited data on the average interanimal distances of nearest neighbors was about three times as great, or 20 meters, in these streamside trees as compared to larger trees more typical of the primary forest where all the animals could feed together in the same tree.

Riverbanks were popular areas for feeding on various grasses, especially *Calamagrostis* sp. Feeding bouts were short, averaging .38 min. (23 sec) per clump.

Foraging Patterns and Fruit Availability in Nonstream Primary Forest

Nonstream areas were less predictable in their food supply. Fruit was significantly rarer on those forest trails away from streams. Large fruiting trees in those areas were more patchy and less rapidly renewing food sources, compared to such pioneer species as *C. farinosa.* For example, only one *Ficus* tree in the sampled hectare was over 30 meters tall or over 110 centimeters in diameter at breast height, and it fruited only once during the study.

The monkeys appeared to monitor nonstream areas for large trees in fruit.

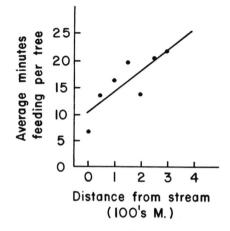

Fig. 9-4. Feeding durations per tree in relation to distance from a stream. Data for this figure are given in Table 9-2.

These trees contained a large number of feeding sites and the troop repeatedly exploited them for several consecutive weeks, until the crop was exhausted. When large trees such as *Koordersiodendron pinnatum* produced fruit, all troop members would feed in the same tree. Feeding bouts were longer (20 min. or more, Table 9-2) in these large trees than in the smaller trees of streamside areas.

As the distance of food source from the river increased, the monkeys tended to feed longer per tree, on the average (Fig. 9-4). This general relationship was probably due to several factors: (1) trees farther from the river were generally larger in size than riverine trees. They were taller, had a greater diameter at breast height and probably produced greater numbers of fruit at one time; (2) the fruits from trees farther from the river were probably larger, on the average, than the fruits from trees in the riverine areas.

To test the latter explanation, the dimensions of 26 fruits were randomly collected and measured at the study site. The length, width and height of each fruit and its seed were measured in centimeters with calipers. The weight of each fruit and seed was measured in grams with a postage scale. Table 9-7 tentatively classifies the habitat of each of these fruit trees. The fruit from trees generally within 100 meters of a stream was labeled a streamside fruit, while fruit from trees generally farther than 100 meters of a stream was labeled nonstream fruit. The classification is based on my work in the sample hectare, experience in the study area, and conversation with native assistants. Table 9-7 shows that the average of all nine dimensions of fruits from nonstreamside trees are larger than the average of all nine dimensions of fruit from streamside trees.

Table 9-7.   Fruit Dimensions.

LOCAL NAME	STREAMSIDE FRUITS								N(SEEDS)
	LENGTH		WIDTH		HEIGHT		WEIGHT		
	F[1]	S	F	S	F	S	F	S	
Bayur	6.0	.5	2.5	.5	3.0	.1	15.0	.1	9
Basa	1.0		.7		.7		.2		
Bungkal	.5		.5		.5		.04		9
Jambu air	1.5	1.0	1.5	1.0	1.2	1.0	1.0	.8	1
Kuranya	1.2	.8	1.0	.6	.8	.2	1.0		5
Leban	.7	1.2	.6	.8	.6	.5	.2		1
Murop	1.5	.4	1.5	.2	1.3	.2	.5		5
Singkuang	1.9	1.3	2.0	1.1	2.0	.7	2.0	1.0	1
Akar mintau	9.5	.4	4.0	.3	4.0	.3	50.0		24
Ulin	9.0		2.7		2.7		33.0		1
Average	3.28	.8	1.7	.64	1.68	.43	10.29	.63	6.2

NON-STREAMSIDE FRUITS

LOCAL NAME	LENGTH		WIDTH		HEIGHT		WEIGHT		N(SEEDS)
	F	S	F	S	F	S	F	S	
Akar kumbayau burung	1.7	.5	1.1	.2	1.1	.2	1.0		1
Akar busur	1.5	1.0	1.5	.8	1.5	.3	.3	.1	1
Akar pisang punan	7.0	2.0	3.5	1.5	3.5	.5	48.0	1.0	8
Asam	6.5	3.5	5.5	2.7	6.0	3.5	125.0	20.0	1
Belau	6.0	3.5	5.0	1.5	5.0	1.5	76.0	4.0	3
Emus	1.0		1.0		1.0		.3		24
Golongan	5.8	4.0	3.1	2.4	3.8	1.3	31.0	8.0	2
Kendis hutan	3.0		2.3		2.1		6.0		
Marsesat	1.0	.7	.6	.4	.6	.2	.2		1
Medang klema	2.5	1.5	2.5	1.1	2.5	1.2	5.0	1.0	1
Medang lelan	2.5	1.5	2.5	1.2	2.5	1.2	7.0	2.0	
Pahung	4.5		4.5		4.5		65.0		3
Pumadam larak	3.5		3.5		3.5		15.0		100
Simpur	5.0		4.0		4.0				
Tebu hitam	3.0	2.8	2.0	1.4	2.0	1.0	8.0	3.0	1
?"Tennis Ball"	6.5	1.7	6.3	1.0	6.0	.5	150.0	1.0	14
Average	3.81	2.06	3.06	1.29	3.1	1.04	36.03	4.46	12.57

[1] F = fruit, S = seed

Maximum length of visits to a single fruit source was a *Koordersiodendron* tree which was utilized continuously between June 8 and August 2, 1975. A Mocong burung tree was exploited for three weeks between February 23 and March 16. Individuals of these tree species, as well as Beringin (*Ficus* sp.), Nyatoh, Putat, Pumadam larak, Tincau hari, and others were systematically visited while they were fruiting. On one unusual day the troop traveled directly to 11 *Koordersiodendron* trees with hardly a pause on the way, and fed in most of them for periods of only 10 to 20 minutes.

These data suggest that the energetic cost of more distant dispersal from the river may have been offset by feeding on food items of greater energy value, occurring in more concentrated packages, than was available in streamside trees.

RANGING BEHAVIOR

Range

The home range of the study troop was about 1.25 km². This figure was obtained by counting all the different hectares that monkeys were ever seen to enter (Fig. 9-3). Since the estimation of home range is somewhat dependent on the number of observation hours spent with the troop, an observation curve is presented in Fig. 9-5. This figure shows several things. Even after 200 hours with the troop, the animals had entered only half the total number of hectares they eventually entered. The presence of several plateaus in the graph is also interesting. It is assumed that as the number of observation hours increases, the addition of a new area to the range will eventually decrease to the point where additional observation results in diminishing returns. This point is conventionally set at the 1 percent level, i.e., each additional observation produces less than 1 percent increase in the total area (Odum and Kuenzler, 1955). Fig. 9-5

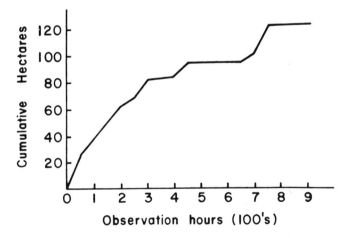

Fig. 9-5. Cumulative hectares entered as a function of the cumulative number of observation hours spent with the troop.

shows that if an observation is defined in terms of 10, 50, or even 100 hours, then one might greatly underestimate home-range size for this troop. The increase in size after 700 hours is due to the utilization of a new area of young secondary forest. This utilization is probably related to increased habituation to the observer such that the animals fed in even shorter trees in more open area, and to a decreased fruit abundance in other parts of the range as determined from fruit surveys.

The average daily distance traveled was about 1,900 meters. This estimate was obtained by averaging the number of hectares entered each day over a sample of 35 days. Fig. 9-3 gives the rank order of hectare use on a per-time basis. The troop, for example, spent *most* of its observation time in the hectare containing its sleeping tree. The troop spent 50 percent of its time in the first 12 ranked hectares and 75 percent of its time in all 27 of the ranked hectares (Table 9-8). These hectares are useful in delineating a core area. The core area was not continuous along the river. This was probably due to the avoidance of this area between cores due to the observer's frequent presence. The monkeys were unhabituated in this area because they had to locomote on the ground in my presence.

Ranging Behavior. The study troop visited some hectares much more than others. In analyzing troop and hectare distribution, the expected frequencies for each were calculated using the Poisson distribution. A Chi-Square Test was then used to show that the troop was significantly nonrandomly distributed (Table 9-9). This table shows that the troop visited 27 hectares only once each in the 40-day sample period. Calculation of the ratio of variance to mean from Table 9-9 yields the large value of 154. This value clearly indicates the troops' clumped or underdispersed distribution. Fig. 9-3 indicates that the troop spent most of its time near a stream or river. In fact, the troop spent about 62 percent of its time between 6 a.m. and 1 p.m. in a hectare containing a river or stream. This percentage would be much greater if afternoon data could be used, since the troop usually spent between 4 p.m. and 6 p.m. in a streamside hectare before they went to sleep. Nevertheless, this pattern of streamside preference may have been slightly influenced by the animals' dispersal from a riverine refuge each morning.

A significant, linear, negative correlation exists between the troop's distance from a stream and its utilization of a hectare. As the distance from a stream increased, the average time per hectare spent by the animals decreased ($r = -.74$, $t = 4.2$, $p < .001$, Fig. 9-6). If we examine the data on frequency of entering a hectare as a function of increasing distance from a stream, the same pattern emerges. As distance from a stream increased, the average frequency of visiting a hectare decreased ($r = -.64$, $t = 3.2$, $p < .005$, Fig. 9-7). These linear relationships would be much stronger if afternoon data were included,

Table 9-8.   Hectare Utilization By Time.

RANK	TOTAL MINUTES	CUMULATIVE FREQUENCY
1	1550	9.1
2	1170	16.0
3	1095	22.4
4	665	26.4
5	622	30.0
6	610	33.6
7	570	37.0
8	515	40.0
9	505	43.0
10	495	45.9
11	430	48.4
12	420	50.9
13	365	53.0
14	345	55.0
15	340	57.0
16	325	59.0
17	322	60.8
18	305	63.0
19	285	64.3
20	250	65.8
21.5	225	67.1
21.5	225	68.4
23	220	69.7
24	210	70.9
25	205	72.2
26	200	73.3
27	195	74.5

Table 9-9.   Troop Visitations/Hectare.

	TROOP VISITATIONS			
	0	1	2 OR MORE	TOTAL HECTARES
Frequency observed	22	27	73	122
Frequency expected	105	16	1	122

since the animals always traveled to a riverine tree, usually about 4 p.m. to 5 p.m.

The attractiveness of the streamside areas was probably related to the monkeys' ability to exploit the predictable and abundant food supplies contained in them. More direct evidence that streamside utilization was related to foraging is seen in Fig. 9-8. As distance from a stream increased, feeding per hectare was less frequent. The relationship is again a linear one ($r = -.75$, $t = 12.7$, $p < .001$, Fig. 9-8). Feeding durations on a per tree basis, however, increased as the distance from the stream increased. This relationship is also linear ($r = .82$, $t = 3.2$, $p < .05$, Fig. 9-4).

Another ranging pattern of these animals was the central refuge pattern. This pattern is the rhythmical dispersal of groups of animals from and their return to, a fixed point in space (Hamilton and Watt, 1970; Fittinghoff and Lindburg, this volume). The study troop slept in their refuge tree on the Sengata River on 217 out of 305 nights, or 71 percent of the time. The troop also spent more time in the half-hectare containing the refuge tree than in any other hectare (see Chapter 8, this volume, for further discussion of riverine refuging for *Macaca fascicularis*).

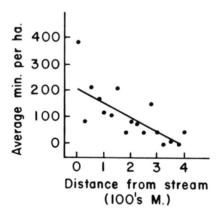

Fig. 9-6. Average time spent in a hectare in relation to its distance from a stream, based on 40 days of observation between the hours of 6 a.m. and 1 p.m.

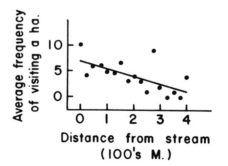

Fig. 9-7. Average frequency of visiting a hectare in relation to its distance from a stream, based on 580 visitations to 100 different hectares.

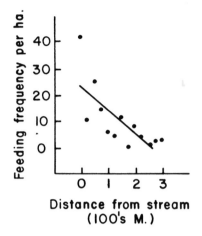

Fig. 9-8. The frequency of feeding in a hectare in relation to its distance from a stream, based on feeding observations collected between the hours of 6 a.m. and 1 p.m. for 40 days.

Arboreality.   The study troop spent more than 97 percent of their time in the trees. To some extent this is an overestimate, since it can probably be assumed that the animals were reluctant to come to the ground in my presence. To test this observational bias, the minutes spent on the ground were plotted as a function of increasing observation time. This technique gave 10 balanced intervals of 40 hr observation blocks and an average observation time of 9.45 hrs/day. A significant positive correlation was obtained, indicating that terrestriality increased as observation hours increased ($r = .55, t = 1.87, p < .05$, Fig. 9-9). It is very doubtful, however, that terrestriality ever exceeded 10 percent of the day in this troop in the absence of the observer since, even with "maximum" habituation, the troop still spent greater than 95 percent of its time each day in the trees. When terrestrial, these macaques spent most of the time along the Sengata River, especially when it was low and the sandy riverbanks were exposed. For the duration of the study, the animals were seen on the ground in the forest on 29 occasions. They usually descended to the ground in areas where the trees were small, and often would flee from the observer, on the ground, in such areas.

DISCUSSION

The crab-eating macaque, *Macaca fascicularis*, is a very common and widely distributed monkey throughout Southeast Asia. It can exploit a variety of environments, including ones made available by man. This species is found, for example, along the seacliffs of Bali, in its temple sanctuaries, and in other rural and urban environments. Systematic surveys, however, report that the

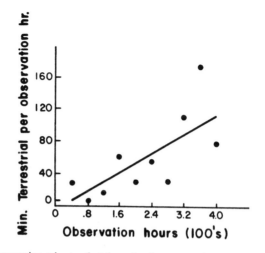

Fig. 9-9. Number of observation minutes that the animals spent on the ground between the hours of 6 a.m. and 6 p.m for 42 days between April 16, 1975 and May 29, 1976.

preferred habitat of this macaque is secondary forest, and particularly riverine secondary habitats (Crockett and Wilson, this volume; Rijksen, 1978; Rodman, 1978a; Southwick and Cadigan, 1972; Wilson and Wilson, 1975, 1977). This study supports the results of such surveywork.

One possible reason for the success of *M. fascicularis* is that its population expansion may be relatively recent. (See discussion by Fooden and Eudey, this volume, but see Delson, this volume, for a dissenting view). Its present-day distribution and its evolutionary history suggest a relatively recent invasion of Southeast Asia. It may have met relatively little competition during its Pleistocene colonization throughout Southeast Asia. Its present-day preference for littoral, riverine, and disturbed habitats may have been a very useful adaptation in the past as it colonized one of the most geologically volatile areas of the world.

Another reason for the success of this primate is its efficient reproductive biology. In comparison to pongids and certain other cercopithecoids, it reproduces at a younger age and has a relatively shorter birth interval. Other aspects of its behavior are also important for colonizing new areas, such as male immigration and troop fission (Wheatley, 1978a).

The close association of man with this primate also contributed to its evolutionary success. One benefit was its utilization by man as a pet, resulting in protection and introduction to new areas. One of the greatest benefits to this monkey may have been the human practice of slash-and-burn agriculture. This ancient type of agriculture is still the most common cause of secondary forests today. The clearing of forests along rivers for the planting of crops and their eventual abandonment by man led to an increase in the type of habitat favored by this monkey. This pattern continues today. Eventually, however, factors related to increased human population and the clearing of secondary forest may lead to a population decline for this macaque just as appears to be the case for the declining population of *Macaca mulatta*. (Southwick and Siddiqi, 1977).

Troop Size and Composition

The average troop size of about 30 individuals in the study troop and 26 individuals in three other troops in the study area may be large in comparison to many other troops in the Kutai Nature Reserve (Fittinghoff, 1978; Kurland, 1973). Wilson and Wilson (1975) found that *M. fascicularis* had larger group sizes in the more disturbed forest on the lower Sengata River than along the smaller rivers deeper in the primary forest.

Troop size data from the literature exist for over 100 troops of *M. fascicularis* (Angst, 1975; Bernstein, 1966, 1967; Chiang, 1968; Fittinghoff, 1978; Fooden, 1971; Furuya, 1965; Kurland, 1973; Rijksen, 1978; Rodman, 1973; Shirek-Ellefson, 1967; Southwick and Cadigan, 1972; Wilson and Wilson, 1975, 1977). The most extensive data are from the primate surveywork of Crockett and Wilson (this volume) in Sumatra. These authors indicated that the mean troop size for *M. fascicularis* in 34 troops in 22 different habitat types was 18.6 individuals, and that larger troops were found in secondary forest.

Data from the literature on group composition show that *M. fascicularis* live in multimale troops (Angst, 1975; Chiang, 1968; Fittinghoff, 1978; Furuya, 1965; Kurland, 1973; Ridley, 1906; Rodman, 1973; Shirek-Ellefson, 1967; Southwick and Cadigan, 1972; Wilson and Wilson, 1977). These results contrast with the predictions of various socioecological models such as those of Crook and Gartlan (1966) and Eisenberg *et al.* (1972) which state that forest species should be in unimale or age-graded troops and not have large-sized troops.

Feeding Behavior

What possible explanations of the heavy utilization of streamside areas are there? The most obvious explanation is that they do not seem to be exploited as much by sympatric primates. The orang-utan, *Pongo pygmaeus*, the gibbon, *Hylobates muelleri*, and the Sunda Island leaf monkey, *Presbytis aygula*, were infrequently sighted during surveywork in streamside areas, and their use of these areas was not great. The smaller body size and greater agility of *M. fascicularis* is an advantage in exploiting streamside areas and riverine refuging (see Chapter 8, this volume).

Another explanation of this animal's utilization of streamside areas is that they contain predictable and abundant sources of food. The fruit survey results showed that the fruit supply in secondary, streamside forest was more abundant and predictable than the fruit supply in old secondary and in primary, nonstreamside forest. In a nearby area, Rodman (1978a) reported the same pattern of fruit abundance, using the same survey methods. The important characteristic of the various pioneer tree species found in riverine areas is that they typically flower and fruit year round (Whitmore, 1975). Over half of the trees found in these areas contain fruit for these macaques. The idea that the food supply of frugivores is temporally and spatially unpredictable as mentioned by Brown and Orians (1970) needs to be carefully qualified.

Diet The crab-eating macaque is highly frugivorous in its diet. Other researchers have also noted this (Fittinghoff, 1978; Fooden, 1971; Furuya, 1962; Kurland, 1973), but only Rodman (1973) reported systematic data when he showed that over 90 percent of their diet was fruit.

The macaques as a genus are supposed to be vegetarian and omnivorous in their diets according to some socioecological models. Recent field studies on other macaques, however, have also noted their predominant frugivory (*M. mulatta*, Lindburg, 1976, 1977; *M. nemestrina*, Bernstein, 1967; Rodman, 1978b; *M. radiata*, Simonds, 1965; Sugiyama, 1971; *M. silenus*, Green and Minkowski, 1977; *M. sinica*, Dittus, 1977; Hladik, 1975). As Hladik (1975) proposed, the term omnivore should be dropped, and the macaques should be considered as frugivores.

Soil ingestion by macaques was noted by Eudey (1978) and Lindburg (1977), and may provide nutritious elements to the diet and influence digestion.

Foraging The crab-eating macaque is unusual in its activity profile as compared to many other species of primates. The only species which appear to spend more time in travel and less time in feeding and resting are *M. mulatta* (Neville, 1968) and *M. nemestrina* (Rodman, 1973). Only those two species have a similar activity profile to *M. fascicularis*. The differentiation between macaques and other primates supports the contention by Bernstein (1970) that activity budgets are sometimes useful in indicating degrees of evolutionary relatedness of species.

A number of important principles relate an animal's behavior to resource distribution and abundance (Altmann, S. 1974; Brown, 1964; Brown and Orians, 1970; Horn, 1968; Kummer, 1971). Altmann, for example, points out the relationship between resource distribution, density, and group size. Resources containing a small number of feeding sites are characteristically exploited by small groups, and those containing a large number can be exploited by large groups. The study troop broke up into small, dispersed groups in areas where fruit had a low density on a per tree basis. The small size of the foraging units in these trees was probably due to the limited number of feeding sites. More aggression occurred in these trees as the animals chased each other out of a tree. The less concentrated food sources on a per tree basis reduced the benefit gained per individual, given the energetic cost of locomotion and intratroop competition. These trees, however, have a high streamside density and their fruits are rapidly renewing. Consequently, the monkeys ranged through these areas as a group and, in addition, monitored fruiting trees away from the stream where resources were more patchy and highly concentrated in so-called 'supermarkets'. As mentioned previously, all of the troop members could feed in one of these trees for longer periods of time.

Similar but more extreme grouping patterns relating to small, widely dispersed trees on the one hand and large clumped trees on the other are reported for another arboreal frugivore, *Ateles belzebuth* (Klein and Klein, 1975, 1977). For the former trees, subgroupings of spider monkeys were small and stable and for the latter, subgroups were larger. Two of the most important palm species in their diet were not spatially clumped and bore ripe, localized fruits over long periods of time. These characteristics of palms are similar to the important food trees of the secondary forest at Kutai, and it is interesting to note that both areas where such trees grow are heavily flooded regions. A similar relationship of party size to the number of feeding sites has also been pointed out by Wrangham (1977) for chimpanzees.

A monitoring strategy for the discovery and utilization of large fruiting trees was noted in several other arboreal primates such as *Cercocebus albigena* and *Cercopithecus mitis* (Struhsaker, 1975; Waser, 1976, 1977). These primates had larger home ranges, diverse daily ranging patterns, and greater overall group spread as an adaptation to locating and exploiting unpredictable and super-abundant fruit sources.

The crab-eating macaque is a terminal branch feeder (Kurland, 1973; Shirek-Ellefson, 1967). The importance of cheek pouches in more fully utilizing the terminal branch area should be mentioned. Although mechan-ically its feeding sphere is not as great as the gibbon (Grand, 1972), extended periods of time hanging below a branch are not necessary to the macaque as it can simply grab what it wants and stuff its cheek pouches. The pattern of retrieving food items, placing them in cheek pouches, and retreating to a higher area for mastication was also noted by Lindburg (1971) and Murray (1975) for *M. mulatta*. These researchers said that such a retrieve and retreat pattern could enhance safety during feeding and outcompeting conspecifics. Another benefit to this pattern might be to scan the distribution and abundance of brightly colored fruits from a vantage point and make another foray into those areas to feed. The idea that mammals can hardly see the brightly colored fruits among the thick foliage as stated by Ridley (1930), for example, seems weak since primates have excellent color vision and fruits are colorful. Conspicuous fruits may be selected for and their seeds dispersed by mammals as well as by birds.

Ranging An animal's distribution in space and time indicates its adaptation to exploitation of essential resources. The distribution of food resources may directly affect such things as the home range size of an animal, its pattern of movement within its range, and other aspects of its ranging behavior. These small-bodied, arboreal frugivores presumably need to complement their diet with animal protein for certain essential amino acids. Hladik (1975), for example, indicated that such a demand for animal protein in *M. sinica* was difficult to meet and necessitated large home ranges in their search for animal prey. The home range size for the study troop is larger than the areas reported by other researchers such as Fittinghoff (1978) and Kurland (1973). Kurland (1973) gave an estimated .8 km² for a *M. fascicularis* troop at the Mentoko. Some other arboreal forest animals have even greater home range sizes. Waser (1976), for example, gave a figure of 4 km² for *Cercocebus albigena*. Green and Minkowski (1977) gave a figure of about 5 km² for *Macaca silenus*. The average length of day range was also larger than previously thought for *M. fascicularis*. Data on the study troop showed a day range of 1,900 meters which is about the same as the average for forest *M. mulatta*, whose day range was reported to be 1,428 meters (Lindburg, 1971), and about 2.6 kilometers (Neville, 1968).

The degree of arboreality in the study troop is one of the highest yet reported for any macaque species. Rodman (1973) reported that a nearby troop of *M. fascicularis* was never contacted on the ground and that during extensive tracking, they never came to the ground except within 5 meters of the edge of the river. The only other study that indicates such a high degree of arboreality in macaques is that of Green and Minkowski (1977) who said that *M. silenus* was on the ground less than 1 percent of the time. These reports of highly arboreal macaques are in contrast to some of the socioecological models that depict macaques as terrestrial.

Concluding Remarks The task of predicting and explaining a troop's or a species' social organization is much more challenging than originally stated by those researchers who proposed models of socioecology. Nevertheless, such a broad task is a worthy one, since prediction and explanation are at the core of science. The study of socioecology stresses adaptation such that various features of social systems are a product of certain selective factors. The various approaches to socioecology can propose multiple explanations of the adaptive features of social organization, but they do not seem to be able to predict very much. For example, small group size can be as adaptive in avoiding predators as large group size is in fighting off predators. This study on an arboreal macaque is especially useful in pointing out the typological nature of some of the socioecological predictions. *Macaca fascicularis* is neither "typical" of all macaques, nor is it "typical" of forest monkeys.

If our models are to have predictability, then this functional aspect of the relation between behavior and ecology is only half the story. Since the social organization of an animal is the product of individual interactions between its members, it is equally important to undertake a causal analysis on the behavioral sources of the troop's social organization. The more we know about a species' or a troop's or an individual's history of its behavioral sources of social organization under different ecological conditions, the better our predictions and explanations of social organization will be. Relative degrees of prediction and explanation may appear too frustrating and too difficult for some researchers, but, after all, simple and complete predictions about life are not possible.

ACKNOWLEDGMENTS

This research was supported by National Science Foundation Grant BMS 74-14190; The Explorers Club and a Sigma Xi Grant-in-Aid of Research.

I am indebted to my sponsor, the Indonesian Institute of Sciences, and also to the Department of Nature and Wildlife Conservation and Pertamina Oil Co. for their kind assistance. Use of the Hilmi Oesman Memorial Research Station and additional assistance was provided by Dr. Smith from the University of Washington. I am especially grateful for the kind assistance and advice of Dr. Rodman, Dr. Lindburg, Dr. Becking, and Mr. Academia. I thank Dr. Williams for the nutritional analyses of fruits, and Dr. Whittig and Dr. Quick for examining the soil sample.

REFERENCES

Altmann, J. Observational study of behaviour: Sampling methods. *Behaviour*, **49**(304): 227-267 (1974).
Altmann, S. A. Baboons, space, time and energy. *Amer. Zool.*, **14**: 221-248 (1974).
Angst, W. Pilot experiments to test group tolerance to a stranger in wild *Macaca fascicularis. Am. J. Phys. Anthrop.*, **38**(2): 625-630 (1973).
_____. Basic data and concepts on the social organization of *Macaca fascicularis*. In L. A. Rosenblum, (ed.), *Primate Behavior: Developments in Field and Laboratory Research.* IV., New York: Academic Press, pp. 325-388 (1975).
Bernstein, I. S. Naturally occurring primate hybrid. *Science*, **154**: 1559-1560 (1966).
_____. A field study of the pigtail monkey (*Macaca nemestrina*). *Primates*, **8**: 217-228 (1967).
_____. Primate status hierarchies. In L. A. Rosenblum, (ed.), *Primate Behavior: Developments in Field and Laboratory Research.* I., New York: Academic Press, pp. 71-109 (1970).
Brown, J. L. The evolution of diversity in avian territorial systems. *The Wilson Bulletin*, **76**(2): 160-169 (1964).
Brown, J. L. and G. H. Orians. Spacing patterns in mobile animals. *Annual Review of Ecology and Systematics*, **1**: 239-262 (1970).

Chiang, M. Use of tools by wild macaque monkeys in Singapore. *Nature*, **214**: 1258–1259 (1967).

_____. *The Annual Reproductive Cycle of a Free-Living Population of Long-Tailed Macaques in Singapore*. M.Sc. thesis, University of Singapore, (1968).

Clutton-Brock, T. H. Primate social organization and ecology. *Nature*, **250**: 539–542 (1974).

Crook, J. H. and Gartlan, J. S. Evolution of primate societies. *Nature*, **210**: 1200–1203 (1966).

Dittus, W. P. J. The socioecological basis for the conservation of the Toque monkey (*Macaca sinica*) of Sri Lanka (Ceylon). In Prince Rainer III and G. Bourne, (eds.), *Primate Conservation*, New York: Academic Press, pp. 237–265 (1977).

Eisenberg, J. F., Muckenhirn, N. A. and Rudran, R. The relation between ecology and social structure in primates. *Science*, **176**: 863–874 (1972).

Eudey, A. Earth-eating by macaques in Western Thailand: a preliminary analysis. In D. J. Chivers and J. Herbert, (eds.), *Recent Advances in Primatology*, I., New York: Academic Press, 351–353 (1978).

Fittinghoff, N. A., Jr. *Macaca fascicularis of Eastern Borneo: Ecology, Demography, Social Behavior*, and *Social Organization in Relation to a Refuging Habitus*. Unpublished Dissertation Ph. D., University of California, Davis (1978).

Fooden, J. Report on Primates collected in western Thailand, January-April, 1967. *Fieldiana Zool.*, **59**: 1–62 (1971).

Furuya, Y. On the ecological survey of wild crab-eating monkey in Malaya. *Primates*, **3**: 75–76 (1962).

_____. Social organization of the crab-eating monkey. *Primates*, **6**: 285–336 (1965).

Grand, T. I. A mechanical interpretation of terminal branch feeding. *J. of Mammalogy*, **53**(1): 198–201 (1972).

Green, S. and K. Minkowski. The lion-tailed monkey and its south Indian rain forest habitat. In Prince Rainier III and G. Bourne, (eds.), *Primate Conservation*, New York: Academic Press, pp. 289–337 (1977).

Hamilton, W. J. III and Watt, K. E. F. Refuging. *Annual Review of Ecology and Systematics*, **1**: 263–286 (1970).

Hill, W. C. O. *Primates. Comparative Anatomy and Taxonomy*, **VII**. *Cynopithecinae*. New York: John Wiley, (1974).

Hladik, C. M. Ecology, diet, and social patterning in old and new world primates. In R. H. Tuttle, (ed.), *Primate Functional Morphology and Evolution*, Chicago: Aldine Press, pp. 3–35 (1975).

_____. A comparative study of the feeding strategies of two sympatric species of leaf monkeys *Presbytis senex* and *Presbytis entellus*. In T. H. Clutton-Brock, (ed.), *Primate Ecology: Feeding and Ranging Behaviour of Monkeys, Lemurs and Apes*, London: Academic Press, pp. 323–353 (1977).

Hoogerwerf, A. *Udjung Kulon*. Leiden: E. J. Brill (1970).

Horn, H. S. The adaptive significance of colonial nesting in the brewer's blackbird (*Euphagus cyanocephalus*). *Ecology*, **49**(4): 682–694 (1968).

Jessen, R. J. Determining the fruit count on a tree by randomized branch sampling. *Biometrics*, **11** (1-4): 99–109 (1955).

Kelly, B. W. Objective methods for forecasting Florida citrus production. *IASI, Estadistica, Marzo*, pp. 56–64 (1958).

Klein, L. L. and Klein, D. J. Social and ecological contrasts between four taxa of neotropical primates. In R. Tuttle, (ed.), *Socioecology and Psychology* of Primates, The Hague: Mouton, pp. 59–85 (1975).

_____. Feeding behaviour of the Colombian spider monkey. In T. H. Clutton-Brock, (ed.), *Primate Ecology: Feeding and Ranging Behaviour of Monkeys, Lemurs and Apes*, London: Academic Press, pp. 153–181 (1977).

Kummer, H. *Primate Societies*. Chicago: Aldine Press, (1971).

Kurland, J. A. A natural history of kra macaques (*Macaca fascicularis* Raffles, 1821) at the Kutai Reserve, Kalimantan Timur, Indonesia. *Primates*, **14**: 245–262 (1973).

Lindburg, D. G. The rhesus monkey in North India: An ecological and behavioral study. In L. A. Rosenblum, (ed.), *Primate Behavior: Developments in Field and Laboratory Research, Vol. 2*, New York: Academic Press, pp. 1–106 (1971).

_____. Dietary habits of rhesus monkeys (*Macaca mulatta* ZIMMERMAN) in Indian forests. *J. Bombay Nat. Hist. Soc.* **73**(2): 261–279 (1976).

_____. Feeding behaviour and diet of rhesus monkeys (*Macaca mulatta*) in a Siwalik forest in North India. In T. H. Clutton-Brock, (ed.), *Primate Ecology: Feeding and Ranging Behaviour of Monkeys, Lemurs and Apes*. London: Academic Press, pp. 223–249 (1977).

Medway, L. The monkeys of Sundaland: Ecology and systematics of the cercopithecids of a humid equatorial environment. In J. R. Napier and P. H. Napier, (eds.), *Old World Monkeys: Evolution, Systematics and Behavior*, New York: Academic Press, pp. 513–553 (1970).

Murray, P. The role of cheek pouches in cercopithecine monkey adaptive strategy. In R. H. Tuttle, (ed.), *Primate Functional Morphology and Evolution*, Chicago: Aldine Press, pp. 151–194 (1975).

Neville, M. K. Ecology and activity of Himalayan foothill rhesus monkeys (*Macaca mulatta*). *Ecology*, **49**(1): 110–123 (1968).

Oates, J. F. The Guereza and its food. In T. H. Clutton-Brock (ed.), *Primate Ecology Feeding and Ranging Behaviour of Monkeys, Lemurs and Apes*, London: Academic Press, pp. 275–321 (1977).

Odum, E. P. and Kuenzler, E. J. Measurement of territory and home range size in birds. *Auk*, **72**: 128–137 (1955).

Richards, P. W. *The Tropical Rainforest*. Cambridge: Cambridge University Press (1966).

Ridley, H. N. The menagerie at the botanic gardens. *Royal Asiatic Soc. of Great Britain and Ireland, Straits Branch, Singapore, Journal*, **46**: 133–194 (1906).

_____. *The Dispersal of Plants Throughout the World*. London: Clowes and Sons (1930).

Rijksen, H. D. A Fieldstudy on Sumatran Orang Utans (*Pongo Pygmaeus Abelii* Lesson 1827). *Mededelingen Lanbouwhogeschool Wageningen, Nederland,* 78-2, H. Veenman & Zonen B. V. 420 pp., (1978).

Rodman, P. S. *Synecology of Bornean Primates*. Ph. D. Dissertation, Harvard University, Cambridge, Mass., (1973).

_____. Feeding behaviour of orang-utans of the Kutai Nature Reserve, East Kalimantan. In T. H. Clutton-Brock, (ed.), *Primate Ecology: Feeding and Ranging Behaviour of Monkeys, Lemurs and Apes*, London: Academic Press, pp. 383–413 (1977).

_____. Diets, densities and distributions of Bornean primates. In G. G. Montgomery and J. F. Eisenberg, (eds.), *Arboreal Folivores*, (1978a).

_____. Synecology of Bornean primates. II: Ecological segregation of sympatric *Macaca fascicularis and Macaca nemestrina*. Manuscript (1978b).

Rowell, T. E. Variability in the social organization of primates. In D. Morris, (ed.), *Primate Ethology*, Garden City, New York: Anchor, pp. 283–305 (1967).

Shirek-Ellefson, J. Visual Communication in *Macaca irus*. Ph. D. thesis, University of California, Berkeley, California, (1967).

_____. Social communication in some old world monkeys and gibbons. In P. C. J. Dolhinow, (ed.), *Primate Patterns*, New York: Holt, Rinehart and Winston, pp. 297–311 (1972).

Simonds, P. E. The bonnet macaque in south India. In I. DeVore, (ed.), *Primate Behavior: Field Studies of Monekys and Apes,* New York: Holt, Rinehart and Winston, pp. 175–196 (1965).

Southwick, C. H. and Cadigan, F. C. Jr. Population studies of Malaysian primates. *Primates*, **13**(1): 1–18 (1972).

Southwick, C. H. and Siddiqi, M. F. Demographic characteristics of semi-protected rhesus groups in India. *Yearbook of Physical Anthropology*, 1976, **20**: 242–252 (1977).

Struhsaker, T. T. Correlates of ecology and social organization among African cercopithecines. *Folia primatol.* **11**: 80–118 (1969).

_____. Correlates of ranging behavior in a group of red colobus monkeys, *Colobus badius tephrosceles. Amer. Zool.,* **14:** 177–184 (1974).

_____. *The Red Colobus Monkey.* Chicago: The University of Chicago Press, (1975).

Sugiyama, Y. Characteristics of the social life of bonnet macaques (*Macaca radiata*). *Primates,* **12**(3–4): 247–266 (1971).

Waser, P. M. *Cercocebus albigena:* Site attachment, avoidance, and intergroup spacing. *Amer. Natur.,* **110**: 911–935 (1976).

_____. Feeding, ranging, and group size in the mangabey, *Cercocebus albigena.* In T. H. Clutton-Brock, (ed.), *Primate Ecology: Feeding and Ranging Behaviour of Monkeys, Lemurs and Apes,* London: Academic Press, pp. 183–222 (1977).

Wheatley, B. P. *The Behavior and Ecology of the Crab-eating Macaque (Macaca fascicularis) in the Kutai Nature Reserve, East Kalimantan, Indonesia.* Ph. D. Dissertation, University of California, Davis (1978a).

_____. Riverine secondary forest in the Kutai Nature Reserve, East Kalimantan, Indonesia. *Malayan Nature Journal* **30**(4): (1978b).

_____. Foraging patterns in a group of longtailed macaques in Kalimantan Timur, Indonesia. In D. J. Chivers and J. Herbert, (eds.), *Recent Advances in Primatology,* **I.**, London: Academic Press, pp. 347–349 (1978c).

Whitmore, T. C. *Tropical Rainforests of the Far East.* Oxford: Clarendon Press (1975).

Wilson, C. C. and Wilson, W. L. The influence of selective logging on primates and some other animals in East Kalimantan. *Folia primatol.,* **23**: 245–274 (1975).

_____. Behavioral and morphological variation among primate populations in Sumatra. *Yearbook of Phys. Anthrop.,* **20**: 207–233 (1977).

Wrangham, R. W. Feeding behaviour of chimpanzees in Gombe National Park, Tanzania. In T. H. Clutton-Brock, (ed.), *Primate Ecology: Feeding and Ranging Behaviour of Monkeys, Lemurs and Apes.* London: Academic Press, pp. 503–538 (1977).

Chapter 10
Population Patterns and Behavioral Ecology of Rhesus Monkeys (Macaca mulatta) in Nepal

Jane Teas,
Thomas Richie,
Henry Taylor
and *Charles Southwick*

INTRODUCTION

Natural populations of rhesus monkeys in Nepal were studied between June 1974 and August 1978 in three areas of Nepal (Fig. 10-1): (1) temple and parkland habitats in Kathmandu valley at an altitude of 1,340 meters; (2) lowland terai jungle in the Karnali-Bardia reserve of southwestern Nepal at an altitude of 180 meters; and (3) pine-oak-spruce forests in the mountains around Rara Lake in northwestern Nepal at an altitude of 3,050. meters. These areas represent three basic habitats which rhesus monkeys occupy in Nepal; temples and parklands surrounded by intensive agriculture and high human population density, lowland monsoon forest between the Gangetic basin and Himalayan foothills, and upper montane forest in the central Himalayas.

The Kathmandu rhesus populations live on the grounds of ancient temples, and interact closely with the human community in and around these temples. One population of approximately 300 monkeys in five social groups lives in Swayambhu, a Buddhist temple site at least 2,000 years old, on the western side

Fig. 10-1   Locations of study areas.

of Kathmandu city (Fig. 10-2). The other population of about 320 monkeys ranges over a complex of parks, residences and temples at Pashupati, a Hindu temple site also more than 2,000 years old, located on the eastern side of Kathmandu city (Fig. 10-3).

Fig. 10-2   Swayambhu Temple, Kathmandu.

Fig. 10-3   Pashupati Temple.

Both temples include open and wooded parklands, and Pashupati contains small tracts of forest. The rhesus monkeys in these locations provide ideal opportunities for studies of primate population dynamics, home-range patterns, and the utilization of time, space, and habitat by rhesus monkeys. They are also ideal for the analysis of social behavior in different environments, and interactions between monkeys and people.

In contrast, the Karnali-Bardia and Rara Lake areas offer opportunities to study rhesus in forest habitats with less frequent human contact. Even in these areas, however, rhesus monkeys interacted with people and agricultural activities at times.

The terai forests of Karnali-Bardia represent the typical "jungle" and wildlife areas of Nepal and northern India. These are ecotonal zones between the Gangetic plain and the first range of Himalayan foothills, characterized by deciduous forest, primarily sal (*Shorea robusta*), terminalia (*Terminalia tomentosa*), sheesham (*Dalbergia sissoo*), acacia (*Acacia catechu*), banyan (*Ficus bengalensis*), pipal (*F. religiosa*), and more than 100 other species of trees, shrubs, and vines. Rainfall is 125 to 200 centimeters during the monsoon season of June to September, with the remainder of the year quite dry. The climate is similar to that of the Gangetic plain, and the area is subject to increasing agricultural invasion.

The montane forests of the Rara Lake area in northwestern Nepal at altitudes around 3,050 meters are dominated by blue pine (*Pinus excelsa* or *Pinus wallichiana*), western Himalayan spruce (*Picea smithiana*), fir (*Abies spectabilis*), birch (*Betula utilis*), and several species of oaks (*Quercus* spp.). Small cultivated fields also occur around the lake, with crops of barley, wheat, corn, millet, potatoes, apples, pulses, and spinach. The climate is cool and moist with winter snows.

Most field studies of rhesus macaques have been undertaken in India, where the monkeys have been subjected to considerable trapping pressure (until the recent export ban beginning April, 1978) with the result that both their population ecology and behavior has been affected (Southwick, Beg and Siddiqi, 1961a, 1961b, 1965; Southwick and Siddiqi, 1977). In Nepal, rhesus monkeys have not been trapped, nor had they been studied prior to our own investigations. They afford substantial advantages for both behavioral and ecological studies of macaques in natural settings.

The diverse rhesus populations of Nepal illustrate the wide range of ecological and behavioral adaptability of the rhesus monkey, probably the greatest of any species of nonhuman primate. Nepal offers an outstanding opportunity to study some of the extremes of rhesus adaptability.

RESULTS

Population Ecology

In the terai forests of Karnali-Bardia, 27.5 km² of forest were surveyed in February and March, 1976, by foot transects, with the sighting of eight groups of rhesus monkeys and 21 groups of langurs (*Presbytis entellus*). This provided a population estimate of 0.29 rhesus groups per square kilometer, and 0.76 langur groups per square kilometer.

Rhesus groups varied from 20 to 51 individuals, averaging 32 per group. This is very similar to the average group size of rhesus monkeys in forests around Dehra Dun (30.3) as observed by Lindburg (1971). Langur groups in Karnali-Bardia varied from 12 to 40 and averaged 26.2 individuals per group.

The rhesus group density of Karnali-Bardia was very similar to that observed in Corbett National Park, India, in the same type of forest habitat (Southwick, Beg and Siddiqi, 1961b) where 0.27 groups per square kilometer were found. Group size in Corbett was higher, however, averaging 50 individuals per group, so overall population density in Corbett was 13.5 individual rhesus per square kilometer compared to only 9.3 rhesus per square kilometer in Karnali-Bardia.

The Karnali-Bardia rhesus population had a low ratio of immature individuals to total population (0.44), whereas that in Corbett had a higher ratio (0.50) indicating better reproduction and survival of young in Corbett. By way of comparison, the Dehra Dun forest populations studied by Lindburg (1971) had a ratio of immatures to total population of 0.50 to 0.54. These and other macaque population studies have shown that an immature to adult ratio of 0.50 is necessary for long-term population maintenance (Southwick and Siddiqi, 1977), suggesting that the Karnali-Bardia population was not in a strong position in 1976.

In the pine-spruce-oak forests of northwestern Nepal in the vicinity of Rara Lake rhesus monkeys were found in October of 1975 at an altitude of 3,050 m and they were reliably reported to ascend a ridge at 3,658 m. This probably represents a high altitude record for rhesus monkeys (Richie, *et al.*, 1978). Only one group was found, consisting of 39 individuals, including 3 adult males, 17 adult females, 8 infants, and 11 juveniles. The immature ratio was 0.49, indicating that this group was near the edge of long-term maintenance. Obviously the survival or mortality of each immature individual is of rather critical importance to this small isolated population.

Several features of the Rara Lake environment aided the survival of rhesus monkeys during harsh winters at this altitude: (1) the southern exposure of the forest which formed their primary home range; (2) the climate-moderating influence of the lake which does not freeze; (3) the presence of two villages with agricultural crops; and (4) the easy access to lower altitudes in the Mugu Karnali Valley near the forest. If two or more of these favorable ecological conditions had been absent, we doubt if the rhesus could permanently survive at this altitude and latitude. Although seasonal migration could not be documented at the time of our study, villagers did report seasonal changes in the monkey population.

Rhesus monkeys have the widest range of habitats among all species of macaques. Not only distributionally, ranging from Afghanistan eastward throughout South and Southeast Asia, they are also found in a wide variety of environments ranging from mangrove swamps to the high mountain forests of Rara Lake at 3,050 meters. Everywhere they have adapted well to the presence of man, and have found viable habitats in villages, towns, cities, railway stations, along roadsides and canals, temples and, of course, forests. Their adaptability is expressed primarily in behavior. In the Kathmandu population, for instance, fully 20 percent of the aggression observed during the study involved other species. Of this, 80 percent was initiated by other species and directed towards the monkeys and the remaining 20 percent was initiated by the monkeys and directed towards other species. Most of the other species involved were people, although dogs, birds, goats, and cows were also involved. This provides some indication of the competitive environment to which rhesus can adapt. The important fact is that rhesus can survive and reproduce well under adverse conditions of crowding and competition. Unlike other congenerics such as the lion-tailed macaque (*Macaca silenus*), rhesus monkeys have sufficient plasticity of temperament and behavioural variability to accommodate close association with man.

Various studies on food habits also show the behavioral adaptability of rhesus monkeys. In the forests around Dehra Dun, Lindburg (1976) found that rhesus relied almost exclusively on natural forest vegetation, consuming over 100 species of plants in their diet. A preliminary study of food habits of the Pashupati monkeys by Marriott (1978) showed that 80 percent of the monkeys' diet came from natural sources. In contrast, a study of rhesus food habits in an agricultural area of north central India showed that only 7 percent of their diet came from natural sources, 93 percent from human sources (Siddiqi and Southwick, 1980). Another indication of the willingness of rhesus monkeys to adapt to the presence of man was recorded in the Karnali-Bardia survey. Even in the terai forests, rhusus troops averaged only 0.25 kilometers from the nearest field edge or village. Our survey was conducted in February, and hence was not a period of harvest, still the rhesus groups appeared to be concentrated along the jungle/field edge. Certainly the success of rhesus in adapting to man's presence is beneficial only as long as man remains tolerant of the monkeys' presence.

More detailed population studies have been conducted on the Kathmandu populations at Swayambhu and Pashupati over a period of five years, from 1974 to 1978. The last two years of 1977 and 1978 have been supported by an Earthwatch program under the Center for Field Research in Belmont, Massachusetts.

The total populations of Swayambhu and Pashupati have been relatively stable, varying from 266 to 327 monkeys for Swayambhu and from 300 to 332 monkeys for Pashupati. The number of groups, however, have not remained constant. At Swayambhu, there has been an increase in the number of groups from five in 1974 to six in 1978, and at Pashupati there has been an increase from seven to ten troops.

All the behavioral studies on the Kathmandu populations have focused on multimale, multifemale troops. On an average, a troop was composed of 8 percent male, 35 percent female, 30 percent juvenile, 15 percent yearling, and 13 percent infants.

Both of the Kathmandu populations have had a very favorable age ratio, with an average of 57.5 percent of the total population immature. Normally, this demographic pattern should produce population growth in the order of 10 to 15 percent per year, yet these populations have apparently been intrinsically regulated without trapping or external removal.

The mechanisms for intrinsic population regulation seem to be low natality (an average of 63 percent of adult females produced one young per year from 1974 through 1978, compared to 78 percent to 90 percent in comparable Indian populations), and high adult mortality (25.8 percent per year in Kathmandu, compared to 12.5 to 17.5 percent per year in comparable Indian populations, see Southwick, *et al.*, 1977).

Infant mortality in Kathmandu (21.5 percent per year) was slightly higher than in India (15.5 percent to 17.7 percent), but juvenile loss rate was much lower (17.0 percent in Kathmandu compared to 31.5 percent to 53.9 percent in India). The high juvenile loss rate in India reflects trapping loss, since this age class was extensively trapped prior to April, 1978, when India issued an export ban on all rhesus monkeys.

All of the above data indicate that the primary mechanisms of population regulation in the Kathmandu rhesus are reduced birth rates and high adult mortality. To a lesser extent, higher infant mortality may also be considered a mechanism of population regulation. Hence these data show considerable areas of agreement with those of Dittus (1975, and this volume). The real causes for these demographic characteristics of the Kathmandu rhesus are unknown; we do not know, for example, the relative roles of disease, nutrition, or behavior in determining these patterns. These unknowns represent challenges for future research.

Home Ranges and the Utilization of Space

The rhesus groups of Swayambhu had overlapping home ranges varying in size from 6 to 15 hectares. The degree of overlap was only 10 to 20 percent however, whereas in the more crowded Aligarh temple monkeys in India it was observed to be 80 percent (Southwick, *et al.*, 1965). Overlap in Kathmandu occurred primarily around the temples, garbage dumps, water holes and major access paths. Each group also had a core area of utilization which was fairly exclusively occupied; these tended to be places for resting, grooming, and play, and were located away from major food resources.

Similar patterns of home range occurred at Pashupati, except the spatial dimensions varied more, from 2.5 to 24 hectares. In general the largest groups had the largest home ranges with the most favorable habitat, though many specific exceptions occurred. At Swayambhu, the two largest groups, Rex's (138 monkeys) and Omshalla's (64 monkeys), had the largest home ranges, whereas Falstaff's (29) and Cyan's (30) had the smallest. The home ranges of Rex's group and Omshalla's were similar in size, but Rex's group had the most favored habitat in terms of access to the temple, and an eastern exposure which gave them early morning sunshine (advantageous after cold nights), and afternoon shade (advantageous during very hot days).

At Pashupati, the two largest groups, Burger's (71 monkeys) and Gandalf's (65 monkeys) also had the largest home ranges and best access to the main temple. Home ranges were not entirely stable from year to year, and often changed dramatically as the size, leadership or dominance relationships of the groups changed.

The Utilization of Time

In the study of behavior, emphasis was placed on the quantitative analysis of behavior profiles (that is, the relative number of monkeys engaged in various activities), and on various ecological factors (season, habitat, time of day) associated with changes in behavioral profiles. For this analysis, we used four

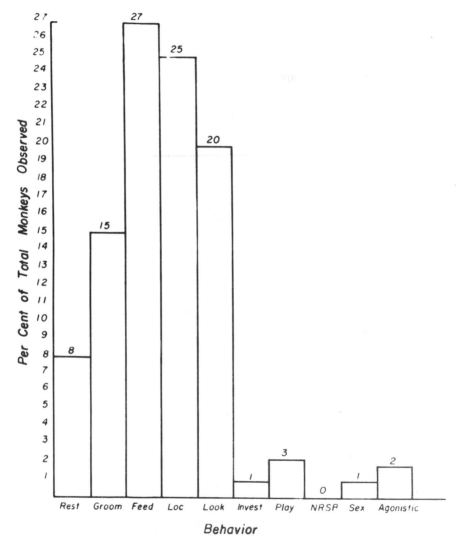

Fig. 10-4   Behavior profiles for 12 months, 1974-75.

and five-way analysis of variance and Duncan's new multiple-range test. For all comparisons, a "p" value of 0.05 was considered to be significant.

Behavioral profile data were obtained by counting the numbers of observed monkeys engaged in each of 10 basic behaviors every 10 minutes during each hour and a half observation period. The data were then averaged to obtain the mean numbers of monkeys engaging in each of the ten types of behavior throughout the observation period. The basic behavioral profile data involved a total of 1,506 hours of observation over a 12-month period. The study was divided equally between four habitats, two times of day and almost equally between the four seasons. Typically 15 to 20 individual monkeys were under surveillance at any given time. Special behavioral data were obtained on aggressive behaviors, grooming, feeding, and maternal-infant relationships since these represent social interactions of particular interest in understanding group dynamics and behavioral ecology.

For all individuals throughout the entire year, feeding and locomotion were the most important behavioral activities, accounting for 52 percent of the monkeys' time. If looking behavior and grooming are added, fully 87 percent of their total behavior is accounted for (Fig. 10-4). Play activities, mostly by juveniles, occupied only 3 percent of the total activity budget; agonistic behavior, 2 percent; and sexual behavior, 1 percent.

Behavioral profiles varied according to age class, season, habitat, and time of day, and the data yielded a great many details on these variations. In general, adults spent significantly more time grooming, resting and looking. Sexual and agonistic behaviors were also more frequently observed among adults. Juveniles spent significantly more time in feeding on natural vegetation and in play (Fig. 10-5). Both age classes spent comparable amounts of time in feeding from human sources and locomotion.

Seasonal changes in behavior were most conspicuous during the fall, when mating occurred, and the spring and early summer when births occurred. The fall mating season was accompanied by a significant increase in grooming, sexual behavior, and feeding from human sources. Agonistic behavior also showed an increase in the fall, but the increase was statistically significant only for adult males when considered alone.

The spring birth season showed remarkably few significant changes in behavioral profile. There were only decreases in locomotion and looking.

Habitat had a number of significant influences on behavior. In the Swayambhu temple grounds, the monkeys were most actively engaged in feeding from human sources, looking, and locomotion (Fig. 10-6). The parklands were characterized by significantly higher percentages of resting and grooming, both quiet activities. The forest habitat was characterized by moderation in all behaviors; that is, no behavior was significantly the most or the least, compared to other habitats. We had anticipated that the forest

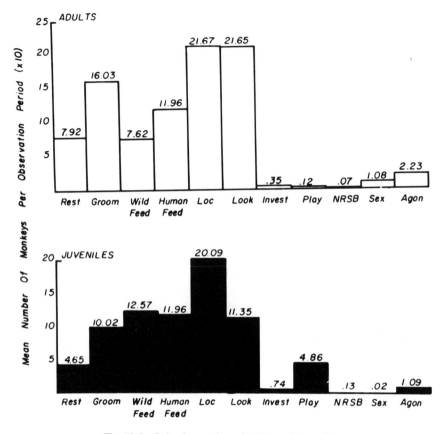

Fig. 10-5   Behavior profiles of adults and juveniles.

environment would result in significantly less agonistic behavior, as was shown in studies of forest-dwelling rhesus in India in comparison with temple, village and urban monkeys (Southwick, 1972), but this was not the case. The rhesus in Pashupati forest showed only slightly less aggression than those in Swayambhu temple, and they actually showed more than those in Pashupati and Swayambhu parklands. None of these differences in agonistic behavior was significant, however. Our only explanation is that the Kathmandu rhesus are basically temple monkeys, even though they venture into parklands and forest patches. Apparently their brief sojourns in the forest did not modify their aggressive behavior significantly; at least they did not reduce its frequency as measured in this study.

The time of day modified behavior by producing a higher frequency of resting, grooming, and feeding from human sources during the morning, and more feeding from natural vegetation in the afternoon. There was not

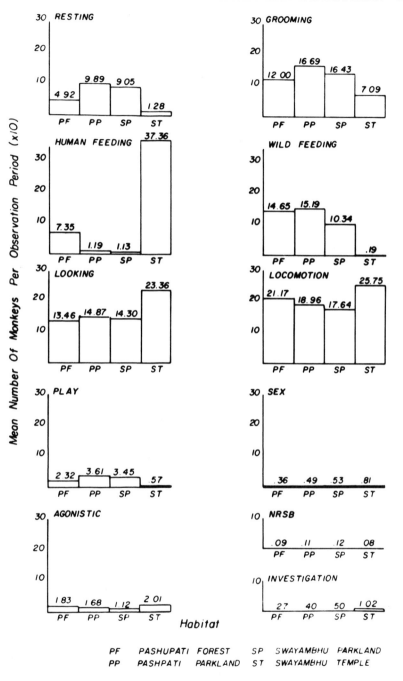

Fig. 10-6. Behavior profiles compared by habitat.

significant variation between morning and afternoon behaviors in other categories.

The variables of season, habitat, and time of day did not intereact in any significant way on the monkeys' behavior; that is, the same general seasonal changes occurred independently in each habitat and in both morning and afternoon time periods. Complete data and more detailed analyses of all of the these topics are presented in other papers and publications (Taylor, 1976; Teas, 1975, 1978, 1979).

Grooming

More detailed data on grooming showed that adult females were twice as active in grooming as juveniles, and three times more active than males (Fig. 10-7). Males received about 30 percent more grooming than they initiated, females initiated about 10 percent more than they received, and juveniles groomed and were groomed approximately equally. There were no significant

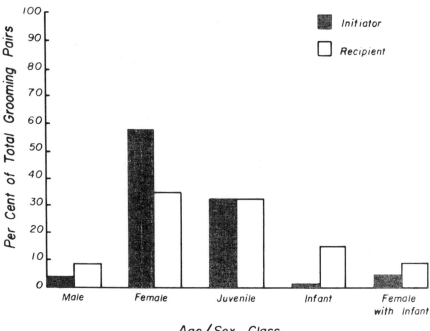

Fig. 10-7   Initiate and recipient relationships in grooming pairs compared by age/sex class.

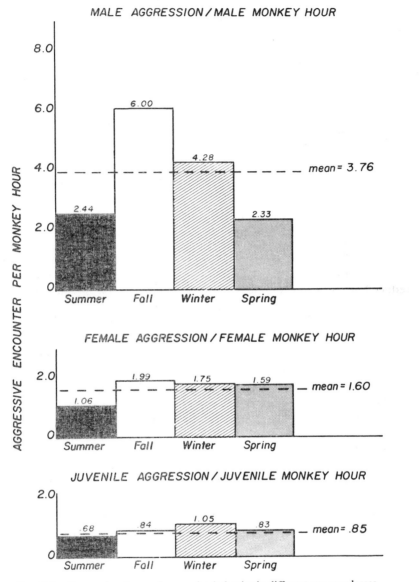

Fig. 10-8. Seasonal patterns of aggressive behavior in different age-sex classes.

seasonal variations in the mean number of each age/sex class engaged in grooming. Habitat influenced grooming significantly, with lowest grooming frequencies in the temple, moderate levels in the forest, and highest frequencies in the parklands.

Agonistic Behavior

The Kathmandu rhesus had high levels of aggression (averaging 1.71 aggressive encounters per monkey per hour), three to four times higher than comparable groups in India. Despite the high level of aggression, 94 percent of all aggressive conflicts in Kathmandu were threat and chase interactions only, and did not involve physical contact.

Males initiated more than twice as many aggressive interactions as females, and four times more than juveniles. The seasonal increase in aggression in the fall was mainly due to an increase in male aggression (Fig. 10-8).

Although we lack direct proof, we feel that the high levels of aggressive behavior may have population significance by acting as a general stressor agent, and thereby increasing adult mortality and reducing natality.

Parental Care

Maternal care in the Kathmandu rhesus was typical of that described in other studies of rhesus monkeys elsewhere, with the exception that mothers showed reduced grooming of infants in the spring and summer. Female rhesus with infants showed appropriate protective and restrictive behavior but only 0.13 grooming bouts per hour were directed toward their infants. This was generally lower than grooming levels recorded for any other age/sex class.

Other studies of rhesus maternal care do not quantify grooming, but they imply that substantially more mother-infant grooming than we saw in Kathmandu is typical. If there is indeed less maternal grooming of infants in the Kathmandu rhesus, this may contribute to the relatively high levels of infant mortality, though we cannot assess the role of infectious disease. We feel the Kathmandu rhesus were also subject to high levels of infectious disease, especially respiratory and enteric infections such as tuberculosis and dysentery, and these would impact most seriously on infants.

Several cases of the care of infants by adult males were observed, and these have been described in detail in a separate publication (Taylor, *et al.*, 1978). For three years in a row, one dominant male adopted an orphaned infant and attempted in each case to care for the infant, but in all cases the infant died of starvation.

SUMMARY

A field study of rhesus monkey ecology and behavior was undertaken in Nepal from June 1974 through August 1978. Population counts were conducted in lowland terai forest in southwestern Nepal, upland montane forest in northwestern Nepal, and in Kathmandu valley in central Nepal. Intensive behavioral studies were done in Kathmandu.

The Kathmandu populations were relatively stable in numbers over the study period, and showed intrinsic biosocial mechanisms of population regulation. These were manifested primarily by reduced birth rates and high adult mortality rates. To a lesser extent, they also showed high infant mortality rates. Population trends of the forest groups are unknown, but age structure data indicated that these populations may be declining or are barely holding their own.

Behavioral studies focused on home ranges, behavioral profiles, and environmental variables. Home ranges of different groups overlapped, but virtually all groups had core areas of exclusive use. In behavioral profiles, the monkeys spent 87 percent of their observed daytime activities in feeding, locomotion, looking behavior, and grooming. Specific time budgets varied between sex and age groups, season of the year, time of day, and habitat.

Grooming was the most prevalent direct social interaction and was dominated by adult females. The Kathmandu rhesus were highly aggressive, although 94 percent of their aggressive behavior was noncontact. The high levels of aggressive behavior may have contributed indirectly to increased mortality and reduced natality. They may also have been related in some way to reduced maternal care. These topics require further study.

The entire study emphasizes the ways in which environmental factors may influence behavior, and the importance of behavior in the ecological adaptations of rhesus monkeys.

ACKNOWLEDGMENTS

This research was supported by the National Geographic Society, and the Center for Field Research in Belmont, Massachusetts. We are indebted to Drs. Barry Bishop, Edwin Snyder, Paul Oehser, and Ms. Mary G. Smith and Ms. Joanne Hess of the National Geographic Society and Ms. Elizabeth Caney and Ms. Pamela Pierce of the Center for Field Research for guidance and encouragement at important phases of the study. In addition, we are grateful for the contributions made by the 17 Earthwatch participants who assisted in data collection during 1977 and 1978. In Nepal, we would like to thank Professors D. R. Uprety, and D. D. Bhatt, Dr. K. R. Pandey, and Mrs. Bina Pradhan of Tribhuvan University; Dr. B. N. Uprety and Mr. H. R. Mishra of the Office of National Parks and Conservation; Dr. Thomas Acker and Gabriel Campbell of the U.S. Educational Foundation in Kathmandu. Dr. B. Marriott, Dr. P. Vandegrift, Mr. Ram Shrestha, Rakesh Shrestha, F. M. Burhans, and R. T. Tashi all assisted with the field work. George and Charlotte Whitesides participated in the Karnali-Bardia survey, and we are very grateful for their contributions.

REFERENCES

Dittus, W. P. J. Population dynamics of the toque monkey, *Macaca sinica.* In, R. H. Tuttle, (ed.), *Socioecology and Psychology of Primates,* Paris: Mouton, pp. 125–151 (1975).

Lindburg, D. G. The rhesus monkey in north India: an ecological and behavioral study. In L. A. Rosenblum, (ed.), *Primate Behavior: Developments in Field and Laboratory Research*, New York: Academic Press, pp. 1–106, (1971).

_____.Dietary habits of rhesus monkeys (*Macaca mulatta* ZIMMERMAN) in Indian forests. *J. Bombay Nat. Hist. Soc.*, **73**: 261–279 (1976).

Marriott, B. Feeding patterns of wild rhesus monkeys (*Macaca mulatta*) in Kathmandu, Nepal. *Fed. Proc.*, 37(3): 759 (1978).

Richie, T., Shrestha, R., Teas, J., Taylor, H. and Southwick, C. Rhesus monkeys at high altitudes in northwestern Nepal. *Jour. Mammalogy*, **59** 43–444, (1978).

Siddiqi, M. F. and Southwick, C. H. Food habits of rhesus monkeys in north central India. Zool. Record (Zoological Survey of India), Calcutta (in press, 1980).

Southwick, C. H. Aggression among nonhuman primates. Module in Anthropology, No. **23**, Reading, Mass: Addison-Wesley, pp. 1–23 (1972).

Southwick, C. H., Beg, M. A., and Siddiqi, M. R. A population survey of rhesus monkeys in villages, towns and temples of northern India. *Ecology*, **42**: 538–547 (1961a).

Southwick, C. H., Beg, M. A., and Siddiqi, M. R. A population survey of rhesus monkeys in northern India: II. Transportation routes and forest areas. *Ecology*, **42**: 698–710 (1961b).

Southwick, C. H., Beg, M. A. and Siddiqi, M. R. Rhesus monkeys in north India. In, I. DeVore, (ed.), *Primate Behavior: Field Studies of Monkeys and Apes*, New York: Holt, Rinehart and Winston, pp. 111–159 (1965).

Southwick, C. H., and Siddiqi, M. F., Farooqui, M. Y., and Pal, B. C. Xenophobia among free-ranging rhesus groups in India. In R. L. Hollaway, (ed.), *Primate Aggression, Territoriality, and Xenophobia*, New York: Academic Press, pp. 185–209 (1974).

Southwick, C. H., Siddiqi, M. F. Population dynamics of rhesus monkeys in northern India. In, Prince Rainier III and G. Bourne, (eds.), *Primate Conservation*, New York: Academic Press, pp. 339–362 (1977).

Taylor, H. Rhesus monkeys of Kathmandu Valley. *Nepal Nature Conservation Society Newletter*, No. 33, (1976).

Taylor, H., Teas, J., Richie, T., Southwick, C., and Shrestha, R. Social interactions between adult male and infant rhesus monkeys in Nepal. *Primates: J. of Prim.* **19**: 343–351 (1978).

Teas, J. The rhesus monkey in Kathmandu, Nepal. *Nepal Nature Conservation Society Special Coronation Issue*, No. **28**, pp. 2–4 (1975).

Teas, J. Behavioral Ecology of Rhesus Monkeys (*Macaca mulatta*) in Kathmandu Nepal. Ph.D. Thesis, Baltimore: The Johns Hopkins University, 131 pp. illus., (1978).

Teas, J., Taylor, H. and Richie, T. Environmental influences on aggression. *Sixth Congress of the Internat. Primatological Society, Proc.* (in press), (1979).

Chapter 11
The Social Regulation of
Primate Populations:
A Synthesis

Wolfgang P. J. Dittus

INTRODUCTION

A long-term study of a population of toque monkeys (*Macaca sinica*) of Sri Lanka suggests that animals indirectly kill one another through social behaviors that involve access to resources and mates. Such mortality is not imposed randomly, but follows differences in social dominance. As dominance differs according to age and sex (and kinship), so too do the socially imposed mortalities. The net result of these relationships is that the density as well as the age-sex distribution of the society (and hence the population) are determined through social means and are regulated towards an equilibrium with the available resources. The evolution of such behaviors is discussed elsewhere (Dittus, 1979). The aim of this paper is to examine the ubiquity of this phenomenon among other primates, paticularly among other macaque species.

For the sake of clarity, and to establish a framework for comparison, I will present first (after the next section) a more detailed account of the relationships between behavior, ecology and demography as observed in the toque monkey. Then I will review published data from other primate species that are relevant to the present topic.

GENERAL INFORMATION CONCERNING *MACACA SINICA*

The Taxonomy, Morphology and Distribution of *Macaca sinica*

Macaca sinica is endemic to the island of Sri Lanka (formerly Ceylon) and belongs to the subfamily of Old World monkeys, Cercopithecinae, which also includes other macaques, baboons (*Papio, Theropithecus*), vervet monkeys or guenons (*Cercopithecus*), and others. Toque monkeys share the subgenus *zati* with the bonnet macaques (*Macaca radiata*) of southern India, and possibly with *Macaca assamensis* of northeastern India and Assam (Hill and Bernstein, 1969).

The toque monkey is a long-tailed macaque and is agile both on the ground and in the trees. It is about one-half the size of the commonly known rhesus macaque (*Macaca mulatta*); the average weight of wild adults is 5.7 kilograms for males and 3.6 kilograms for females. Its common name refers to a caplike whorl of hair radiating symmetrically outward from a central point on the crown of its head (Fig. 11-1). Its nearest relatives, the bonnet macaques (*M.*

Fig. 11-1. Photograph of a group of toque monkeys (*Macaca sinica*). The adult male, sitting in the foreground, was cut on the side of the face while fighting with another male over access to an estrous female. Three adult females and a nursing infant are sitting behind the male.

radiata), have a similar arrangement of head hair. However, relative to *M. radiata*, *M. sinica* is somewhat smaller, possesses a brighter pelage with orange and reddish hair; and its ears, lower lips and border of the eyelids are pigmented black instead of pink. Adult females of both species frequently have red faces. Hill (1932) and Pocock (1932) describe the morphology of these species in greater detail.

The distribution of *M. sinica* is confined to areas of natural forest that have an ample water supply. Groups have not been observed living in towns, as is often reported for *M. mulatta* of India (Southwick, *et al.*, 1961). Estimates of population size for the species as a whole and for each of its three subspecies have been published (Dittus, 1977b).

The Study Area and Schedule

The study site was located in a semievergreen forest area of the Polonnaruwa Sanctuary in the northeastern dry zone of Sri Lanka. Demographic, ecological and certain behavioral data were recorded for 18 troops of nearly 450 toque monkeys. Detailed social and ecological data were recorded for two troops. The study was continuous between September 1968 and May 1972 and from March 1975 to the present. Intermittent observations were taken between May 1972 and March 1975.

Age Classification

The ages of infants and juveniles in general were estimated by comparison to infants and juveniles whose exact ages and morphological development were known. Females were considered adult with the first pregnancy, on the average at approximately five years of age. Males were not adult until seven to eight years old as judged by body size, testes, canine teeth and muscular develop- ment. Subadult males from five to seven years old were as large as, or larger than, adult females, but were smaller and physically less developed than adult males. Adults were classified according to five broad categories; young, young to middle age, middle age, old and senile. A host of morphological changes, similar to those criteria one might use in subjectively assessing the ages of humans, formed the basis of this classification. With increasing age there occurred erosion and tartarization of the teeth, wrinkling of the facial skin, especially near the eyelids, lips and cheek pouches, and loosening of folds of body skin. The degree of facial pigmentation often intensified with age, and facial hair became more prominent. The pelage of old or senile individuals was frequently dull, and some hair-loss occurred, especially on the tail. Older inidividuals frequently were more scarred than younger ones, and crippling and stiffness of the joints was evident among very old animals. Longevity in the toque monkey was thought to be approximately 30 years (Dittus, 1975) which is supported by records of capitve animals (Hill, 1937; Jones, 1962). Young adult males (7 to 10 years old) were larger than subadult males, and as large as most adult males, but their muscular development was less than in males in their prime (young to middle age class). The numerical age ranges (e.g., Fig. 11-2) attributed to adult classes were determined by dividing the number of unassigned adult years (25 for females, 20 for males) by the number of unassigned adult age classes (5 for females, 4 for males). A more detailed and systematic scheme of age class determination among adults was presented in Appendix 1 of Dittus (1974).

THE BEHAVIORAL REGULATION OF POPULATION DENSITY AND AGE-SEX DISTRIBUTION IN *MACACA SINICA*

The original data which support the conclusions and observations abstracted below have been presented and discussed in greater detail elsewhere (Dittus, 1975, 1977a).

Population Structure and Demography

In order to assess population processes it is useful to transform census information into vital statistics. The theory and practical methods for life-table analyses as applied to mammalian populations have been established (Deevey, 1947; Quick, 1963; Caughley, 1966, 1977). Basically, an accurate knowledge of the age and sex distributions of populations under equilibrium conditions is requisite. Mortality is calculated by assuming that the decrease in the number of individuals in successive age classes is the result of mortality, once the effects of natality, immigration and emigration have been taken into account.

   Demographic changes were traced in the study population of approximately 450 macaques from 18 troops. Troop size increased through birth and immigration and decreased through death and emigration. Under relatively stable ecological conditions troops fluctuated in size within limits (the fluctuation corresponding to birth seasons and subsequent mortality); but the net growth of most troops and of the population as a whole was zero ($R_0 = 1$). This was further confirmed by tracing the histories of 131 females, over three to nine years: the 33 females beginning to reproduce balanced the 33 reproducing females that died. The average birth rate of 0.688 infants per adult female per year was sufficiently high to permit, theoretically, rapid population growth ($R_0 > 1$). The overall equilibrium state was achieved mainly through a balance between natality and mortality. Of all macaques born, 90 percent of males and 85 percent of females died prior to adulthood (Dittus, 1975).

The age and sex composition of the study population is given in Table 11-1. The total number of males nearly equalled that of females. There were 2.3 times as many adult females as adult males, but among the juveniles and infants there were more males than females. The sex ratio at birth, 1.042 (290 births) for the population did not differ significantly from 1.000 ($z = 0.294$, $p = 0.772$, two-tailed binomial test). Females do not migrate, and solitary males and all-male groups are included in the census. The ages of juveniles were known or estimated according to the morphology of juveniles with known ages. The sex ratios among the juveniles and adults therefore are not attributable to sex differences in the rates of maturation or migration; rather, they reflect differences in sex-specific rates of mortality (Fig. 11-2) as outlined in a life-table (Dittus, 1975).

The Causes of Mortality

A major question concerns the causation of the observed mortality. In the population of toque monkeys at Polonnaruwa, congenital or other disease was not apparent, and predation, mainly from feral dogs, was not a major cause of mortality. Although disease may have contributed to mortality, particularly among infants where mortality was extreme, there is no evidence to suggest that disease or predation by themselves underlie the observed pattern of

Table 11-1. The Age-Sex Distribution of *Macaca Sinica* at Polonnaruwa, Sri Lanka, in October, 1971[1].

| AGE CLASS | NUMBER OF ANIMALS PER AGE-SEX CLASS | | |
	AGE IN YEARS	MALES	FEMALES
Adult Female	5 - 30	--	111
Adult Male	7 - 30	48	--
Subadult	5 - 7	23	--
Juvenile	1 - 5	83	54
Infant	0 - 1	30	21
Total		184	186

[1] After Dittus, 1975.

mortality (Fig. 11-2). More likely disease and predation act in concert with more determinate phenomena, perhaps those which arise through food related behavioral-ecological interrelationships. The latter were examined more closely.

Mortality in Relation to Differential Access to Resources

An individual was considered dominant if it had priority of access to food, mates, and other resources. Records of agonistic behavior permitted the ordering of all animals in a troop according to a linear hierarchy of dominance, with few departures from linearity. A threat was defined as any behavior that caused the respondent to alter its spatial position or behavior in avoidance. The frequency of agonistic behavior varied according to the context of activity (e.g., moving, resting, playing, grooming); and 82 percent of all threats occurred while foraging—searching for and consuming food. Subordinate individuals, when threatened, were prevented from feeding, and foods that they had found were usurped in 36 percent of all threat interactions. In some instances dominant individuals removed and consumed food that was already in the cheek pouches of subordinates. Since access to food is crucial to growth and

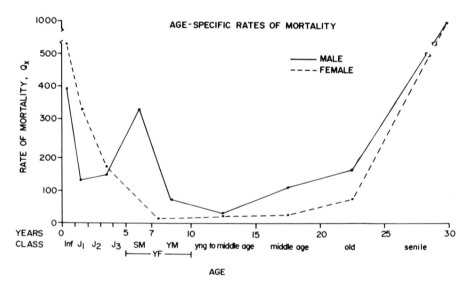

Fig. 11-2. The age-specific rates of mortality (the proportion of animals dying (q_x) per 1,000 entering each age class) for a population of *Macaca sinica*. The method of age classification is described in the text. Infants (Inf.) are 0 to 1 year old. The ages of juveniles are (J_1) 1 to 2 years; (J_2) 2 to 3½ years; (J_3) 3½ to 5 years. Young adult females (YF) span the subadult male (SM) and young adult male (YM) age classes. (From Dittus 1975).

survivorship, and mortality differed by age and sex, behavioral data were summarized according to the frequency of threats occurring between age-sex classes during foraging. The ratio of the frequencies with which individuals threatened and were threatened during foraging, directly measures their relative access to contested resources. Such ratios (Table 11-2) indicate that adult males had greatest relative access and infant females the least.

To test for the possibility of agonistic discrimination between animals of different age and sex, the observed frequencies of threat were statistically tested against several progressively stronger null hypotheses. The latter assumed random threatening between animals of different age and sex according to first, the age-sex distribution of the society; and second, to the age-sex distribution plus the frequencies with which animals of different age and sex were closely associated in space. It was found that: (1) adult males were dominant to animals of all other age-sex classes and threatened them more frequently than expected if threats had occurred randomly; (2) dominant adult males displaced subordinate adult males, subadult males and old juvenile males to the troop's periphery away from favored feeding areas; (3) subadult males and adult females exploited the juveniles which in turn exploited the infants; (4) among the juveniles and infants the males dominated their female peers; (5) overall, female juveniles and infants were threatened by all members of the society (including each other) about twice as often as the males of these age classes; (6) subadult males were peripheral to the troop and interacted minimally with the remainder of the troop.

Affiliative behaviors during foraging reflected the agonistic discrimination against infant and juvenile females. For example, during foraging adult and older juvenile males frequently approached and hugged infant and younger juvenile males. Such favoritism was never accorded young females. Juvenile females associated much less with adult males during foraging than juvenile males. As these males monopolized the favored foods, the young females were in effect avoiding these foods.

The behavioral relationships during foraging apparently influences the foraging efficiency of animals. Adult males were assumed to be the most efficient foragers because they spent the least amount of time (i.e., effort) in foraging, and the most in resting, but had the highest feeding rates and consumed the greatest proportion of foods that are high in protein and calories. By these same measures adult females were next in foraging efficiency, then juvenile males, and juvenile females were least efficient.

I have proposed (Dittus, 1977a) that mortality is socially imposed through direct competition for vital resources, which is most acute for animals prior to adulthood. The greater mortality manifest by the youngest animals in general, and by the females relative to the males amongst the juveniles and infants in particular, is probably the result of the much greater frequency with which they

Table 11-2. Ratios of Threats Given To Those Received Per Age-Sex Class During Foraging in Two Different Sized Troops[1].

	ADULT MALE	SUBADULT MALE	ADULT FEMALE	JUVENILE MALE	JUVENILE FEMALE	INFANT MALE	INFANT FEMALE	TOTAL
Number of animals in troop B	4	3	8	9	9	3	1	37
Frequency of threats:								
given	344	28	355	355	333	52	13	1480
received	33	22	299	350	563	126	87	1480
Ratio of threats given:received	10.4	1.3	1.2	1.0	0.6	0.4	0.1	
Number of animals in troop A	2	3	4	3	3	1	2	18
Ratio of threats given:received (N = 229 threats)	8.4	2.4	1.1	0.8	0.4	0.0[2]	0.3	

[1] Adapted after Dittus 1977a. The magnitude of the ratio correlates directly with social rank and reflects relative access to food resources by age and sex.
[2] The infant male in Troop A was a neonate, mostly carried by the mother and foraging very little.

were exploited for, and prevented access to, resources. Whether death is through starvation alone, or through an agent secondary to food deprivation (e.g., physiological stress) has not been distinguished. Adult and subadult males had additional sources of mortality.

Mortality in Relation to Mating and Migration

During the mating season, rates of threatening increased two-fold, but were confined primarily to adult males, estrous females and subadult and older juvenile male "followers" that were directly involved in mating activity. Most threats were between adult and subadult males as they competed for access to estrous females. The incidence of wounding increased, especially among males during the mating season.

Adult males recently deposed from high status in the dominance hierarchy, and subadult males, wandered between troops during the mating season. Rates of migration were highest in subadult males, and, on the average, all males left their natal troop prior to adulthood. Migration was a stressful period because the subadult immigrants were relegated to low priority of access to resources by the new troop, and they were frequently wounded from fights with established males. The peak in mortality among subadult males (Fig. 11-2) coincided with the peak of migration in males, which suggested that the rigors of migration underlie the observed mortality. Adult males migrated less frequently than the subadults, and had better chances of access to resources. Their rates of mortality were less than those of subadults, but higher than those of adult females which never shifted between troops (Dittus, 1975).

Behavioral Regulation of Social Structure and Density in Relation to Changes in the Food Supply

Under relatively stable ecological conditions at Polonnaruwa up to 1974, the net growth rate of the population was zero ($R_O = 1$). A decrease in the food supply, caused mainly by a prolonged drought in 1974, coincided with a 13.5 percent decrease in population size. One troop, however, had access to a garbage dump bordering the study area, where food scraps were deposited daily. This troop grew at an annual rate of 12.5 percent. These observations suggested that a decrease in food supply caused the observed reduction in population size, whereas the abundant food supply stimulated population growth (Dittus, 1977a).

As might be expected from the age-sex differences in access to food resources, changes in the food supply affected different age-sex classes differently. Thus the decrease in population size resulted mainly from increased mortality among the nonadults, particularly among the youngest juveniles, and more among the female infants and juveniles than among their male peers. In contrast, the increase in population size under an abundant food supply resulted mainly from greater survivorship among the juveniles, and particularly among the female infants and juveniles (*ibid.*).

Population Regulation: Some Hypotheses

The limits to population growth appear to be set mainly by the availability of food and water. However, subordinate animals have lowest priority of access to other resources as well, such as favored sleeping places and refugia from predators. Under conditions where these resources might be in short supply it is conceivable that their availability may also influence populaton growth. Predation and disease appear to have no major or long-term impact on population growth. However, such nonsocially imposed mortality alters the age-sex composition of troops, and hence, alters the behavioral forces which determine social density and age-sex composition.

Competition between troops, in part, determines the amount of food available per troop, and thereby contributes to setting the upper limit to troop size. The molding of the age-sex structure is then determined by social processes within the troop.

I hypothesized (Dittus, 1977a) that the effects of social behavior (mainly aggressive and affiliative behaviors) and possibly its nature, change in accordance with the age-sex distribution of the society, its density, and available resources. Social behavior determines, through socially imposed mortality, the age and sex distribution of the society (and hence of the population) in such a way as to maximize the reproductive success of some of its members, and results in bringing the society (and population) towards an equilibrium state ($R_0 = 1$) with the resources and nonsocially imposed mortality.

THE BEHAVIORAL REGULATION OF POPULATION DENSITY AND AGE-SEX DISTRIBUTION IN OTHER PRIMATES

To my knowledge no other studies have as yet dealt with the social regulation of population size or age-sex composition among primates. Therefore such comparisons will be incomplete for any one species. However, there exist relevant data from many studies, and my aim here is to assess them collectively with the view of examining the generality of the data and hypotheses presented on the basis of the data for *M. sinica*. By virtue of the emphasis in primate research, most of these comparisons will concern the Old World Cercopithecinae, particularly those of the genus *Macaca*, and the New World howler monkeys, genus *Alouatta*.

Demographic Comparisons

An assessment of mortality by age and sex necessitates accurate census data of animals of all ages by sex. Unfortunately, most field studies of primates are incomplete in this regard. Data from the few reports where sex classification among juveniles has been achieved or attempted are listed in Tables 11-3 to 11-5. In the original publications, census data often were presented for each troop in the population studied. I have totaled data for all troops per population or study, omitting troops that were incompletely censused. Contiguous troops which constitute a subpopulation in each area were studied in all cases except Simonds' (1973) study of *M. radiata* (Table 11-3). Data were tallied according to equal-age classes between the sexes for juveniles and infants. In some instances, this meant combining the age categories the original authors had assigned them. For example, young juveniles aged one to two years old, and old juveniles aged two to three years old, are tabulated here as juveniles aged one to three years old. The ages of adulthood indicated by the original authors were adhered to in all cases.

Solitary or all-male groups are reported to be either absent or extremely rare among most species listed in Tables 11-3 to 11-5. Where they do occur they have been included in the population census: *M. sinica, M. radiata, M. fuscata, M. mulatta,* and *Alouatta caraya.*

Without exception all populations showed a prevalence of females among the adults, but males outnumbered females among the juveniles. Except for the data of *Papio anubis* of Kenya (Table 11-3), addition of the subadult to the adult males, such that the post-juvenile age classes are equal-aged between the sexes, does not alter the sex ratio favoring adult females. Hence, this difference cannot be ascribed to the greater maturation time males require to reach adulthood. Similarly, adding all the unsexed juveniles to the juvenile females does not alter the sex ratio favoring males (except in *P. anubis* from Uganda where many juveniles are not sexed). The data are less clear among infants because too many remain unsexed.

Hall and DeVore (1965) do not distinguish subadult males in *P. anubis* (Table 11-3), hence the sex ratio among old juveniles may be confounded by unequal age classes. However, young juveniles (one to two years old) were aged with greater confidence (Hall and DeVore, 1965), and the males clearly outnumber the females. In his census of *Alouatta seniculus,* Neville (1976) attributes one to two years duration to the subadult male phase and one year to subadult females. Therefore, the numerical differences between subadult males and females of equal age are less pronounced than indicated in Table 11-4. Pope (1968) does not recognize a subadult phase in *A. caraya,* however, she distinguishes young adult males and females of equal age on the basis of dental wear (Table 11-4).

Table 11-3. The Demography of Several Wild Populations of Cercopithecinae[1].

SPECIES AND LOCATION OF STUDY	ADULT MALE	ADULT FEMALE	SUB-ADULT MALE	JUVENILE MALE	JUVENILE FEMALE	INFANT MALE	INFANT FEMALE	JUVENILE SEX UNKNOWN	INFANT SEX UNKNOWN	TOTAL	SEX RATIO NEONATE	SEX RATIO ALL AGES (MALE:FEMALE)
Macaca sinica Anuradhapura Sri Lanka	48 (>7)	117 (>5)	20 (5-7)	106 (1-5)	71	29 (0-1)	26	0	0	417	18:18	203:214
Macaca mulatta North India (1)	7	19 (>3)	2 (3-?)	15 (1-3)	7	10 (0-1)	8	0	0	68	10:10	34:34
Macaca radiata South India (2)	37 (5+)	61-62 (4+)	18 (3-4)	23 (1-3)	17	? (0-1)	?	0	50-51	206-208	?	78+:78+
Papio cynocephalus Kenya (3,4)	44 (6)	65 (4)	7 (4-6)	34 (1-4)	20	4 (0-1)	5	5	12	196	3:3	89+:90+
Papio anubis Uganda (5)	19	21	3	13	6	?	?	8	19	90	?	35+:27+
Papio anubis Kenya (6,7)	31 (>8)	55 (5)	? (4-8)	18 old 32 yng (1-4)	14 old 15 yng	7 (0-1)	7	2	12	193	?	88+:91+
Cercopithecus aethiops Kenya (8)	8.34 (≥4)	17.44 (≥4)	?	11.13 (1.5-4)	6	? (0-1.5)	?	0.65	24.69	68.25	?	19.47+:23.44+
Cercopithecus aethiops Loloui Island Lake Victoria (9)	53	74		43	35	?	?	0	24	229	?	96+:109+

Sources: 1. Lindburg 1973. 2. Simonds 1973. 3. Altmann and Altmann 1970. 4. Hausfater 1975. 5. Rowell 1966. 6. Hall and DeVore 1965. 7. DeVore and Hall 1965. 8. Struhsaker 1967b. 9. Hall and Gartlan 1965.

[1] The numbers in parentheses indicate the age ranges (in years) attributed to the various age-sex classes.

Table 11-4. The Demography of Wild Populations of South American Howler Monkeys[1].

SPECIES AND LOCATION OF STUDY	ADULT		SUBADULT		JUVENILE		INFANT		UNSEXED		TOTAL	SEX RATIO (MALE:FEMALE) ALL AGES
	MALE	FEMALE	MALE	FEMALE	MALE	FEMALE	MALE	FEMALE	JUV.	INF.		
Alouatta seniculus Venezuela (1)	30 (4.5+)	47 (3.5+)	13 (2.5-4.5)	6 (2.5-3.5) Young adult male female	23 (1-2.5)	18	13 (0-1)	10	0	3	163	79:81
Alouatta caraya Argentina (2)	8	20	46 (unspecified equal age)	28	16 (1-3)	11	6 (0-1)	5	0	0	135	71:64

NUMBER OF ANIMALS PER AGE-SEX CLASS

Sources: 1. Neville 1976.   2. Pope 1968.

[1] The numbers in parantheses indicate the age ranges (in years) attributed to the various age-sex classes.

275

The neonatal sex ratio is known for only a few species. For *M. mulatta* raised in captivity, van Wagenen (1954) reports a neonatal sex ratio of 103 males: 105 females. Drickamer (1974) reports a similar ratio of 179 males: 183 females for rhesus born in the provisioned colony at La Parguera. Gilbert and Gillman (1951) found seven male and seven female births among captive *P. ursinus.* These data support the trend towards an equal sex ratio at birth among the Cercopithecinae listed in Tables 11-3 and 11-5.

Assuming an approximately equal neonatal sex ratio among the populations considered here, the census data suggest that (1) the sex ratio favoring males among juveniles is the result of greater mortality among juvenile females; (2) mortality in males is extreme during the subadult or adolescent phase such that (3) the sex ratio among adults favors females. Pope (1968) had attributed the age-sex composition of *A. caraya* (Table 11-4) to similar age and sex specific differences in survivorship. It is of particular interest that most populations tend towards an equal total sex ratio. This suggests that, although mortality is distributed differently by age between the sexes, the total or average survivorship for the male and female cohorts is the same. The theoretical implications of this are discussed by Dittus (1979). In the absence of more substantial demographic data and statistical scrutiny, these conclusions remain tentative.

The ages of wild primates, particularly among juveniles, have generally been estimated, at least in part, on the basis of body size. Among laboratory and colony reared *M. mulatta* (Gavan and Hutchinson, 1973) and in the wild population of *M. sinica* (Dittus, 1977a) juvenile males grow faster than juvenile females. Conceivably, such growth differences may occur also among other wild primate populations. The use of body size for estimating juvenile ages, therefore, harbors the risk of overestimating the ages of juvenile males relative to juvenile females. Consequently, large-sized juvenile males may be classed as subadult rather than juvenile. In life-table analyses such classification would tend to overestimate the mortality among juvenile males relative to the juvenile females. The possible error in age estimation due to growth differences by sex among juveniles, therefore, would tend to underestimate rather than overestimate the difference in mortality between male and female juveniles and infants.

Population Growth in Relation to Food Availability

The provisioning of food to colonies of *M. mulatta* and *M. fuscata* stimulated an increase in the growth rate of these populations by 16 percent (Koford, 1965) and 8.9 percent (based on data in Itani, 1975), respectively. The converse situation prevails in the Masai-Amboseli Game Reserve of Kenya where recent ecological and edaphic changes have caused the reduction of natural

Table 11-5. The Demography of Macaque Populations That Are Provisioned With Food By Man[1].

SPECIES AND LOCATION OF STUDY	ADULT MALE	ADULT FEMALE	SUB-ADULT MALE	JUVENILE MALE	JUVENILE FEMALE	INFANT MALE	INFANT FEMALE	JUVENILE SEX UNKNOWN	INFANT SEX UNKNOWN	TOTAL	SEX RATIO (MALE:FEMALE) NEONATE	SEX RATIO ALL AGES
Macaca fuscata Takasakiyama (1962) (1)	156 (>4)	228 (>3)	47 (3–4)	118 (1–3)	89	42 (0–1)	44	2	18	744	182:163	363+:361+
Macaca fuscata Takasakiyama (1965) (2)	245 (>4)	321 (>3)	57 (3–4)	140 (1–3)	118	67 (0–1)	72	0	0	1020	?	509:511
Macaca mulatta Cayo Santiago (1967) (3,4)	96 (>4.5)	197 (3.5)	18 (3.5–4.5)	153 (0.5–3.5)	143	—	—	0	0	607	143:149	267:340

Sources: 1. Itani et al. 1963. 2. Carpenter and Nishimura 1969. 3. Wilson 1968. 4. Koford 1965.
[1] The numbers in parentheses indicate the age ranges (in years) attributed to the various age-sex classes.

277

vegetation (Western and Van Praet, 1973) which constitutes the main diet of *P. cynocephalus* (Altmann and Altmann, 1970) and of *Cercopithecus aethiops* (Struhsaker, 1967a, 1973). The population of *C. aethiops* declined by 33.3 percent between 1964 and 1971, and by 43 percent between 1964 and 1975; Struhsaker (1973, 1976) suggests that this resulted from the reduction in food supply. The decline in the population of *P. cynocephalus* from 2,600 in 1963 to approximately 200 in 1971 is attributed to the same cause (Hausfater, 1975). Food shortage in a troop of *C. aethiops* in Senegal resulted in a 21 percent reduction in troop size, which Galat and Galat-Luong (1977) attribute to increased mortality through malnutrition. Glander (1975, 1977) suggests that the population size of *Alouatta palliata* of Costa Rica is limited by the availability of palatable foods.

Predation has generally been de-emphasized as a major factor limiting primate populations (e.g., Rowell, 1969; Hall and DeVore, 1965; Hall and Gartlan, 1965; Pope, 1968; Galat and Galat-Luong, 1977; Glander, 1975), and it is likely that food availability determines the sizes of primate populations in general, as previously suggested by Crook (1972).

Behaviorally Induced Starvation According to Dominance by Age and Sex

Dominance hierarchies are frequently described in field studies of primates, yet few investigators examine dominance relations between animals of different age and sex. Where such analyses have been attempted, adult males are described as dominant to all other troop members, the youngest animals being the most subordinate; e.g., *C. aethiops* (Struhsaker, 1967b), *P. anubis* (Rowell, 1967), *M. mulatta* (Southwick, 1969), *P. ursinus* (Saayman, 1971), *M. radiata* (Sugiyama, 1971) and *A. palliata* (Glander, 1975). Usually the most dominant male(s) forces the most subordinate adult males and the subadult males to assume a position peripheral to the troop; e.g., *M. mulatta* (Southwick, Beg and Siddiqi, 1965; Kaufmann, 1967; Lindburg, 1971), *M. radiata* (Sugiyama, 1971), *M. fuscata* (Imanishi, 1957), *P. anubis* (DeVore and Hall, 1965), *C. aethiops* (Struhsaker, 1967b) and *A. palliata* (Glander, 1975). In the provisioned colonies of *M. mulatta* and *M. fuscata*, where long-term genealogical records have been maintained, Kawamura (1958), Koford (1963), Sade (1967), Koyama (1967) and Norikoshi (1974) have stressed the importance of the mother's rank in determining that of her offspring. Thus, juveniles of high-ranking mothers dominate adult females and others that rank below the mother. Such social inheritance of rank is consistent with the hypothesis that mothers should invest most in their own young, a topic that is examined in greater detail elsewhere (Dittus, 1979).

Sex differences in agonistic behavior have been investigated by Altmann

(1968) among the colony of *M. mulatta* at Cayo Santiago. As in *M. sinica*, adult male rhesus were more aggressive than adult females, and juvenile males and females did not differ in their total frequency of aggressive behavior. However, unlike *M. sinica* the frequency of submissive behaviors between juvenile males and females did not differ among the rhesus. As Altmann pointed out, however, no systematic sampling technique was used in collection of his data, nor did he specify the context of activity in which the behavior occurred. In a controlled study of *M. mulatta* in captivity, juvenile males showed more dominant behaviors and fewer fear behaviors than juvenile females (Møller *et al.*, 1968). Likwise, among juvenile peers of *M. fuscata* males show more dominant behavior than females (Norikoshi, 1974). Hausfater (1975) indicated that in wild *P. cynocephalus* adult males dominate all other troop members, adult females dominate most young, and among the juveniles and infants, older ones dominate the younger ones, but among peers males dominate females. These data are consistent with those for *M. sinica*.

Aggressive discrimination by adult males against young females has been observed in a colony of crab-eating macaques, *Macaca fascicularis*. Newly dominant adult males frequently attacked and killed infants in the social group. Notably, they attacked females more than males, killing only two of the 18 available male infants, but 11 of the 16 available female infants (Angst and Thommen, 1977). Evidence from colony raised pig-tailed macaques, *M. nemestrina*, suggests that aggressive discrimination against female infants may begin even before birth. Sackett *et al.* (1975) report that pregnant females carrying female fetuses were attacked and bitten by other group members more frequently than females carrying male fetuses. This discrimination occurred only during the latter half of the pregnancy term when male hormones from the fetus begin circulating in the mother's blood—presumably communicating her infant's sex to other group members. Other forms of behavioral discrimination have been observed by Rowell, Din and Omar (1968) among *Papio anubis*, where adult males interacted more with male than female infants. Similarly, adult male *M. mulatta* showed more physical contact with male than female infants (Redican, 1975), and adult and juvenile male *M. radiata* preferred to sleep and groom with juvenile and infant males (Koyama, 1973).

The studies quoted above demonstrate differences, by age and sex, in dominance and other behavioral relations without considering the involvement of resource competition. The latter, of course, need not be evident in all behavioral manifestations, and any one behavior may have several effects. The importance of rank to feeding ecology, however, is demonstrated by the fact that dominant-subordinate relationships are defined frequently and most reliably on the basis of priority of access to food or other contested items among the Cercopithecinae (Bernstein and Sharpe, 1966; Richards, 1974; Farres and

Haude, 1976) and *Alouatta* (Glander, 1975). In addition, many investigators report heightened frequency and intensity of aggression, usually with an adult male emerging as a despot, during artificial feeding, e.g., *P. anubis* (Hall and DeVore, 1965), *M.mulatta* (Kaufmann, 1967; Koford, 1963; Lindburg, 1971; Southwick, 1972), and *M. speciosa* (Bertrand, 1969). Southwick, Beg and Siddiqi (1965:p. 153) report that during feeding adult male *M. mulatta* were most aggressive towards infants and juveniles, especially the former. When food was very limited, caged rhesus mothers aggressively prevented their own infants from feeding and some infants were killed (Carpenter, 1942).

In an experimental study of agonistic behavior in captive *M. mulatta*, Southwick (1967) reported that under restricted food supply "the level of food intake of each monkey was directly related to its position in the dominance hierarchy. The highest ranking animals had first access to food, and apparently took as much as they wanted. . . The mid ranking animals then consumed the remaining food. . . the lowest half of the dominance hierarchy. . . was under severe starvation." The two lowest ranking subadults had to be removed from the social group to prevent their death through starvation. Similar behaviorally induced starvation was noted by Bertrand (1969) among captive *M. mulatta*, *M. fuscata*, and *M. speciosa*, and by Bernstein (1969) among captive *M. nemestrina*.

These data, then, indicate that dominant animals impose mortality upon subordinates through competition for resources among other Cercopithecinae. Conceivably, such mortality may vary according to the differences in rank by age and sex, similar to the pattern found in *M. sinica*.

Agonistic Behavior in Relation to Food Availability

The above observations might lead one to suspect that food shortage should increase aggression. Behavioral observations of *M. sinica* (unpublished data) and of free-ranging *M. mulatta* under conditions of food shortage on Descheo Island, Puerto Rico (Morrisson and Menzel, 1972), and of a natural population of *P. ursinus*, facing starvation (Hall, 1963), fail to support this contention. Instead aggression decreased, animals became lethargic, and spatially more dispersed and spent most of the time searching for food. Southwick (1967) confirmed these field data by experimentally creating a food shortage among captive *M. mulatta*.

A decrease in overt aggression under food shortage is consistent with Lack's (1966) hypothesis that subordinate animals should not contest the priority of access to resources by dominant animals in order to conserve energy in a conflict that would be lost anyway. Instead all energy is devoted to the acquisition of food by other means, and any mortality that results is owed to imposed starvation rather than increased fighting.

Socially Imposed Mortality in Relation to Mating and Migration

An increase in wounding and mortality from fighting between males during the mating season is documented for *M. mulatta* of Cayo Santiago by Vandenbergh and Vessey (1968) and Wilson and Boelkins (1970). Among wild *M. mulatta*, Lindburg (1971) reports increased aggression among males, and cites several cases of severe wounding and two deaths that resulted from such male competition for mates. Males of *A. seniculus* (Neville, 1972b) and of *A. palliata* (Glander, 1975) harrass one another in a sexual context. Neville (1972a) and J. F. Eisenberg (personal communication) have frequently observed scars and wounds among subadult and adult males of *A. seniculus*, and Eisenberg attributes such injuries to male-male fighting.

The migration of males between troops has been noted in several species: *M. mulatta* (Koford, 1963, 1966; Lindburg, 1969; Neville, 1968), *M. fuscata* (Kawai and Yoshiba, 1968a, b; Kawanaka, 1973), *M. radiata* (Sugiyama, 1971; Simonds, 1973), *P. anubis* (DeVore and Hall, 1965; Rowell, 1969), *P. cynocephalus* (Altmann and Altmann, 1970), *C. aethiops* (Struhsaker, 1973), and *A. seniculus* (Neville, 1972a). Where noted, all reports indicate that rates of migration are highest during the mating season, and detailed studies of *M. mulatta* (Boelkins and Wilson, 1972) and of *M. fuscata* (Kawai and Yoshiba, 1968a, b) indicate that the highest rates of migration occur in subadult males.

The link between mortality and migration has not been established in the studies of these species. However, I have shown earlier (Dittus, 1975) that the peak in mortality in males of *M. mulatta* at Cayo Santiago and of *M. fuscata* at Takasakiyama corresponds to the ages where migration rates are the highest. Also, experimental studies that simulate migration (changing group membership) indicate that such change is socially stressful and can lead to socially imposed mortality. Strangers introduced to established captive groups of *M. mulatta* were attacked 4 to 10 times more often than established group members (Southwick, 1969). Newcomers were attacked similarly in captive groups of *M. mulatta, M. fuscata, M. speciosa* and *M. silenus* (Bertrand, 1969), and *M. nemestrina* (Bernstein, 1969). Altmann and Altmann (1970) indicate that an emigrant male *P. cynocephalus* had lost status recently through fighting with other males, and Kawanaka (1973) suggests that emigrated male *M. fuscata* remain solitary because of their inability to achieve dominant rank in a new troop.

Deaths among experimentally introduced strangers that resulted from bite wounds they received from established members has been noted in *M. speciosa* (Bertrand, 1969), *M. nemestrina* (Bernstein, 1969) and *Papio* (Hall, 1964). In *M. nemestrina* death was also imminent among introduced strangers because established members prevented their access to food and water. In wild populations, Saayman (1971) reports an adult male *P. ursinus* killing two young males on the periphery of a troop.

Adult males compete for mates much more frequently and vigorously than adult females; and males, particularly the subadults, most frequently change troop membership. Both male behaviors involve greater risks of mortality for the males than for the females. Therefore, the sex ratio favoring females among the adults may be the result of the greater behaviorally mediated mortality among adult and especially subadult males.

SUMMARY AND CONCLUSIONS

Demographic data from several species of macaques (*Macaca*), baboons (*Papio*), vervet monkeys (*Cercopithecus*) and the South American howler monkeys (*Alouatta*), indicate a common pattern of age and sex-specific mortality whereby: (1) among the juveniles and infants females die at a greater rate than their male peers; (2) mortality in males surpasses that of females with the advent of adulthood; (3) mortality in males is extreme during the subadult phase; and (4) males continue to be subject to greater mortality than females in adulthood. Population growth among many primate species is closely linked with the availability of palatable food. Predation or disease cannot adequately account for the maintenance of troop size nor for the pattern of age and sex specific mortality evident in the various species of Cercopithecinae or *Alouatta*. Behavioral data, particularly for the Cercopithecinae, indicate that subordinate animals frequently die as a result of being prevented access to food and water by the dominant individuals in their societies. In those species for which adequate records exist, dominance relationships differ by age and sex as follows: adult males dominate all other age and sex classes; adult females dominate juveniles and infants; juveniles dominate infants; and among the infants and juveniles the males dominate their female peers. Together these data suggest that the greater mortality of young juveniles versus older ones and adults, and particularly of females versus males among the infants and juveniles, is a direct result of their limited access to vital resources, though other forms of discrimination against them may also be involved.

Socially restricted access to resources may also contribute to mortality among subadult and adult males, and is most acute for subordinate ones which are forced to remain peripheral to preferred feeding sites by the dominant adult males. In addition, subadult and adult males are subject to mortality which results from competition for mates. Such competition frequently leads to wounding and emigration of males. The extreme mortality in subadult males is linked to their high rates of migration, that is, to the rigors (frequent wounding and low priority of access to resources) that migration invites.

These comparative data are in direct accord with those from a more detailed study of the relationships between demography, ecology and behavior in *Macaca sinica*. Several hypotheses of population regulation through socially imposed mortality were presented on the basis of that study. It is possible that these hypotheses also pertain to most Cercopithecinae, *Alouatta* and other mammals whose social organization is similar to that of *Macaca sinica.*

Mortality through competition for resources and mates does not, of course, preclude other kinds of socially imposed mortality. It may represent only one important manifestation of a behavioral syndrome which favors differential survivorship by age and sex; other manifestations including the direct harassment and killing of individuals. Socially imposed mortality and the resulting regulation of population size and age-sex composition, might be considered as the outcome of behaviors whose function, in an evolutionary sense, is to maximize the fitness of animals (see Dittus, 1979).

ACKNOWLEDGMENTS

This paper was prepared with financial support by National Science Foundation grant BNS 76-19740; Wenner-Gren Foundation for Anthropological Research grant 3199; and Deutsche Forschungsgemeinschaft grant Dil73/1, awarded to the author. I thank Ms. A. Baker-Dittus for critical reading and typing of the manuscript.

REFERENCES

Altmann, S. A. Sociobiology of rhesus monkeys. IV. Testing Mason's hypothesis of sex differences in affective behavior. *Behaviour* **32**: 50–70 (1968).

Altmann, Stuart A. and Altmann, Jeanne, *Baboon Ecology.* Chicago: University of Chicago Press (1970).

Angst, W. and Thommen, D. New data and a discussion of infant killing in Old World monkeys and apes. *Folia primatol.,* **27**: 198–229 (1977).

Bernstein, I. S. Introductory techniques in the formation of pigtail monkey troops. *Folia primatol.,* **10**: 1–19 (1969).

——————. and Sharpe, L. G. Social roles in a rhesus monkey group. *Behaviour* **26**: 91–104 (1966).

Bertrand, M. The behavioral repertoire of the stumptail macaque. *Bibliotheca Primatologica,* No. 11. New York: S. Karger (1969).

Boelkins, C. R. and Wilson, A. P. Intergroup social dynamics of the Cayo Santiago rhesus (*Macaca mulatta*) with special reference to changes in group membership by males. *Primates,* **13**: 125–140 (1972).

Carpenter, C. R. Societies of monkeys and apes. *Biol. Symp.,* **8**: 177–204 (1942).

——————. and Nishimura, A. 1969. Takasakiyama colony of Japanese macaques (*Macaca fuscata*). In *Proc. 2nd Int. Congr. Primat.,* pp. 16–30, Basel and New York: S. Karger (1969).

Caughley, G. Mortality patterns in mammals. *Ecology,* **47**: 906–918 (1966).

Caughley, Graeme. *Analysis of Vertebrate Populations.* New York: John Wiley (1977).

Crook, J. H. Sexual selection, dimorphism, and social organization in the primates. In B. Campbell (ed.), *Sexual Selection and the Descent of Man,* pp. 231–281. Chicago: Aldine (1972).

Deevey, E. S. Life tables for natural populations of animals. *Quart. Rev. Biol.,* **22**: 283–314 (1947).

DeVore, I. and Hall, K. R. L. Baboon Ecology, In I. DeVore, (ed.), *Primate Behavior,* pp. 20–52. New York: Holt, Rinehart and Winston (1965).

Dittus, W. P. J. The ecology and behavior of the toque monkey, *Macaca sinica.* Ph. D. Dissertation, University of Maryland (1974).

_____. Population dynamics of the toque monkey, *Macaca sinica.* in R. H. Tuttle, (ed.), *Socioecology and Psychology of Primates*, pp. 125–152. The Hague: Mouton (1975).

_____. The social regulation of population density and age-sex distribution in the toque monkey. *Behaviour*, **63**: 281–322 (1977a).

_____. The socioecological basis for the conservation of the toque monkey (*Macaca sinica*) of Sri Lanka (Ceylon). In Prince Ranier III and G. H. Bourne, (eds.), *Primate Conservation*, pp. 237–265. New York: Academic Press (1977b).

_____. The evolution of behaviors regulating density and age-specific sex ratios in a primate population. *Behaviour*, **69**: 265–302 (1979).

Drickamer, L. C. A ten-year summary of reproductive data for free-ranging *Macaca mulatta. Folia primatol.*, **21**: 61–80 (1974).

Farres, A. G. and Haude, R. H. Dominance testing in rhesus monkeys: comparison of competitive food getting, competitive avoidance and competitive drinking procedures. *Psychol. Rep.* **38**: 127–134 (1976).

Galat, G. and Galat-Luong, A. Demographie et regime alimentaire d'une troupe de *Cercopithecus aethiops sabaeus* en habitat marginal au Nord Sénègal. *La Terre et la Vie*, **31**: 557–577 (1977).

Gavan, J. A. and Hutchinson, T. C. The problem of age estimation: a study using rhesus monkeys (*Macaca mulatta*). *Am. J. Phys. Anthrop.*, **38**: 69–82 (1973).

Gilbert, C. and Gillman, J. Pregnancy in the baboon (*Papio ursinus*). *So. Afr. J. Med. Sci.*, **16**: 115–124 (1951).

Glander, K. E. Habitat and resource utilization: an ecological view of social organization in mantled howler monkeys. Ph. D. Dissertation, University of Chicago, (1975).

_____. Poison in a monkey's Garden of Eden. *Natural History*, **86**: 34–41 (1977).

Hall, K. R. L. Variations in the ecology of the chacma baboon, *Papio ursinus. Symp. Zool. Soc. Lond.*, **10**: 1–28 (1963).

_____. Aggression in monkey and ape societies. In J. D. Carthy, (ed.), *The Natural History of Aggression*, New York: Academic Press (1964).

_____. and DeVore, I. Baboon social behavior. In I. DeVore, (ed.), *Primate Behavior*, pp. 53–110. New York: Holt, Rinehart and Winston (1965).

_____. and Gartlan, J. S. Ecology and behaviour of the vervet monkey, *Cercopithecus aethiops*, Lolui Island, Lake Victoria. *Proc. Zool. Soc. Lond.*, **145**: 37–57 (1965).

Hausfater, G. Dominance and reproduction in baboons (*Papio cynocephalus*). *Contrib. Primatol.* **7**: 1–150 (1975).

Hill, W. C. O. The external characters of the bonnet monkeys of India and Ceylon. *Ceylon J. Sci.*, **(B)16**: 311–322 (1932).

_____. Longevity in a macaque. *Ceylon J. Sci.* **20**: 255–256 (1937).

_____. and Bernstein, I. S. On the morphology, behavior and systematic status of the Assam macaque (*Macaca assamensis* McClelland 1839). *Primates*, **10**: 1–17 (1969).

Imanishi, K. Social behavior in Japanese monkeys, *Macaca fuscata. Psychologia* **1**: 47–54 (1957).

Itani, J. Twenty years with Mount Takasaki monkeys. In G. Bermant and D. G. Lindburg, (eds.), *Primate Utilization and Conservation*, pp. 101–125. New York: Wiley-Interscience (1975).

_____, Tokuda, K., Furuya, Y., Kano, K. and Shin, Y. The social construction of natural troops of Japanese monkeys in Takasakiyama. *Primates*, **4**: 1–42 (1963).

Jones, M. L. Mammals in captivity—primate longevity. *Laboratory Primate Newsletter*, **1**: 3–13 (1962).

Kaufmann, J. H. Social relations of adult males in a free-ranging band of rhesus. In S. A. Altmann, (ed.). *Social Communication Among Primates*, pp. 73–98. Chicago: University of Chicago Press (1967).

Kawai, M. and Yoshiba, K. Some observations on the solitary male among Japanese monkeys. *Primates*, **9**: 1–12 (1968a).

_____. and_____. Sociological study of solitary males in Japanese monkeys. *Proc. VIIIth Internat. Congr. of Anthrop. and Ethnological Sciences*, **1**: 259–260 (1968b).

Kawamura, S. Matriarchal social ranks in the Minoo-B troop: a study of the rank system of Japanese monkeys. *Primates*, **1**: 149–156 (1958).

Kawanaka, K. Intertroop relationships among Japanese monkeys. *Primates*, **14(2-3)**: 113–160 (1973).

Koford, C. B. Rank of mothers and sons in bands of rhesus monkeys. *Science*, **144**: 356–357 (1963).

_____. Population dynamics of rhesus monkeys on Cayo Santiago. In I. DeVore, (ed.), *Primate Behavior*, pp. 160–174. New York: Holt, Rinehart and Winston (1965).

_____. Population changes in rhesus monkeys: Cayo Santiago, 1960–1965. *Tulane Studies in Zool.*, **13(1)**: 1–7 (1966).

Koyama, N. On dominance rank and kinship of a wild Japanese monkey troop in Arashiyama. *Primates*, **8**: 189–216 (1967).

_____. Dominance, grooming, and clasped-sleeping relationships among bonnet monkeys in India. *Primates*, **14**: 225–244 (1973).

Lack, David. *Population Studies of Birds*. Oxford: Clarendon Press (1966).

Lindburg, D. G. Rhesus monkeys: mating season mobility of adult males. *Science*, **166**: 1176–1178 (1969).

_____. The rhesus monkey in north India: An ecological and behavioral study. V. Characteristics of the study population. In L. A. Rosenblum, (ed.), *Primate Behavior: Developments in Field and Laboratory Research*, pp. 1–106, New York: Academic Press (1971).

_____. Grooming behavior as a regulator of social interactions in rhesus monkeys. In C. R. Carpenter, (ed.), *Behavioral Regulators of Behavior*, pp. 124–148. Lewisburg: Bucknell University Press (1973).

Møller, G. W., Harlow, H. E. and Mitchell, G. D. Factors affecting agonistic communication in rhesus monkeys (*Macaca mulatta*). *Behaviour*, **31**: 339–357 (1968).

Morrison, J. A. and Menzel, E. W., Jr. Adaptation of a free-ranging rhesus monkey group to division and transplantation. *Wildlife Monographs*, **31**: 5–78 (1972).

Neville, M. K. Male leadership change in a free-ranging troop of Indian rhesus monkeys (*Macaca mulatta*). *Primates*, **9**: 13–27 (1968).

_____. The population structure of red howler monkeys (*Alouatta seniculus*) in Trinidad and Venezuela. *Folia primatol.*, **17**: 56–86 (1972a).

_____. Social relations within troops of red howler monkeys (*Alouatta seniculus*). *Folia primatol.*, **18**: 47–77 (1972b).

_____. The population and conservation of howler monkeys in Venezuela and Trinidad. In R. W. Thorington, Jr., and P. G. Heltne, (eds.), *Neotropical Primates, Field Studies and Conservation*, pp. 101–109. Washington, D. C.: National Acad. Sci. (1976).

Norikoshi, K. The development of peer-mate relationships in Japanese macaque infants. *Primates* **15**: 39–46 (1974).

Pocock, R. I. The long-tailed macaque monkeys (*Macaca radiata* and *M. sinica*) of southern India and Ceylon. *J. Bombay Natur. Hist. Soc.*, **35**: 267–288 (1932).

Pope, B. L. Population characteristics. In M. R. Malinow, (ed.), *Biology of the howler monkey (Alouatta caraya) in northern Argentina*, Bibliotheca Primatologica, No. 7: 13–20. Basel: S. Karger (1968).

Redican, W. K. A longitudinal study of behavioral interactions between adult male and infant rhesus monkeys (*Macaca mulatta*). Ph. D. Dissertation, University of California at Davis, (1975).

Richards, S. M. The concept of dominance and methods of assessment. *Anim. Behav.*, **22**: 914–930 (1974).

Quick, H. F. Animal population analysis. In H. S. Mosby and O. H. Hewitt, (eds.), *Wildlife*

Investigational Techniques, pp. 190–228. Washington, D. C.: The Wildlife Society (1963).

Rowell, T. E. Forest living baboons in Uganda. *J. Zool. Lond.,* **149**: 344–364 (1966).

_____.A quantitative comparison of the behaviour of a wild and a caged baboon group. *Anim. Behav.,* **15**: 499–509 (1967).

_____.Long term changes in a population of Uganda baboons. *Folia primatol.* **11**: 241–254 (1969).

_____, Din, N. A. and Omar, A. The social development of baboons in their first three months. *J. Zool. Lond.* **155**: 461–483 (1968).

Saayman, G. S. Behaviour of the adult males in a troop of free-ranging chacma baboons (*Papio ursinus*). *Folia primatol.,* **15**: 36–57 (1971).

Sackett, G., Hohn, R. and Landesman-Droyer, S. Vulnerability for abnormal development: pregnancy outcomes and sex differences in macaque monkeys. In N. R. Ellis, (ed.), *Aberrant Development in Infancy, Human and Animal Studies,* pp. 59–76. New York: John Wiley (1975).

Sade, D. S. Determinants of dominance in a group of free-ranging rhesus monkeys. In S. Altmann, (ed.), *Social Communication among Primates,* pp. 99–122. Chicago: University of Chicago Press (1967).

Simonds, P. E. Outcast males and social structure among bonnet macaques (*Macaca radiata*). *Am. J. Phys. Anthrop.,* **38**: 599–604 (1973).

Southwick, C. H. An experimental study of intergroup agonistic behavior in rhesus monkeys (*Macaca mulatta*). *Behaviour,* **28**: 182–209 (1967).

_____. Aggressive behavior of rhesus monkeys in natural and captive groups. In S. Garattini and E. B. Sigg, (eds.), *Aggressive Behavior (Proc. Symp. on Biology of Aggressive Behavior, Milan, 1968),* pp. 32–43. Amsterdam: Excerpta Medica (1969).

_____. Aggression among non-human primates. *Addison-Wesley Module,* **23**: 1–23 (1972).

_____, Beg, M. A. and Siddiqi, M. R. A population survey of rhesus monkeys in villages, towns and temples of northern India. *Ecology,* **42**: 538–547 (1961).

_____, and _____. Rhesus monkeys in North India. In I. DeVore, (ed.), *Primate Behavior,* pp. 111–159. New York: Holt, Rinehart and Winston (1965).

Struhsaker, T. T. Ecology of vervet monkeys (*Cercopithecus aethiops*) in the Masai-Amboseli Game Reserve, Kenya. *Ecology,* **48**: 891–904 (1967a).

_____. Social structure among vervet monkeys (*Cercopithecus aethiops*). *Behaviour,* **29**: 83–121 (1967b).

_____. A recensus of vervet monkeys in the Masai-Amboseli Game Reserve, Kenya. *Ecology,* **54**:930–932 (1973).

_____. A further decline in numbers of Amboseli vervet monkeys. *Biotropica,* **8(3)**: 211–214 (1976).

Sugiyama, Y. Characteristics of the social life of bonnet macaques (*Macaca radiata*). *Primates,* **12(3-4)**: 247–266 (1971).

Vandenberg, J. G. and Vessey, S. Seasonal breeding of free-ranging rhesus monkeys and related ecological factors. *J. Reprod. Fert.,* **15**: 71–79 (1968).

van Wagenen, G. Body weight and length of the newborn laboratory rhesus monkey (*Macaca mulatta*). *Fed. Proc.,* **13**: 157 (1954).

Western, D. and Van Praet, C. Cyclical changes in the habitat and climate of an East African ecosystem. *Nature (Lond.),* **241**: 104–106 (1973).

Wilson, A. P. Social behavior of free-ranging rhesus monkeys with an emphasis on aggression. Ph. D. Dissertation, University of California, Berkeley (1968).

_____. and Boelkins, R. C. Evidence for seasonal variation in aggressive behavior by *Macaca mulatta. Anim. Behav.,* **18**: 719–724 (1970).

Chapter 12
Female Choice and Mating Strategies Among Wild Barbary Macaques (Macaca sylvanus L.)

David Milton Taub, Ph. D.

INTRODUCTION

Among primates, sexual selection has been envisioned as operating principally among the males as they compete intrasexually for access to receptive, but passive, females (Goss-Custard, *et al.*, 1972). Perhaps as a result, most primate studies of reproduction have focused on the male's role in mating, especially the pattern and outcome of intermale competition for receptive females. Thus primate mating systems have usually been described as being male dominated or controlled, wherein the females are envisioned as passive and inactive recipients of "male choice." Recently, however, several investigators have reported partner "preferences" among female primates in their mating activities (Loy, 1970; Stephenson, 1974), and Lindburg (1975) has suggested that choice of mating partners among rehesus monkeys (*Macaca mulatta*) may be a female prerogative. Thus it appears that, in some instances at least, females do play a more active role in determining the outcome of their mating activities than has been thought previously. Theoretical models of mating strategies have always emphasized that male and female reproductive "interests" may be in opposition (Trivers, 1972). Hence there is usually a dynamic, synthetic inter-action between the mating strategies used by males and females to maximize their different, often conflcting, interests. On theoretical grounds alone one would expect that primate males and females both would be active participants in their mating systems. For the most part, a clear understanding of the interplay of male *and* female roles and differential mating strategies among primates has not been developed.

Early studies of nonwild groups of Barbary macaques (*Macaca sylvanus*) suggested that the males of this species showed intense interest in infants (Lahiri and Southwick, 1966; MacRoberts, 1970; Burton, 1972). When first studied in the wild condition some 10 years ago, Deag, and Crook (1971) were able to confirm that there were elaborate interactions between Barbary macaque males and infants, and they were immediately impressed with the magnitude, intensity and multiplicity of relationships between males and infants. Taub's study (1978) amplified and clarified some aspects of the Barbary macaques' male care-taking system. While it had been known for some time that males of monogamous species, including primates, developed intensive care-taking relationships with their offspring, it was remarkable to discover analogous behavior among males of a social, polygamous mammal.

The role of the parents in the care and rearing of the offspring has been envisioned as a critical factor in the evolution of mating systems (Orians, 1969; Trivers, 1972). For example, when males play little or no role in the care and feeding of the young, or do not maintain resources essential to female and/or offspring survival (e.g. defending a territory rich in resources), selective forces have tended to favor a polygamous mating system. On the other hand, when the male does participate in the care and feeding of the young, monogamy has been favored (Brown, 1975). It is therefore puzzling to find a highly developed system of male interactions with infants in a social primate that lives in groups of multiple males and females among whom mating is polygamous. If the occurrence of this male-infant system is true "investment," then it has not been predicted by, nor accounted for in any of the recent models concerning "parental investment" (Trivers, 1972; Maynard-Smith, 1977). Although the evolutionary pressures that originally favored the evolution of male care-taking behavior among Barbary macaques are conjectural, field studies (Deag, 1974; Taub, 1975, 1978) have verified that the males of this species do, in fact, show pronounced interactions with infants. Once this behavior pattern had become established, its existence must have created other selective forces that further influenced the patterns of mating in this species. How then do Barbary macaque male and female sexual roles differ? What are the possible consequences of, and what is the relationship among, these strategies and the unusual phenomenon of male care-taking activities that are so well developed in this species?

METHODS AND MATERIALS

Study Site and Study Group

This study was conducted in the Middle Atlas region of Morocco, North Africa. The primary study site was located at Ain Kahla, about 5° 13′ West by

33° 15′ North, 30 kilometers southwest of the city of Azrou. Ain Kahla lies within the Sidi M'Guild forest, which exemplifies the high mixed cedar and oak forest that is typical of the Middle Atlas region, and which is primary habitat for the Barbary macaque (Taub, 1977). Elevation there averages 2,000 meters and there are extensive snowfalls of up to two meters from December to April (see Taub, 1978 and Deag, 1974, for further details of the Ain Kahla region).

A study group, "S," was located and habituated to close approaches during the Fall months of 1973, and all behavioral data of this project were collected on this group. After the birth of all infants of the 1974 season, "S" group contained 39 animals (Table 12-1). No strange animals joined "S" group during the 15 months over which the study was conducted. With the exception of the temporary move by adult male BB and juvenile male DOF, no juvenile, subadult or adult male (n = 16) of "S" group became peripheral to the group or emigrated out of "S" group during the course of the study. No solitary animals or all-male "bachelor" groups were ever seen during this study, which is consistent with Deag's observations (Deag, 1974). "S" group was relocated in 1977 and at that time several previous members of the group were missing, but there were no animals present in the summer of 1977 that had not previously resided in "S" group. With the possible exception of juvenile male DOF, it is probable that the loss of animals was due to mortality and not emigration. It is typical for subadult male Japanese and rhesus macaques to become peripheral to the group as they mature, and then to emigrate out of their natal groups (Koyama, 1970; Kurland, 1977; Boelkins and Wilson, 1972; Lindburg, 1969). In contrast, Barbary macaque subadult and juvenile males do not peripheralize; rather they appear to mature in, and become functioning adults in, their natal groups. These data suggest that the pattern of subadult males becoming peripheral and emigrating to nonnatal groups—common among some macaques—is not typical of them all, as has been widely supposed. Intertroop male mobility among Barbary macaques appears to be greatly reduced.

Methods[1]

Observation Protocols. Focal female protocols were done (Altmann, 1974); however, each female was sampled only during her period of estrus rather than throughout the entire menstrual cycle. This procedure allowed the collection of the greatest amounts of mating data, but at the expense of systematically establishing female behavioral profiles during nonfertile portions of the cycle.

---

[1] See Taub, 1978, for details of methodology.

Table 12-1. Composition of the Study Group During 1974.

ADULT MALE	ADULT FEMALE	SUBADULT MALE	SUBADULT FEMALE	JUVENILE MALE	JUVENILE FEMALE	YEARLING MALE	YEARLING[a] FEMALE	INFANT MALE	INFANT FEMALE
Mo	LP	4SA	F	LJM	DOE	YM1	YF1	iLN	iLP
RD	Bu	Ro		MJ		YM2	YF2	iRSE	iBu
CM	LN	BL		Ba		YM3	YF3	iPP[b]	
WN	RSE	BS		DOF[c]		YM4			
V	LP			nj					
NM	MM								
BB[d]	SG								
	PP								
	SS								

[a] It was extremely difficult to judge differences between a late yearling female and a young juvenile one; it is possible that at least one of the three yearling females could have belonged to the juvenile class.

[b] Disappeared from "S" group between July 28 and August 1, 1974. No carcass found.

[c] First noted absent from "S" group on November 8, 1974, and observed resident in a neighboring group on December 4. 1974.

[d] First noted absent from "S" group on August 18, 1974, and observed resident in a neighboring group on September 25, 1974. Rejoined "S" group permanently on October 24. 1974.

In conventional usage, the period of behavioral and sexual receptivity in the female mammal has been referred to as "estrus" (Cockrum, 1962). This term has been used widely to describe the sexual cycles of many nonhuman primates. Some investigators have recently challenged the validity of this usage (Rowell, 1970; Loy, 1970; Hanby, *et al.*, 1971), suggesting that among many nonhuman primates behavioral receptivity may not be closely synchronized with ovulation. Data from field studies on wild primates in nonprovisioned habitats indicate that nonhuman primates do have behavioral homologues of true mammalian estrus (Petter-Rousseaux, 1964; DeVore and Hall, 1965; Saayman, 1970; Lindburg, 1971; Ransom, 1971; Hausfater, 1975; Kummer, 1968; Dunbar and Dunbar, 1975; Taub, 1978). In this paper, estrus among Barbary macaques is defined as the several days when the perineum reaches maximum turgescence, and when the frequency of sexual behaviors, especially consort formation and ejaculatory copulations is highest; "estrous female" refers to a female at this time in her sexual cycle.

As with baboons (*Papio* spp.) and pigtail macaques (*M. nemestrina*), the sexual skin of Barbary macaques goes through marked sequential changes; these changes correspond to a stage of perineal turgescence or inflation, followed by a period when the perineum and labium are at maximum swelling, after which a period of rapid deturgescence or deflation occurs, followed by perineal inactivity or arrest, which separates the cycles from one another (Taub, 1978). As with their close relatives, it is probable that ovulation in Barbary macaques also occurs after maximum perineal turgescence is achieved (Hendrickx and Kraemer, 1971; Bullock, *et al.*, 1972; White, *et al.*, 1973; Graham, *et al.*, 1973; Blakley, *et al.*, 1977). During this study however, it was unknown on exactly which day ovulation occurred of the several days when the perineum was maximally swollen.

The descriptions of perineal morphology had the practical value of providing reasonably objective, external indicators of estrus. Morphological criteria alone were sometimes insufficient, however, to allow selection of a female for focal sampling. Consequently, at times it was necessary to use other indicators of estrus, such as the behavior of adult females and males, in order to select from among several swollen females. Hausfater (1975) found that while maximum perineal turgescence was an excellent predictor of ovulation in yellow baboons, behavioral changes, such as consort formation, were also reliable indicators of estrus. Wild female Barbary macaques also show unambiguous behaviors indicating sexual receptivity, such as sexual solicitations and presentations and the formation of consort partnerships, and these behaviors coincide with the several days when the sex skin has reached maximum turgescence.

Portions of nine estrous periods of six adult females were sampled in 1974 (Table 12-2). To assess rates of copulation, of consort exchange and of other

Table 12-2. Focal-Female and *Ad Libitum* Sample Minutes Accumulated During Observations On Estrous Females.

FEMALE	ESTRUS	TOTAL FOCAL-FEMALE SAMPLES (N)	AD LIBITUM SAMPLES	AVERAGE FOCAL-FEMALE SAMPLES PER DAY OF OBSERVATION
LP	1	418 minutes (2)	55 minutes	209 minutes
	2	322 minutes (1)	130 minutes	322 minutes
LN	1	980 minutes (3)		327 minutes
PL	1	376 minutes (2)	50 minutes	188 minutes
MM	1	304 minutes (1)	62 minutes	304 minutes
	2	1672 minutes (3)		557 minutes
PP	1	424 minutes (2)	90 minutes	212 minutes
	2	1289 minutes (4)		322 minutes
SS	1	340 minutes (2)	243 minutes	170 minutes
All female averages		6125 minutes (20)	630 minutes	306.25 minutes

related components of sexual activity, it was necessary that the focal samples be lengthy. Thus focal-female samples were not fixed; rather, once selected, the focal-female was observed until contact with her was lost for over one hour. Each subject female was observed for at least 130 consecutive minutes during each focal period. For all females, the average length of a daily focal sample was 306 minutes, with a range of 130 to 441 minutes. Considering the three females whose estrous periods were most intensively sampled—LN #1, MM #2 and PP #2—the length of daily focal-female samples ranged from 190 to 441 minutes, with a three-female pooled average of 394 focal minutes per observation day. Most focal samples were made without losing contact with the focal subject, but on a few occasions short interruptions were inevitable. At this time of the year there are about 12 hours of daylight at Ain Kahla, which means that the focal-female samples obtained in this study covered from one-third to one-half the day. Because substantial portions of the day were covered, the frequencies of sexual activities derived from these samples may be considered estimates of absolute frequencies of male and female sexual behavior.

Terminology, Computations and Indices. The sexual associations of Barbary macaques appear to be unique (Taub, 1978), and I propose the term "consociation"—a transitory but exclusive sexual association during estrus between a Barbary macaque female and a single male—to refer to this species-typical courtship pattern.

The "originator" of a consociation was defined as the animal that both initiated and pursued the approach to the other resulting in the establishment of a consociation. In some cases, when both partners began to approach each other simultaneously, a determination of who was the originator was not possible, and these consociations were considered to be "mutually" formed.

Focal observations on a female ended when males no longer associated with them. Although this usually coincided with the onset of deturgescence, sometimes a female's perineal swelling did not show observable signs of deturgescence for a day or two (particularly among younger females) after this loss of male interst. Hence the sample periods in this study cannot be correlated in relation to the onset of deturgescence, as has been done so successfully with baboons (Hausfater, 1975). Therefore, the data have been presented in terms of several days of maximum swelling rather than in relation to each day prior to deturgescence.

The absolute time when each mount and copulation occurred and when each consociation started and terminated was recorded during each focal sample. The length of mounts was timed with a stop watch and the length of the ejaculatory pause of each mount was estimated by counting off seconds (e.g. "one-thousand and one," "one-thousand and two," etc.). Pelvic thrusts were counted off into a tape recorder as they occurred. Female refractory periods—the intervals between successive ejaculatory copulations for a single female—were computed from focal periods during which occurrences of all copulations were known. Copulation frequencies are expressed as a rate—i.e. "x" copulations per sample hour—which were computed by dividing the total number of copulations occurring during a focal sample by the number of observation minutes from the onset to the end of that sample period. Likewise, consociation formation rates were calculated by dividing the number of consociations formed during a focal sample period by the length of that period and expressed as a rate per sample hour.

DOMINANCE HIERARCHIES AND AGGRESSION

Dominance or hierarchial status relations have been reported for many nonhuman primates, including the Barbary macaque (Deag, 1977). Although a diversity of criteria have been used to define primate status hierarchies (Bernstein, 1970), the least ambiguous ranking usually derives from aggressive interactions (Sade, 1967; Bernstein, 1970; see Rowell, 1966 for an opposite view), primarily because dominance among primates is usually conceived as an aspect of agonistic behavior.

The behaviors of Barbary macaques considered for this study were contact and noncontact aggression (see Taub, 1978, for details). When the direction of all dyadic aggressive encounters recorded during *ad libitum* and focal-female

Table 12-3. Matrix of Dyadic Agonistic Encounters.

ACTOR \ RECIPIENT	Mo	RD	CM	WN	V	NM	BB	4SA	RO	BL	BS	LP	Bu	LN	RSE	MM	PL	SG	PP	SS	LJM	MJ	Ba	DOF	nj	F	DOE
Mo		31	15	5	5	7	1	3	2	11	4	8	3	6	2	6	3	2	1	1	8	4	1	1	1	1	1
RD	13		5	7	27	3	3	37	23	11	16	5	4	6	5	7	3	2	1	1	15	1		1	1	1	1
CM		1		1	6	3		22	4	11	10	1	4	4	2	2	3	1	1		7	2	1	1	1	1	
WN		2	1		3	7	3	20	3	14	11	4	8	5	3	5	3	3	2		7	1			1		
V	3	3				1			3	10	13	2	2	2	8			4	2	1		4				1	2
NM	2	2	1		1		4	2	1	5	13	1	4	2	4	7	2	4		2	4	1		1	1	2	1
BB	1	1				3		1	1	1	2					1				3	5	1	1		1	5	1
4SA			2	1		1	1		2	10	5	7	3	2	3	1	3	2	1	3	6	1	2		1		1
Ro	3	3	1	2		1	1	1		7	4	1	2	2	3	5	1	2	1	3	5	1	4	1	1	1	1
BL			2				1				2		1	1	2		1			4	2		2	1	2		
BS			1	1					1			4		3	2	3	3	2		2	2	3	3	2	1	2	3
LP											4		4	3	4	4	3	3	4	4	2	3	4	2	4	4	2
Bu														12	4	3	1	3	7	1	2	2			1	6	3
LN											2				4	10	5	2	1	2	1	1			1	5	1
RSE	1															6	6	1	1	2	2	1	1	1	1	2	1
MM		1															2	4	5	13	2	1	2	2	1	6	1
PL																		13	15	2			1				
SG																			1	4		1	1	1			
PP																				1	1	1	1		1		1
SS																							4			1	
LJM	1		3	1			1					1		3	3	3	2	6	1		1	1	4	2	1	2	2
MJ								1													1		1			3	3
Ba																				4	5			5	4	1	
DOF																			1	3							
nj																				1		1	1				
F		1																		6	1			1			2
DOE												1															

samples is plotted in a matrix so that the lowest numbers of interactions occur below the diagonal (i.e. reversals in the direction of the aggressive act), a linear relationship results among the members of "S" group (Table 12-3). The rank order among individuals was defined by the pattern of dyadic interactions among all participants, and reference to social status is based on the rank orders resulting from the distribution shown in Table 12-3. For descriptive purposes, males Mo, RD, CM and WN are considered "high ranking" males, while the remaining adult and subadult males are considered "low ranking."

Equivalent sampling periods were not accumulated for each member of the troop, thus quantitative analyses of the agonistic patterns themselves are not possible. The sample biases implicit in my aggression data notwithstanding, qualitative assessments of these data suggest that aggression among Barbary macaques is neither frequent nor intensive, which are conclusions also reached by Deag.

MALE STRATEGIES

Because data on the sexual behavior of wild Barbary macaques are almost nonexistent, a qualitative description of some features of male sexual behavior will be presented here as a framework for the quantitative results which follow. Since focal-male samples were not made, these descriptions of male behavior are probably incomplete, but based on observations from *ad libitum* and focal-female samples, three distinct patterns emerge of a male's sexual relations with estrous females. Each "S" group male used only one of these strategies in his interactions with estrous females.

The "Proximity-Possession Principle"

Among some nonhuman primates, it has been observed that proximity to an object is often sufficient to guarantee "ownership" of that object, even against higher-ranking rivals (Van Lawick-Goodall, 1968; Menzel in Kummer, 1973; Kummer, 1968, 1973; Kummer, *et al.*, 1974). A similar phenomenon—referred to here as the "proximity-possession principle"—occurs among Barbary macaques, and operates in relation to "ownership" of estrous females. A male's close proximity to a female prevents rivals from attacking him or otherwise appropriating the female, irrespective of his social status. While an estrous female remains within a consort male's "sphere of influence" (usually less than 15 meters, and most often between 1.5 and 8 meters, rival males will not attempt to displace this consort. In my observations, the only exceptions occurred when a nonconsort male could enlist the support of other males in a coalition, but this was rare. Lowest-ranking subadult males were observed to copulate with an estrous female while a high-ranking rival that had just previously consorted with that female sat passively by, less than 3 meters away. Although most copulations occurred in clear view of other males, and the "copulation calls" of the female always attracted others' attention, harassment and interruption of a copulating pair, frequent among other macaques (Stephenson, 1974; Hanby and Brown, 1974), was never observed. The close physical association between the estrous female and the consorting male may not only inhibit rivals, it may, conversely, disinhibit the "owner." Subadult males threatened or attacked higher-ranking adult males only when the subadults were in consociation with an estrous female.

This pattern of guaranteeing possession of an estrous female by spatial proximity is most effective at the time the female is actively soliciting a consociation with a new partner. When the female begins to terminate the consociation, the male's maintaining close proximity will not prevent her loss. This is because the female and the new prospective consort will maneuver so that the rival himself brings her into his "sphere of influence." Then the "proximity-possession principle" acts to protect this newly formed consociation against disruption by the deposed or other rivals. This principle thus operates as a mechanism which allows an estrous female and any male, regardless of his social rank, to maintain a brief sexual association against rivals, but only so long as the female herself chooses to remain with that male.

In discussing a similar phenomenon whereby hamadryas baboon males maintain harem females, Kummer, *et al.* (1974) have suggested that the inhibiting stimuli of the so-called "triadic differentiation" is the "pair gestalt" (pp. 73), i.e. the "perception of ownership" based on the spatial distance between the pair. They further suggest (pp. 74–75) that this inhibitory system". . . protects existing pair bonds against rivals. . . increases the stability of social structure. . . but does not help an inferior male in obtaining a new consort." It is unknown if this explanation applies to Barbary macaques, but in this species the "proximity-possession principle" appears to be essential in allowing an "inferior" (i.e. low ranking) male to obtain the estrous female. I was unable to identify proximate cues other than the physical closeness of the partners and the solicitation behavior of the female.

The Pertinacious Strategy (PS)

"Pertinacity" is defined as ". . . exhibiting unyielding purpose; perversely resolute; persistent. . .," and this is precisely how the behavior of the males using this approach (Mo and WN) can be characterized. The distinctive feature of this strategy is that the male persistently follows the estrous female regardless of the status of her sexual associations. A PS male will follow 3 to 15 meters behind a female while she is in consociation with another male, and will monitor the behavior of the pair constantly, adjusting his movements relative to theirs so that he always remains close to them. Consequently, when the estrous female terminates that consociation, the PS male is usually the closest male to her and the first one she usually encounters. As the estrous female becomes unattached, the PS male approaches her directly, and by bringing her into his "sphere of influence" forms a *de facto* consociation. If the female is disinterested in consorting with the PS male, she will ignore his approaches and continue her search for another consort, but the PS male will continue to follow diligently after her. When there are several females in estrus simultaneously, the PS male may cease monitoring one consorting pair and seek another estrous female. A PS male will continue this pattern throughout each day of a female's estrus, often alternating every third or fourth consociation with her.

The most common way a PS male attempts to prevent termination of the consociation is to follow the female diligently and interpose himself between her and a rival. These maneuvers are ultimately unsuccessful, but sometimes this tenacity succeeds in preventing, temporarily, the female from joining another male. PS males have the greatest capability for altering the female's control of formation of consociations, and they have more direct control in the formation of their own sexual associations than do males using other strategies.

The Peripheralize and Attract Strategy (PAS)

All subadults and the lowest ranking, adult males of "S" group used the PAS male strategy. PAS males do not approach estrous females directly nor do they remain closely associated with them outside the consociation. PAS males tend to maintain substantial distances from a consorting pair (40 to 100 meters). The PAS males' primary means for establishing consociations with an estrous female is to attract the female's attention by "displaying" to her, but these displays are neither elaborate nor stereotyped. These displays include (1) branch shaking loudly (also seen among Japanese macaques; Modahl and Eaton, 1977); (2) walking along ridges or other conspicuous places; and (3) carrying an infant dorsally in clear view of the female. These displays appear to be designed so that the male can "advertise" his location to the estrous female and thereby draw her attention to his presence. When an estrous female sights a PAS male and decides to join him, she begins to approach him, usually followed diligently by her resident consort. Instead of counter-approaching the estrous female, as a PS male would, the PAS male always moves away from the approaching female, usually climbing high into a tree. If the female continues her pursuit and climbs into the tree after him, the PAS male will either sit and await her approach, or continue retreating, sometimes leaping from tree to tree. The PAS male continues to retreat most often when the resident consort climbs into the tree after the estrous female. The PAS male is very vigorous in this behavior and will continue to alter his positions until the estrous female can approach him and enter his "sphere of influence"; once the female joins the PAS male, the pursuing rival ceases his pursuit.

The PAS male thus plays a rather passive role in the formation of sexual associations. The estrous female is the primary director of consociation formation, although the PAS male may influence the formation of consociaitons in two ways: by his success in attracting the estrous female's attention, and by his effectiveness in maneuvering his position relative to the female and the pursuing consort. In keeping with this passive role, PAS males make little effort to prevent the female from terminating consociations; consequently, they are easily displaced whenever the females come in the vicinity of other males.

The Combination Strategy (CS)

The CS male strategy combines tactics used in each of the other two male strategies. CS males (RD and CM) use the PAS stratagem during consociation formation, since they too will move away from the estrous female and try to induce her to join them by attraction displays. Consequently, the estrous female, rather than the CS male, forms significantly more of their consociations. Occasionally, the CS male will use a less intense version of the PS male tactic of maintaining close proximity to the estrous female while she is consorting with others. Because CS males do not show the same determination that PS males do in pressing proximity to the female, they tend to be easily displaced by PS male consorts. Hence they are often obliged to move away, whereupon they revert to trying to attract the female's attention and induce her to join them. The CS male's behavior is most similar to that of the PS male's at the time the female begins to terminate their consociation, since they too try vigorously to remain close to the female to prevent her from leaving. Thus as with PAS males, it is the female who is most responsible for directing the establishment of consociations with CS males, although unlike the PAS males, CS males sometimes play a direct role in forming their sexual associations. CS males are not as reliant on the female for their consociations as are the PAS males, but neither are they as independent from the female's selective behavior as the PS males appear to be.

QUANTITATIVE RESULTS

Consociations

Formations. Barbary macaques formed sexual associations—which were always fluid, brief and temporary—only during estrus. Each day during estrus, a female solicited, established and terminated numerous consociations, one

Table 12-4. Directionality of Consociation Formation By Individual Estrous Females[a].

FEMALE	CONSOCIATIONS[b]	CONSOCIATION FORMATION					
		FEMALE ORIGINATED		MUTUALLY ORIGINATED		MALE ORIGINATED	
		NUMBER	PERCENT	NUMBER	PERCENT	NUMBER	PERCENT
LP	14	10	(71)	2	(14)	2	(14)
LN	56	27	(48)	12	(21)	17	(30)
PL	16	11	(69)	1	(6)	4	(25)
MM	44	26	(59)	7	(16)	11	(25)
PP	71	38	(54)	8	(11)	25	(35)
SS	26	13	(50)	9	(35)	4	(15)
All females combined	227	125	(55)	39	(17)	63	(28)

[a] Consociations with all males combined.
[b] Only those consociations in which both the originator *and* the terminator of that consociation were known positively.

Table 12-5. Directionality of Consociation Formation by Individual Males[a].

FEMALE	NUMBER OF CONSOCIATIONS[b]	CONSOCIATION FORMATION					
		FEMALE ORIGINATED		MUTUALLY ORIGINATED		MALE ORIGINATED	
		NUMBER	PERCENT	NUMBER	PERCENT	NUMBER	PERCENT
Mo	71	15	(21)	21	(30)	35	(49)
RD	34	16	(47)	10	(29)	8	(24)
CM	30	27	(90)	3	(10)	0	(00)
WN	44	30	(68)	1	(02)	13	(30)
V	7	6	(86)	0	(00)	1	(14)
NM	7	5	(71)	1	(14)	1	(14)
BB	5	3	(60)	2	(40)	0	(00)
4SA	6	6	(100)	0	(00)	0	(00)
Ro	9	7	(78)	0	(00)	2	(22)
BL	6	4	(67)	0	(00)	2	(33)
BS	8	6	(75)	1	(13)	1	(13)
All males combined	227	125	(55)	39	(17)	63	(28)

[a] Consociations with all females combined.
[b] Only those consociations where both the originator and terminator of that consociation were positively known.

after the other in rapid succession, with several different males. Tables 12-4 and 12-5 show the number and percent of consociations formed by the estrous female, by the male consort, or mutually, when considering, first, all females combined and, second, all males combined. As can be seen, it was the estrous female who was most instrumental in determining with whom consociations were formed. The estrous female was the sole originator of the consociation in a majority of all cases (55 percent), and this was significantly different from that portion formed by males or formed mutually ($x^2 = 52.04$, df = 2, p < .001). The percent of consociations that an individual female was exclusively responsible for ranged from a low of 48 percent to a high of 71 percent (Table 12-4). When summing cases in which females were assisted in consociation formation by the male (i.e. those mutually formed) and those instances when the female was the originator, an average of 72 percent (range 69 to 85 percent) of all sexual associations involved the estrous female.

Males were the sole originators of only 28 percent (n = 63) of all consociations. As will be recalled, the PS strategy males played an active role in directing the formation of their own consociations, and to some degree, neutralized the female's role in directing their establishment. This pattern can be seen clearly in the data presented in Table 12-5. The two PS strategy males, Mo and WN, collectively accounted for 77 percent (48 of 63) of all male-originated consociations; Mo accounted for 56 percent (n = 35) and WN for 21 percent (n = 13) of them. Excluding these two males, only 7 percent of all remaining consociations were solely established by males; the female now was the sole originator of 71 percent of all remaining consociations (80 of 112).

These data support convincingly the notion that the estrous female controls the formation of her sexual associations. At the same time, they lend support to the conclusion that PS males are as active as the estrous female in forming their consociations.

Terminations. Consistent with her control of the formation of consociations, the estrous female also controlled the termination of consociations. Shown in Tables 12-6 and 12-7, are the number and percent of consociations that an estrous female, the consort male or others were responsible for terminating, first for all females combined and second for all males combined. The estrous female herself terminated an average of 72 percent of all consociations, which is significantly greater than the number of consociations terminated by either the male or other animals ($x^2 = 155.67$, df = 2, p < .001). The percent of consociations terminated by individual females ranged from a low of 60 percent to a high of 88 percent.

Another competing male or female rival caused the termination of 20 percent of all consociations, while the consort male was responsible for terminating only 8 percent of all consociations. In no case did an individual

Table 12-6. Directionality of Consociation Terminations by Individual Estrous Females[a].

	NUMBER OF CONSOCIATIONS[b]	CONSOCIATION TERMINATION					
		FEMALE TERMINATED		MALE TERMINATED		OTHER TERMINATED	
FEMALES		NUMBER	PERCENT	NUMBER	PERCENT	NUMBER	PERCENT
LP	14	11	(78)	1	(7)	2	(14)
LN	56	40	(71)	0	(00)	16	(29)
PL	16	14	(88)	0	(00)	2	(12)
MM	44	36	(82)	3	(7)	5	(11)
PP	71	43	(60)	11	(15)	17	(24)
SS	26	20	(77)	4	(15)	2	(8)
All females combined	227	163	(72)	19	(8)	45	(20)

[a] Consociations with all males combined.
[b] Only those consociations where both the originator and terminator of that consociation were positively known.

Table 12-7. Directionality of Consociation Terminations by Individual Males[a].

MALES	NUMBER OF CONSOCIATIONS[b]	CONSOCIATION TERMINATION					
		FEMALE TERMINATED		MALE TERMINATED		OTHER TERMINATED	
		NUMBER	PERCENT	NUMBER	PERCENT	NUMBER	PERCENT
Mo	71	61	(86)	6	(8)	4	(6)
RD	34	22	(65)	2	(6)	10	(29)
CM	30	22	(73)	3	(10)	5	(17)
WN	44	31	(70)	6	(14)	7	(16)
V	7	4	(57)	0	(00)	3	(43)
NM	7	6	(86)	0	(00)	1	(14)
BB	5	3	(60)	0	(00)	2	(40)
4SA	6	3	(50)	1	(17)	2	(33)
Ro	9	6	(55)	0	(00)	3	(67)
BL	6	1	(17)	1	(17)	4	(50)
BS	8	4	(50)	0	(00)	4	(50)
All males combined	227	163	(72)	19	(8)	45	(20)

[a] Focal samples for all females combined.
[b] Only those consociations where both the originator and terminator of that consociation were positively known.

303

male direct more than 17 percent of the terminations of his consociations (Table 12-7), and almost all instances of male-directed consociation termination occurred on those days when copulation rates were lowest. Indeed, terminations of consociations by the male consort proved to be a reliable behavioral indicator of the ending of that female's estrus.

These data show that the estrous female played an even more dominating role in terminating her consociations than she did in establishing them. It thus appears that the decision to exchange sexual partners among Barbary macaques was almost exclusively a female prerogative.

Differential Selection of Males for Consociations. As seen in Table 12-5, there are considerable differences among individual males in the degree to which each was involved in establishing his own consociations. Since estrous females select different males for consociations, it may be asked if they prefer certain males more than others. While it is clear that Barbary macaque females play a dominant role in partner selection during estrus, it is impossible, from these data, to assess with confidence whether females selectively chose certain males significantly more often than others. Table 12-8 shows, of all female-

Table 12-8. Proportion of All Female-Originated Consociations That Are Directed Toward Each "S" Group Male.

MALE CONSORT	NUMBER OF CONSOCIATIONS DIRECTED TOWARD EACH MALE[a]	PERCENTAGE OF ALL FEMALE-ORIGINATED CONSOCIATIONS DIRECTED TOWARD MALES
Mo	15	(12)
RD	16	(12)
CM	27	(21)
WN	30	(24)
V	6	(4)
NM	5	(4)
BB	3	(2)
4SA	6	(4)
Ro	7	(5)
BL	4	(3)
BS	6	(4)
	125	(100)

[a] Data pooled from all observation days during estrus: exact day of ovulation was not known.

originated consociations, what percentage was directed toward each sexually mature male of "S" group. Thus 12 percent of all female-directed consociations were directed to male Mo, while 24 percent of them were directed toward male WN, etc. These data suggest that estrous females were choosing males WN and CM proportionally about twice as often as they were selecting any other male. But male Mo actually formed about twice as many consociations (n = 71) as did either of the "preferred" males CM and WN (Table 12-4), while male RD formed more consociations than did the "preferred"male CM. Thus even if females are preferentially selecting these two "preferred" males, this female preference, in itself, is not sufficient to give these males numerical superiority in associating with them. There is always the possibility that specific female preferences do exist and could be restricted to the day when ovulation occurs, although from the data in Tables 12-4, 12-5, and 12-8, which are summed for all days of estrus, this does not appear to be the case. Females appear, instead, to choose to associate with as many males as possible on all days of estrus (see below). It is suggested here that even if estrous females did selectively prefer a few males, even on the day of ovulation, the reproductive advantages of such preferences may be neutralized or negated by those males that actively pursue females, e.g. PS males, whether they themselves are preferred or not.

Serial Consociations with Multiple Males. Closely related to the estrous female's ability to direct her own sexual association, is her proclivity to establish consociations with many different, sexually mature males during estrus. Table 12-9 presents the number of different males consorted with during estrus and the percentage of all the "S" group males represented. These data show that on any single day of estrus, a female will consociate with from 27 percent (n = 3) to 91 percent (n = 10) of all adult and subadult males in the group. On average, a female will consociate with a majority (55 percent) of all group males each day of estrus. For the three females for whom the most consecutive days of estrus were observed—LN # 1, MM # 2 and PP # 2—it can be seen that for the entire period of estrus each female consorted, at least once, with 90 to 100 percent of all group males. For any single day of these estrous periods, these three females associated with from 5 (45 percent) to 10 (91 percent) males per day. It should be emphasized that these data are underestimates, since they are derived from focal-female samples covering only a portion of the entire day.

It has been reported that among other primates, e.g. baboons, (Saayman, 1970), an estrous female may consort with several males; in these cases, serial sexual associations with multiple males appear to be restricted to those segments of the estrous cycle when ovulation is unlikely, whereas the number of males consorted with when ovulation is imminent is quite limited

Table 12-9. Number and Percentage of Sexually Mature Males that Estrous Females Formed Consociations With During Estrus.

FEMALE	FOCAL SAMPLE	TOTAL NUMBER CONSOCIATIONS	MALES CONSORTED WITH PER DAY	PERCENT ALL MALES	TOTAL NUMBER OF MALES CONSORTED WITH DURING ESTRUS			
					ESTRUS #1		ESTRUS #2	
					NUMBER	PERCENT	NUMBER	PERCENT
LP	1	11	4	(36)	9	(87)	4	(36)
	2	11	8	(73)				
	3	11	4	(36)				
LN	1	19	10	(91)	11	(100)		
	2	30	8	(73)				
	3	12	5	(45)				
PL	1	11	5	(45)	6	(55)		
	2	12	5	(45)				
MM	1	12	6	(55)	6	(55)	10	(91)
	2	9	4	(36)				
	3	21	9	(82)				
	4	20	5	(45)				
PP	1	13	3	(27)	4	(36)	11	(100)
	2	8	4	(36)				
	3	12	7	(65)				
	4	19	8	(73)				
	5	24	9	(82)				
	6	17	9	(82)				
SS	1	18	5	(45)	5	(55)		
	2	12	4	(36)				
Average			6.1	(55)	7	(64)	8	(73)

(Hausfater, 1975). As seen above, this is *not* the case for Barbary macaques. During the several days within which ovulation is probable (judged by perineal turgescence and behavior), Barbary macaque females are associating with most—probably all—of the sexually mature males in the group. A select few males cannot monopolize the female since she herself controls consociation formation and termination.

Social Status and Consociation Formation and Termination. The social rank of a male appeared to be immaterial to an estrous female when soliciting or terminating a consociation. For example, the female was responsible for forming 38 percent (n = 84) and terminating 40 percent (n = 24) of all consociations with low ranking males, while the female formed 60 percent (n = 37) and terminated 67 percent (n = 137) of her consociations with high-ranking males. Considering next those cases in which the estrous female both terminated the consociation *and* was then exclusively responsible for forming the next one, estrous females left a higher-ranking male for a lower-ranking one 72 percent (n = 71) of the time, whereas they left a lower-ranking male for a higher ranking one 28 percent (n = 27) of the time. As shown previously, dominant males participate more actively in forming their own consociations. Consequently, the female's role in consociation formation with these males is diminished and tends to correlate inversely with the male's social status. On the other hand, these data do suggest that the social rank of a male is not particularly important in determining whether the estrous female will seek him out, or, in turn, terminate a consociation with him.

Rates of Consort Exchange. A Barbary macaque female regularly changes sexual partners at a remarkably high, sustained rate throughout estrus. Table 12-10 presents summaries of average consociation lengths and the rates of consort exchange for all focal samples. These data show that daily consociation turn-over rates ranged from a high of one new partner every 14 minutes to a low of one every 35 minutes, with an all-female daily average of a new partner every 20 minutes. Eighty-six percent of all consociation exchange rates fell between one every 14 and one every 25 minutes. It seems likely that among Barbary macaques the rate at which consorts are changed during the several days of estrus is a function of several interdependent factors, such as intrinsic female differences, the strategy used by the prospective male consort, and proximity to ovulation. Because sample sizes are too small to allow definitive conclusions, and because the actual day of ovulation was not known, it is impossible to assess the variation in consort exchange rates at the time ovulation was imminent. But, these data do suggest that high consociation formation rates are occurring during the several days within which ovulation is likely to occur.

Table 12-10. Consociation Formation Rates and Durations.

FEMALE	ESTRUS	FOCAL SAMPLE	LENGTHS OF CONSOCIATIONS			RATE OF CONSORT EXCHANGE ONE PER "x" MINUTES
			DURATION (AVERAGE)	NUMBER	SD	
LP	#1	1	15 mins.	10	14.3	16
		2	15 mins.	15	11.4	22
	#2	1	25 mins.	11	15.8	30
LN	#1	1	13 mins.	19	12.1	14
		2	13 mins.	30	11.4	14
		3	21 mins.	12	19.2	23
PL	#1	1	16 mins.	11	11.2	16.5
		2	14 mins.	14	9.4	16
MM	#1	1	24 mins.	12	24.0	25
	#2	1	28 mins.	9	27.6	35
		2	17 mins.	21	14.0	17
		3	19 mins.	20	16.9	20
PP	#1	1	14 mins.	12	12.0	20
		2	16 mins.	10	17.8	21
	#2	1	15 mins.	12	16.4	16
		2	15 mins.	19	14.7	18
		3	17 mins.	23	15.9	18
		4	16 mins.	16	12.1	18.5
SS	#1	1	18 mins.	18	17.9	18.5
		2	15 mins.	12	14.4	20
All Female average			17.3 mins.			19.9

The length of each consociation *per se* was likewise markedly attenuated, reflecting the high rate at which sexual partners are changed in this species. The distribution of the length of 297 consociations is presented in Fig. 12-1. The duration of a consociation ranged from less than one minute (all such cases were rounded off to one minute) to a maximum of 93 minutes. The average length was about 17 minutes, while the median length was 13 minutes. Over one-fourth (26 percent, n = 77) of all consociations lasted five minutes or less; 41 percent (n = 122) lasted 10 minutes or less; 27 percent (n = 80) of all consociations were 11 to 20 minutes long; 15 percent (n = 46) were 21 to 30 minutes long; 7 percent (n = 20) were 31 to 40 minutes long; and 10 percent (n = 29) were longer than 41 minutes. Only 2 percent (n = 5) of all recorded consociations lasted one hour or longer. Since the average length of a con-

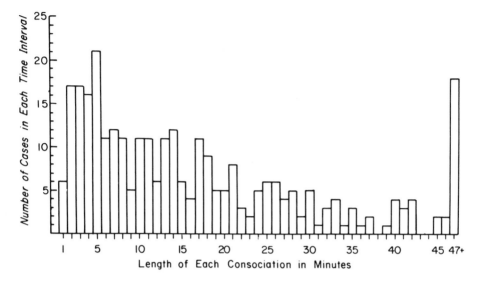

Fig. 12-1. Distribution of the lengths of consociations during estrus.

sociation was about 17 minutes and there are an average of about 12 hours of daylight at this time of year, estrous females could be expected to form up to 40 different consociations in a day. The greatest number of consociations actually observed during any focal-female sample was 30; since this sample was 436 minutes long (about 61 percent of the total daylight hours) it would appear that estrous females do, in fact, form in excess of 40 separate sexual associations per day during estrus.

Distribution of Accumulated Consociation Time. As seen, all sexually mature males have some real access to a female during her estrus. Males do not, however, accumulate equivalent amounts of time with estrous females.[2] The percentage of each female's consociation time accumulated by each male is presented in Table 12-11. Considering first the data for all females combined, it can be seen that the total amount of consociation time a male accrued with an estrous female ranged from a low of only 1 percent to a high of 34 percent. While a single male was thus able to associate with a female for only about a third of all time, the two PS strategy males collectively did accumulate 58 percent of all consociation time. All PAS strategy males collectively accounted for only 13 percent of the total time in association with estrous females.

[2] Although the lengthy focal-female samples have been considered to be relatively unbiased samples of behavior, absolute frequencies are not strictly comparable across estrous periods because unequal focal minutes and unequal numbers of days per estrus were collected for each female.

Table 12-11.   Percent of Consociation Time Accumulated By Each "S" Group Male During a Female's Estrus.

FEMALES AND ESTROUS PERIOD	HIGH RANKING MALES				LOW RANKING MALES						
	MO	RD	CM	WN	V	NM	BB	4SA	RO	BL	BS
LP, Estrus #1	34	16	13	18	5	*	*	2	1	3	8
Estrus #2	57	6	15	23	*	*	*	*	*	*	*
LN, Estrus #1	44	12	8	17	.4	.8	2	2	.8	.9	12
PL, Estrus #1	29	31	5	25	*	*	*	*	2	*	7
MM,Estrus #1	26	19	36	2	*	2	*	*	15	*	*
Estrus #2	27	40	5	19	2	3	.4	.3	3	.3	*
PP, Estrus #1	47	12	3	38	*	*	*	*	*	*	*
Estrus #2	29	8	16	26	4	3	3	4	2	5	.6
SS, Estrus #1	44	6	11	37	*	*	*	2	*	*	*
All Females combined	34	18	12	24	2	1	1	2	2	2	3

* Indicates that no consociation was accumulated by this male with this female.

When next considering each female's estrous period individually, these same patterns are seen: i.e. the majority of consociation time is accumulated by several, specific males, while the remaining time is distributed more or less equally among the other males. Over-all the most successful male, Mo, associated with estrous females (all female averages) about 30 percent more often than did the second most successful male, WN. The third most successful male, RD, associated with females only half as often as did the most successful male, while the fourth most successful male, CM, accumulated 64 percent less consociation time than did male Mo. Each of the remaining males was about equal, and was relatively unsuccessful, in the amount of time accumulated with estrous females. All these data suggest that there are fundamental differences in the abilities of males to associate with estrous females.

The most reliable assessment of intra- and intermale differences in associating with estrous females comes from comparing data for the three most completely sampled females; LN's first estrus, and MM's and PP's second estrus (Fig. 12-2 and Table 12-11). For example, during adult female LN's first estrus, adult male Mo accumulated 44 percent of all her time, adult male RD accrued 12 percent, CM 8 percent, WN 17 percent, subadult male BS 12 percent, etc. (Fig. 12-2). There was as much as a 20-fold difference between the

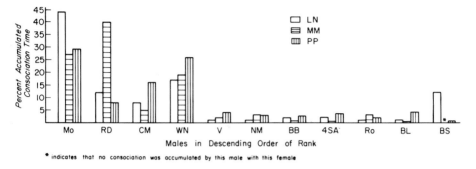

Fig. 12-2. Percentage distribution of accrued consociation time.

most and least successful males in accumulating consort time with this female, and this pattern holds for the other females. A male that is the most successful in accumulating consociation time with one female may not be the most successful male with other females. For example, during LN's first estrus, adult male Mo accumulated the most consort time with her (44 percent), but during MM's second estrus, male RD accumulated the most time (40 percent). Similarly, a male accumulated differing amounts of time with different females—thus male RD accumulated 12 percent of LN's time (first estrus) but he accrued 40 percent of female MM's time and 8 percent of female PP's time during their second estrus.

In general, the majority of all time spent in association with an estrous female was accounted for by a limited number of males in the group. During LN's first estrus, 73 percent of all time spent with her was accounted for by three males, Mo, RD and WN; during MM's second estrus fully 86 percent of consort time with her was accumulated by males Mo, RD and WN; 61 percent of time spent with PP during her second estrus was accounted for by three males, Mo, CM, and WN. The total amount of time a male can sequester any estrous female varies considerably between females and from estrus to estrus. Thus during LN's first estrus, one male, Mo, was able to associate with her for 44 percent of all time, but during PP's second estrus the maximum amount of time any male associated with her was 20 percent.

Social Status and Success in Accumulating Consociation Time. The relationships between the success of a male in associating with estrous females and his social status are ambiguous. For example, the majority of all consociation time was accumulated by the highest-ranking males *collectively*, while the remaining time was distributed about equally among the low-ranking adult and subadult males (Table 12-11). Rank orders by social status and by accumulated consociation time, based on data for each of three females, these

Table 12-12. Comparison of "S" Group Male Rank Orders By Social Status[a] and Accumulated Consociation Time.

RANK ORDER BY SOCIAL STATUS	DESCENDING RANK ORDER BY ACCUMULATED CONSOCIATION TIME FOR:				
	LN. ESTRUS #1	MM. ESTRUS #2	PP. ESTRUS #2	LN. MM. PP POOLED DATA	ALL ESTROUS FEMALES POOLED DATA[b]
#1 Mo	#1 Mo	#1 RD	#1 Mo	#1 Mo	#1 Mo
#2 RD	#2 WN	#2 Mo	#2 WN	#2 WN	#2 WN
#3 CM	#3 RD & BS	#3 WN	#3 CM	#3 RD	#3 RD
#4 WN	#4 CM	#4 CM	#4 RD	#4 CM	#4 CM
#5 V	#5 BB & 4SA	#5 NM & Ro	#5 BL	#5 BS	#5 BS
#6 NM	#6 NM, Ro & BL	#6 V	#6 V & 4SA	#6 V, NM, 4SA & Ro	#6 V, 4SA, Ro & BL
#7 BB	#7 V	#7 BB, 4SA & BL	#7 NM & BB	#7 BB & BL	#7 NM & BB
#8 4SA			#8 Ro		
#9 Ro			#9 BS		
#10 BL					
#11 BS					

a See Table 12-3.
b From Table 12-11.

312

females pooled and all females pooled are compared in Table 12-12. Spearman's rank order correlations between social rank and the rank order for each of these data sets are all significant ($p < .01$, r_s = .89, .91, .88, .95, and .93, respectively), suggesting that high social status is positively correlated with a male's ability to accumulate consociation time with estrous females.

On the other hand, taken individually, the social rank of each male is not consonant with his position in rank order by accumulated consociation time (Table 12-12), as several substantial reversals occur in rank order positions. For example, considering the pooled data for all females, #4 socially ranked male WN ranked second in accumulated consociation time, while #11 socially ranked subadult male BS ranked fifth. Similar reversals may be seen when considering the data for each individual female. These data show that individual male social status *per se* does not correlate linearly with their ranking by accumulated consociation time. Thus, while the majority of all consociation time tends always to be accumulated by the highest-ranking males collectively, high social rank itself does not always guarantee a male success in associating with an estrous female. Neither does being a low-ranking male necessarily mean that this male cannot accumulate as much time with estrous females as can his higher ranking rivals.

Copulations

Components. Barbary macaques are not "multimounters" in the fashion of rhesus or Japanese macaques. Rather, ejaculations occur at the termination of a single mount, which is straightforward and brief (Table 12-13). Considering data for all males combined, the average length of an ejaculatory copulation was 8.7 seconds (SD = 1.1 second, n = 205), with a range of 6 to 14 seconds. Ejaculatory pauses ranged in length from one to four seconds each, with an all-male average of 2.4 seconds each (SD = 0.7, n = 193). On the average, ejaculation occurred approximately 6.3 seconds after the male mounted the female (length of mount minus length of ejaculatory pause). Among individual males, the average number of pelvic thrusts occurring before ejaculation was achieved varied from 8.2 to 9.8, while the all-male combined average pelvic thrusts per copulation was 9.1 (SD = 1.8, n = 207), with an all-male range of 5 to 21. Copulations occurred almost immediately after a consociation was formed! Of the 205 instances when the interval from consociation formation to copulation was positively known, the average delay was only 2.3 minutes. Elapsed time from consociation to copulation ranged from less than one minute to a maximum of only 16 minutes; in 62 percent of the cases (n = 129), the delay was one minute or less; in 77 percent (n = 158) it was two minutes or less; and in 84 percent of all instances (n = 173) only three minutes or less elapsed before copulation. A male never achieved three

Table 12-13. Summary of Male Copulation Components.

MALE	LENGTH OF COPULATION (SECONDS)				PELVIC THRUSTS				EJACULATORY PAUSE (SECONDS)			
	N	X̄	SD	RANGE	N	X̄	SD	RANGE	N	X̄	SD	RANGE
Mo	53	8.4	.9	7 - 11	53	8.7	1.2	5 - 11	51	2.2	.6	1 - 4
RD	34	9.4	2.3	8 - 14	34	9.3	3.6	8 - 21	33	2.7	.6	1 - 4
CM	35	8.4	.9	7 - 11	35	9.2	1.6	6 - 13	32	2.4	.6	2 - 4
WN	39	9.7	1.6	7 - 13	40	9.8	1.9	6 - 14	36	2.4	.6	1 - 3
V	11	7.8	.9	6 - 9	11	9.3	1.4	8 - 11	10	1.9	.6	1 - 3
NM	9	8.6	.7	8 - 10	9	8.9	1.8	6 - 11	8	2.1	.4	2 - 3
BB	4	8.0	1.2	7 - 9	4	8.5	2.5	6 - 12	4	2.3	.5	2 - 3
4SA	6	9.3	1.2	8 - 11	7	9.4	1.9	8 - 13	6	2.7	1.0	2 - 4
Ro	5	7.8	.8	7 - 9	5	8.2	.8	7 - 9	5	2.6	.9	2 - 3
BL	4	7.3	1.0	6 - 8	4	9.5	1.7	8 - 11	4	2.3	1.0	1 - 3
BS	5	7.4	.9	6 - 8	5	8.0	1.2	6 - 9	5	2.0	0.0	2
All males	205	8.7	1.1	6 - 14	207	9.1	1.8	5 - 21	193	2.4	.7	1 - 4

ejaculations with a female during a single consociation. Of 278 consociations recorded during focal-female sampling, two copulations occurred in the same consociation only nine times (3 percent), while only one ejaculatory copulation occurred during 198 consociations (72 percent). There was no ejaculation in 71 consociations (25 percent).[3] In 119 cases (58 percent) the copulation occurred immediately after a solicitation by the female, while the copulation followed a male's initiative in 87 cases (42 percent).

Copulation Rates and Intercopulatory Lengths. Data on average rates of copulation and the lengths of the intercopulatory intervals for estrous females (Table 12-14) show that an estrous female copulates at an extraordinarily high rate. Daily average copulation rates per female ranged from a high of one ejaculation every 12 minutes to a low of one every 41 minutes, while the daily all-female average rate of copulation was one every 27 minutes. Considering the data for the three estrous periods sampled most completely, rates per-day varied from one every 12 minutes to one every 32 minutes, with individual female per-estrus averages of one ejaculation every 18, 27 and 23 minutes. If

[3]Of the 71 nonejaculatory consociations, Mo accounted for over half the cases (56 percent, n = 40). Excluding Mo, only 12 percent of all consociations were nonejaculatory.

the highest rates of copulation occur near ovulation, as seems likely, a copulation rate of about one every 12 to 23 minutes may be diagnostic of proximity to ovulation. The distribution of refractory intervals for all females combined during estrus is presented in Fig. 12-3. Two consecutive copulations might occur within as short a time as two minutes or as long as over 80 minutes, but the combined all-female average interval was 27 minutes. The average per-female intercopulatory intervals ranged from 12 to 49 minutes (Table 12-14), with an all-female daily average of about 27 minutes. The daily average refractory period for the three most completely sampled females ranged from 12 to 38 minutes. Thus, although there might be considerable individual variation in the interval between any two consecutive copulations for each

Table 12-14. Copulation Rates and Refractory Periods For Females During Estrus.

FEMALE	ESTRUS	FOCAL SAMPLE	FEMALE REFRACTORY PERIODS				RATE OF COPULATION
			DURATION (AVERAGE)	NUMBER	SD	RANGE (MINUTES)	ONE PER "X" MINUTES
LP	#1	1	18 mins.	9	16.9	2 - 57	18
		2	27 mins.	3	13.2	12 - 37	40
	#2	1	32 mins.	9	20.9	4 - 62	32
LN	#1	1	12 mins.	21	8.6	2 - 37	12
		2	19 mins.	22	10.2	2 - 36	18
		3	21 mins.	11	17.1	5 - 33	23
PL	#1	1	23 mins.	6	12.6	5 - 43	20
		2	20 mins.	9	7.4	9 - 27	19
MM	#1	1	36 mins.	5	32.9	3 - 85	34
	#2	1	38 mins.	7	26.1	7 - 73	32
		2	20 mins.	18		3 - 65	19
		3	34 mins.	10		10 - 63	30
PP	#1	1	49 mins.	4	25.0	12 - 81	32
		2	38 mins.	3	16.6	19 - 50	41
	#2	1	17 mins.	10	9.9	2 - 39	17
		2	24 mins.	8	30.9	2-111	22
		3	23 mins.	16	19.1	5 - 68	23
		4	32 mins.	9	21.4	5 - 74	32
SS	#1	1	37 mins.	8	27.4	11 - 85	37
		2	33 mins.	6	26.5	7 - 76	33
All female average			27.1 mins.				26.7

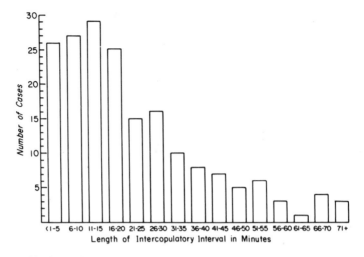

Fig. 12.3. Distribution of intercopulatory intervals for all females.

female, on the average, females consistently copulated at a remarkably high rate during each day of estrus.

Serial Copulations with Multiple Males. Barbary macaque females copulate sequentially with many different males during each day of estrus. Presented in Table 12-15 are data on the number and percentage of "S" group males with whom an estrous female copulated. A female might copulate with 2 to 10 (18 to 91 percent) of the group's sexually mature males on any one day of estrus. The total number of different males copulated with during estrus ranged from 3 (27 percent of all group males) to 11 (100 percent); but when only the three most complete estrous periods are considered, the total number of different males a female copulated with during estrus ranged from 9 (82 percent) to 11 (100 percent). Since these data were collected from focal samples extending over a portion of the entire day, it is probable that these values are underestimates of the actual number of males a female copulated with during any single day of estrus or throughout the entirety of estrus.

This proclivity to copulate with multiple males is not restricted to segments of the estrous cycle when ovulation is least likely, but, on the contrary, females appear to copulate with the greatest number of different males when ovulation is most likely, i.e. when perineal swellings are largest. This pattern can be seen clearly when comparing number of males copulated with to rates of copulation. At those times when copulation rates are highest (at least one every 20 minutes or less), estrous females solicited and copulated with the majority of sexually mature males in the group. For example, from Table 12-14 it can be seen that the highest copulation rate for LN during her first estrus was a

Table 12-15. Number and Percentage of Sexually Mature Males That Estrous Females Copulated With During Estrus.

FEMALE	FOCAL SAMPLE	MALES COPULATED WITH PER DAY	PERCENT ALL MALES	TOTAL NUMBER OF MALES COPULATED WITH DURING ESTRUS			
				ESTRUS #1		ESTRUS #2	
				NUMBER	PERCENT	NUMBER	PERCENT
LP	1	4	(36)	6	(55)	4	(36)
	2	5	(45)				
	3	4	(36)				
LN	1	10	(91)	11	(100)		
	2	8	(73)				
	3	5	(45)				
PL	1	5	(45)	6	(55)		
	2	5	(45)				
MM	1	5	(45)	5	(45)	9	(82)
	2	4	(36)				
	3	9	(82)				
	4	4	(36)				
PP	1	3	(27)	3	(27)	11	(100)
	2	2	(18)				
	3	7	(65)				
	4	8	(73)				
	5	9	(82)				
	6	8	(73)				
SS	1	5	(45)	5	(45)		
	2	4	(36)				
Average		6	(55)	6	(55)	8	(73)

copulation every 12 minutes, and on that day she copulated with 10 different males (91 percent) of the group. Likewise, MM copulated with 9 different males (82 percent of the group) at the time her copulation rate was highest (one every 19 minutes). And when PP's copulation rate was highest (one every 17 minutes) she copulated with 7 different males (65 percent of the group).

It has been shown previously that during estrus the Barbary macaque female herself tends to direct with whom she will consort. It has been shown here that the female is also most active in soliciting copulations, and that she will associate and copulate with a majority of the males of the group during estrus. These patterns support strongly the proposition that the estrous Barbary macaque female purposely directs sexual activity with a large proportion of the group's males. Consequently, most, if not all males, have some *real*, albeit unequal, possibility of impregnating the estrous female. This tends to contrast sharply with most multimale social groups of catarrhine primates, in which some portion of the sexually mature males of the group is excluded from associating with the estrous female at the time most proximate to ovulation. (See "Comparison with Other Species" in Taub, 1978).

The apparent uniqueness of the Barbary macaques' sexual system is probably a real species difference, as seems likely, but it may reflect in part, that few comparable data on other nohuman primates, wild or captive, are available.

Male Sexual Performance

In evolutionary biological models of reproduction, the term "reproductive success" (R.S.) is used in a restricted sense to refer to the real genetic contributions made by an individual to succeeding generations. The assessment of real R.S. has been verified only among some species of fruitflies (*Drosophila*) (Bateman, 1948). Because the direct measurement of R.S. is difficult, the assumption is frequently made that "male reproductive success varies as a function of the number of copulations" (Trivers, 1972; pp. 138), so that the assessment of R.S. is equilibrated with the assessment of copulatory frequencies (e.g. Bertram, 1976). The term "reproductive success" has been used widely in primate studies to refer, in a broad sense, to success in achieving copulations (e.g. Hausfater, 1975), as well as an indirect measurement of real R.S. Among primates, the relationship between frequencies of copulation and the probabilities of fertilizing females is conjectural rather than empirically established. I believe confusion will be minimized if the use of the term "reproductive success" is limited to its sociobiological meaning; i.e. to refer to real individual genetic contributions to succeeding generations. Furthermore, investigators also use such terms as "reproductive behavior" or "reproductive performance" (c.f. Taub, 1978) incorrectly—all such behavior is sexual but only some of it results ultimately in reproduction. In this paper, the terms "sexual performance" (S.P.) or "frequency of copulation" will be used to refer to a male's relative success in achieving copulations.

Distribution of Copulations Among Males. The total number of copulations a male achieved with a female during estrus and the percentage of all copulations represented is summarized in Table 12-16. As can be seen, there are considerable differences among the males of "S" group in copulations achieved. Yet one of the most striking and perhaps fundamental features of these data is that no single male monopolized an estrous female to the exclusion of his rivals. For example, when considering first the data for all females combined, the most successful male, Mo, accounted for 23 percent of all copulations, but the second, third and fourth most successful males accounted for 18, 17, and 18 percent, respectively, differences that are not statistically significant (X^2 = 2.94, df = 3, p > .50). Although the actual percentages change, this same pattern is seen when considering the data for each female individually. For example, for the three most completely sampled estrous periods, the maximum percentage of copulations achieved by a single male ranged from 19 to 30 percent, with a combined average of 23 percent. The proportion of copulations achieved by the most successful males is, in general, fairly equally distributed among them.

Because the data derived from LN's first, and MM and PP's second estrus were the most completely sampled, discussion of male S.P. may be more accurate when based on these data. Hence, intermale differences in copulatory performances will be presented first from data for these three females (Fig. 12-4 and Table 12-16). Because each of these three females was observed for an average of 394 focal minutes per observation day of estrus—about half of the daylight hours at this time of the year—the distribution of copulations derived from these focal samples is considered to be an accurate reflection of a male's sexual activity with these three females during estrus. The addition of more observation hours may well have changed the absolute number of copulations recorded, but likely would not have changed the proportional distribution of them among the males.

Male S.P. and LN's First Estrus. All males of "S" group copulated with LN at least once during the several days of her first estrus. The four highest socially ranked males, Mo (22 percent), RD (14 percent), CM (17 percent) and WN (17 percent) accounted for 70 percent of LN's copulations at this time. The lowest ranking male BS, a subadult, accumulated 10 percent of her copulations, while the remaining 20 percent of LN's copulations were divided about equally

Table 12-16. Male Copulatory Performances With Estrous Females.

MALES	FEMALES AND ESTROUS CYCLE NUMBER																ALL FEMALES COMBINED	
	LP #1		LN #1		PL #1		MM #1		MM #2		SS #1		PP #1		PP #2			
	N	%ª	N	%	N	%	N	%	N	%	N	%	N	%	N	%	N	%b
"High ranking"																		
Mo	5	31	15	22	7	37	1	11	10	25	3	19	4	33	10	19	55	23
RD	2	13	9	14	4	21	3	33	12	30	3	19	3	25	4	7	41	18
CM	4	25	11	17	2	11	2	22	3	8	5	31	0	0	10	19	39	17
WN	3	19	11	17	4	21	0	0	7	18	4	25	5	42	7	13	42	18
"Low ranking"																		
V	1	6	1	2	0	0	0	0	3	8	0	0	0	0	4	7	9	4
NM	0	0	3	5	0	0	1	11	1	3	0	0	0	0	3	6	8	3
BB	0	0	2	3	0	0	0	0	0	0	0	0	0	0	3	6	5	2
4SA	0	0	2	3	0	0	0	0	1	3	1	6	0	0	4	7	8	3
Ro	0	0	1	2	1	5	2	22	2	5	0	0	0	0	4	7	9	4
BL	0	0	2	3	0	0	0	0	1	3	0	0	0	0	3	6	8	3
BS	1	6	6	10	1	5	0	0	0	0	0	0	0	0	2	4	10	4
Sum	16		63		19		9		40		16		12		54		234	

ª Figured as a percentage of each male's copulations achieved with each female.
b Figured as a percentage of each male's copulations achieved with all females combined.

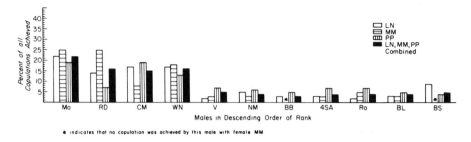

Fig. 12-4. Male copulatory performances with females LN, MM and PP.

among six lower ranking adult and subadult males. If the probability of fertilizing a female is assumed to be a function of the proportion of copulations achieved by each male, the following probability statements follow from these data: The most successful male in achieving copulations, Mo, had about one chance in four (15 of 63 copulations) of impregnating LN. Males CM and WN, second with 11 copulations each, were about equal in their ability to copulate with LN, each having about one chance in six to fertilize her, although Mo is about 30 percent more likely to do so than either of them. RD had the third most copulations, but his chances of impregnating LN—about one in seven—were about 40 percent less than those of the most successful male, Mo. Qualitatively there would appear to be no substantial differences in copulations achieved among these four most successful males (Fig. 12-4). Considering the high-ranking males collectively as a single subgroup, a Chi-square test substantiates that there are no statistically significant differences among Mo, RD, CM and WN in the numbers of copulations each achieved with LN ($x^2 = 1.64$, df = 3, p > .70). The fourth most active male was BS, who had a probability of about one in ten of impregnating LN, but this was 70 percent less than the probabilities for the most successful male, Mo. BS was qualitatively more successful than the other six low-ranking males, who had a small probability—from one in 50 to one in 20—of impregnating LN. Considering the lower-ranking males collectively as another subgroup (including BS), a Chi-square test shows that there are no significant differences among these males in the numbers of copulations each achieved with LN ($x^2 = 7.35$, df = 6, p > .20). A comparison between the high- and low-ranking male subgroups shows a strikingly significant difference in the average numbers of copulations achieved by the males comprising these two subgroups (t = 7.32, df = 9, p < .001).

Based on these data, the following patterns emerge regarding male S.P. with female LN during her first estrus. One or two males did not monopolize sexual access to LN. Four males—45 percent of the sexually mature males in the group—were all about equally successful in copulating with LN. If the assumptions regarding the relationship between copulatory frequency and probabilities of fertilization are valid, none of these males differed significantly in their probabilities of impregnating LN. The other, low-ranking, males were significantly less likely to impregnate LN than were the most successful males, although among themselves none achieved significantly more copulations than the others. On the other hand, if there is no relationship between copulatory frequency and the probabilities of fertilization, the fact that all males copulated with LN during estrus may suggest that all of them may be equally likely to sire her offspring.

Male S.P. and MM's Second Estrus. Males BB and BS did not copulate with MM during the observations of her second estrus. Seventy-seven percent of all MM's copulations were performed with three of the top four ranking males; Mo accumulated 25 percent, RD 30 percent and WN 18 percent. Males CM and V each accumulated 8 percent of all MM's copulations, subadult male Ro, 5 percent, and males NM, 4SA and BL each had 3 percent of the total. Male RD, the most successful, had about one chance in three (12 of 40 copulations) of impregnating MM, assuming a positive correlation between copulatory frequencies and probabilities of fertilization. The second most active male, Mo, had one chance in four (10 of 40) to fertilize MM. The third most successful male, WN, had about one chance in 5.5 to impregnate MM, which was about 30 percent to 40 percent less than the probabilities for males Mo and RD, respectively. With 8 percent each, males CM and V had about one chance in 12 of fertilizing MM. But even though CM had 75 percent fewer copulations than RD, there were no statistically significant differences in the copulations achieved among the high-ranking males as a group ($X^2 = 5.74$, df = 3, p > .10). As with LN, the least successful males all have about an equal, but rather small chance—one in 40 to one in 12—of fertilizing MM. There were also no significant differences among the unsuccessful, low ranking, males in the numbers of copulations each accumulated with MM ($X^2 = 1.98$, df = 6, p > .90). However, as with LN, there was a marked difference in the copulatory performances of the high-ranking males as a group versus the low-ranking males (t = 3.63, df = 7, p < .01).

Whereas 70 percent of LN's copulations were accumulated by four males, 77 percent of MM's copulations were accumulated by three males; but as with LN, male S.P. among these three males did not differ, and all were equally likely to impregnate MM. The remaining males, although not different from one another in their individual S.P., all had a significantly smaller chance of fertilizing MM than did the successful, high-ranking males.

Male S.P. and PP's Second Estrus. All males of "S" group were observed to copulate with PP at least twice during her second estrus. As with LN and MM, the highest-ranking males accumulated the majority of the copulations with her (51 percent by males, Mo, CM and WN). Mo and CM were equally successful in copulating with PP, as each had about one chance in five (10 of 52) of fertilizing her. WN, the next most active, had about one chance in seven to do so. Again there were no statistically significant differences among the higher-ranking males as a group in the numbers of copulations each achieved ($x^2 = 3.18$, df = 3, p > .30). The least successful males achieved from 60 to 80 percent fewer copulations than did the most successful males, but each of these seven low-ranking males was about equal to one another in probabilities of fertilizing PP ($x^2 = 1.06$, df = 6, p > .95). Again, there was a marked difference between the average copulatory performance of the high-ranking male subgroup and that of the low-ranking male subgroup (t = 4.10, df = 9, p < .01).

Male S.P. with All Females. Given the individual male S.P. with females LN, MM and PP, it is not surprising that the same patterns emerge when the data for these females are pooled. That is to say, the four high-ranking males are not significantly different from one another in their copulatory performances ($x^2 = 2.94$, df = 3, p > .30)—neither are the low-ranking males different from one another in the copulations each achieved ($x^2 = .97$, df = 6, p > .98)—but the high-ranking males were significantly more successful in achieving copulations than were the low-ranking subgroup of males (t = 13.11, df = 9, p < .001).

Statistical testing of the pooled data for all females is difficult because unequal numbers of focal samples were accumulated on each of the other females of the group. A qualitative comparison of these data shows concordant patterns. For example, in Fig. 12-4 the results of pooled copulatory performances for LN, MM and PP are shown, and Fig. 12-5 shows the results of pooled copulatory performances for all females. The two distributions are almost identical. These data suggest that total male S.P.—defined as a male's cumulative copulatory performance with all fertile females during their estrus—may remain fairly constant, and that over-all S.P. for high-ranking males did not differ significantly among them.

From the previous analyses of male S.P., the following tentative conclusions regarding Barbary macaque male copulatory performances may be advanced: A few males do not monopolize sexual access to a female during estrus, but instead three or four high-ranking males accumulate most of the copulations. These males are all about equal to one another in their abilities to achieve copulations, and hence have about equal probabilities of impregnating the female. Most of the other sexually mature males of the group do copulate with estrous females, and therefore possess some real, albeit small, possibility of fertilizing them; these relatively unsuccessful males are all about equal to each other in their individual S.P.

As mentioned previously, because the day of ovulation was not known in

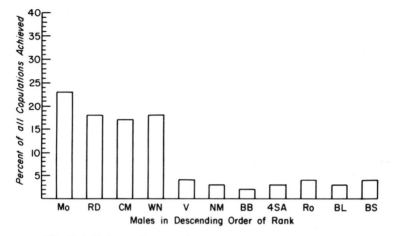

Fig. 12-5. Male copulatory performance with all females combined.

this study, the data have been presented as cumulative sexual performances summed over all sample-days of estrus. It seems important to acknowledge the possibility that on the day of ovulation females restricted copulations to fewer males, so that the probability of fertilization of the female was more limited among the males than the foregoing analyses suggest. It will be recalled that females were selected for focal sampling on the basis of perineal morphology (i.e. when the perineum was at maximum turgescence) and of behavior if the state of the prineum alone was an inadequate criterion. Thus for the three most intensively sampled females—three consecutive days for LN and MM and four consecutive days for PP—it seems probable that ovulation occurred at some point during the period sampled. Table 12-17 shows the daily distribution of copulations on each sample day for males with these three females. The numbers of copulations achieved by each male on a per-day basis are too small to allow statistical tests of significant differences. But, although there is some individual between-male variation, these data do not appear to support the conclusion that one or two males monopolized copulations with the female at ovulation (whichever day that might be) to the exclusion of their rivals. Furthermore, it can be seen that the fewest total number of different males copulated with on any observation day was four (36 percent of the group's males), while on seven of 10 sample-days (70 percent) the female copulated with a minimum of seven sexually mature males (64 percent). If the probability of fertilizing a female is a function of copulating with her on the day of ovulation, then instead of three or four high-ranking males being about equally likely to do so (based on the assumption that probability of fertilization is a function of number of copulations), four to nine high- and low-ranking

Table 12-17. Distribution of Daily Male Copulatory Performances For Females LN, MM and PP.

SAMPLE-DAY	MO	RD	CM	WN	V	NM	BB	4SA	RO	BL	BS
Observation day one											
LN	3	3	4	3	0	2	2	2	1	1	1
MM	3	4	1	0	2	0	0	0	0	0	0
PP	3	1	3	0	1	1	1	0	0	1	0
Observation day two											
LN	6	4	4	4	1	0	1	0	0	1	3
MM	4	4	2	3	1	0	1	1	1	1	0
PP	3	1	2	3	0	0	0	2	1	1	1
Observation day three											
LN	6	2	3	4	0	0	0	0	0	2	2
MM	3	4	0	4	0	0	0	0	1	0	0
PP	4	1	3	3	2	1	0	1	3	0	1
Observation day four											
PP	0	1	2	1	1	2	2	1	0	1	0

males (i.e the number of different males copulating with the female on one day) have equal chances of doing so.

Based on the available evidence, it appears as though the tentative conclusions about Barbary macaque male mating performance derived above are valid. The net result of the female's deliberate choice of multiple male sexual partners during estrus has been to equalize the copulatory performances of "S" group's most sexually active males, as well as to allow all sexually mature males some real probability of siring her offspring.

Dominance and Male S.P. That dominant males mate exclusively or at least significantly more often with receptive females than do low-ranking males has been reported for many nonhuman primate species characterized by multimale breeding groups (Tokuda, 1961; Altmann, 1962; DeVore, 1965; Kaufmann, 1965; Struhsaker, 1967; Hanby, *et al.*, 1971; Dixson, *et al.*, 1973). In contrast, others have reported few or no relationships between male dominance and mating success (Saayman, 1970; Eaton, 1974). For yellow baboons, the so-called "priority of access" model of mating success has been tested recently by Hausfater (1975; pp. 121) who found," . . . first ranking males carried out a far lower proportion of copulations than predicted, and males of all ranks lower than three carried out a larger proportion of copulations than predicted. Second ranking males carried out the highest proportions of copulations during the study period." Thus Hausfater concluded (pp. 109) that the " . . . data do not support the hypothesis that first ranking males have a higher reproductive success than do lower ranking males."[4] On the other hand, as seen in Hausfater's Figure 22 (1975; pp. 111), the number one ranking males accounted for most of the copulations on the estimated day of ovulation, D-3 (i.e. the third day before deturgescence), and in fact ". . . the first ranking males carried out 61 percent of all their copulations on cycle days D-4 through D-2 combined" (pp. 111). Thus despite Hausfater's assertion that during estrus (i.e. D-7 through D-1) the first-ranking males achieved fewer copulations than predicted by the "priority of access" model; in fact, during the critical days surrounding ovulation, they did achieve the majority of the copulations.

As shown for the Barbary macaques in this study, the several highest-ranking males collectively accumulated significantly more copulations than did low-ranking males. Among only these high-ranking males, however, there is no linear relationship between individual male dominance rank *per se* and copulatory success. Since there are no statistically significant differences among these high-ranking males in their individual S.P., it would appear they are all equally likely to fertilize an estrous female—assuming a positive relationship between copulatory frequencies and fertilization—irrespective of

---

[4] "Reproductive success" here referred to copulatory frequency.

their individual positions in the dominance order. The consequences of female choice in forming consociations with these males and in the achievement of copulations, have been to equalize access to her, and by so doing, equalize individual male S.P. among these high-ranking, most successful, males. Likewise, for the seven low-ranking males, there is no linear relationship between their social rank and S.P. While some generalized relationship between social status and S.P. does exist—since collectively high-ranking males are significantly more successful in achieving copulations than are low-ranking males—the exact nature of this relationship remains to be determined. Clearly, however, differential male S.P. depends on factors other than a male's social status.

Male Strategies and S.P. As discussed previously, three different male strategies used by Barbary macaques were identified in this study. These strategies are artificial constructs, based on the similarity of the behaviors of the males in their competition for access to receptive females. Naturally, the success of any male strategy in achieving copulations is a function of the average S.P. of the individual males using each strategy. Consequently an analysis of S.P. by male strategies will reflect the same patterns shown previously when comparing mating data by individual males. Analyzing the data in this manner however, is useful because it underscores the relatively poor S.P. of the low ranking males (i.e. the PAS strategy males), even though, as seen previously, on any given day some low ranking, subadult males, may achieve as many copulations as do high ranking adult males.

For each of the three male strategies, the total number of copulations accumulated, the percentage of the female's copulations represented, and a corrected S.P. mean index (see footnote "a", Table 12-18) are presented in Table 12-18. Whether considering the data for all females combined, for females LN, MM and PP or for these three females combined, the rank order of male strategies by numbers of copulations achieved remains the same. The PS male strategy is always the most successful, reflecting the individual S.P. of males Mo and WN. The CS male strategy always ranks second to the PS male strategy, but the differences between these two are small, and as with individual male S.P., are not significant (t = 1.09, df = 2, p > .20; pooled data for female LN, MM and PP). It will be recalled that the CS males are more dependent on the female's selectivity in forming consociations than are PS males. Thus although the total S.P. of these two strategies are essentially equal, they reflect marked differences in the male's behavior toward the estrous female. As suggested previously, it is the consequence of female choice to equalize male S.P. among the males of these two strategies, and to neutralize the potential effectiveness of the tenacious tactics of the PS male strategy.

The PAS male strategy is the least successful of the three, and these

Table 12-18. Male Strategies and Sexual Performance.

	MALE STRATEGIES								
	PS STRATEGY			CS STRATEGY			PAS STRATEGY		
DATA BASE	MATINGS	PERCENT	MEAN INDIVIDUAL SUCCESS (N=2)[a]	MATINGS	PERCENT	MEAN INDIVIDUAL SUCCESS (N=2)	MATINGS	PERCENT	MEAN INDIVIDUAL SUCCESS (N=7)
LN, Estrus 1	26	(41)	13	20	(32)	10	17	(27)	2.4
MM, Estrus 2	17	(43)	8.5	15	(38)	7.5	8	(20)	1.1
PP, Estrus 2	17	(31)	8.5	14	(26)	7	23	(43)	3.2
LN, MM, PP combined	60	(38)	30	49	(31)	24.5	48	(30)	6.8
All females combined	97	(41)	48.5	80	(34)	40	57	(24)	8.1

[a] Because of differences in sample sizes, to give comparable indices the numbers of accumulated copulations per strategy are divided by the number of males using that strategy.

differences are statistically significant ($t = 12.58$, df $= 9$, $p < .001$). The lack of success in mating shown by this strategy probably reflects the passive nature of the PAS male tactics and their almost total dependence on choice by the estrous females for establishing sexual associations with them. Thus, as with individual male S.P., PS and CS male strategies are the most effective, but they are not significantly different from each other, although the behavior of the males using these two approaches is quite different. The males of either or both of the strategies are significantly more successful in achieving copulations with the estrous female than are the males using the PAS strategy.

DISCUSSION

Sexual selection deals with the evolutionary consequences of competitive mating. As originally conceived by Darwin (1871) and accepted by most subsequent authors, it has been envisioned as operating in two ways: intra-sexually, through competition between members of one sex (usually the males) over sexual access to receptive members of the opposite sex, and intersexually, through differential selection by the limiting sex (usually females) of certain members of the opposite sex over others. It is this latter intersexual process that has been referred to variously as "female choice," "sexual preference" and "epigamic selection." As Ghiselin (1974) has discussed, the concept of female choice has been controversial, in part, due to the difficulty of establishing empirically that female choice has occurred, and partly because of the in-adequacies of theoretical models. It is now generally accepted, however, that females in many species, especially among birds, make a definite choice of the sexual partner (see review by Mayr, 1972). Traditional discussions of female choice have centered on explaining the evolution of secondary sex character-istics and sexual dimorphism. In recent years, aspects of female choice have also been viewed in relation to the evolution of mating systems, and ecological factors influencing female choice (Brown, 1975). For example, the size and quality of the territory defended by a male have been thought to be important factors determining a female's selection of a mating partner. Thus, it is argued that females may have greater reproductive success by "choosing" to mate polygamously with a male who possesses a rich territory than they would if they mated monogamously with a male in a poorer territory (Orians, 1969). In addition, other aspects of male behavior, such as the degree of male care invested in offspring (Trivers, 1972), in combination with ecological factors, have been viewed as determining factors for female choice and in the evolution of mating systems.

The data from this study establish convincingly the existence of "female choice" of mating partners among wild Barbary macaques. The estrous Barbary macaque female plays an active and predominant role in her reproductive activities by purposefully selecting multiple, serial consort partners. In its most elementary form, female choice among Barbary macaques occurs when the female continually seeks out many different males of the group with whom to associate and copulate during the several days of estrus. After a brief association with one partner, during which one copulation typically occurs, the female deliberately abandons this male and immediately solicits another. This routine continues throughout estrus. In effect the female is "choosing" to be a resource that cannot be monopolized, and the consequence of this form of mate selection has been to limit the amount of time any single male can sequester the estrous female and the numbers of copulations he can achieve. Thus all males have some potential for fertilizing the receptive female because she herself chooses to associate with many males. This strategy, used by all females, has effectively equalized the mating performances of the most sexually active males of the group.

How does this form of female choice among Barbary macaques fit with the conventional concepts of female choice and sexual selection? The results of recent studies on wild lions (*Panthera leo*) (Schaller, 1972; Bertram, 1973, 1975a, 1975b, 1976) may provide some insights into the Barbary macaques' mating system. Bertram has summarized the mating system of lions thusly (1975a; pp. 58):

> "Lionesses do not appear to have a regular estrous cycle in the wild; they come into heat at variable intervals of between three weeks and a few months. A lioness usually remains in heat for a few days, and during that time she will mate on the average of every 15 minutes. The first male to encounter a female coming into estrus temporarily gains dominance over all other males. He mates with her and by his presence keeps other males 20 yards or more away. Competition among males for an estrous female is rare. A female sometimes changes mates while she is in heat, but she seldom does so more than once a day. Occasionally, if several females are in heat at the same time, a male may mate with more than one female."

Characteristically, there are large numbers of matings (about 300) during each female lion's estrous period; fertilization rates are low (about one pregnancy per five estrous periods); litter size is small (2 to 3 cubs) and infant mortality is high (70–80 percent). Consequently, a very large number of matings per female is required for each cub reared in the next generation. Bertram (1975a) has calculated that it may take 3,000 copulations to produce one sexually mature progeny. Because it requires such large numbers of copulations to produce viable offspring, Bertram suggests that the "value" of each copulation to a male is low and hence selective pressures favoring intermale competition are low. Thus ". . . 'mating inefficiency' in lions could well be an adaptation by the female serving to reduce competition among the pride males" (1975b; pp. 479).

The mating system of lions is analogous to that of the Barbary macaques in that during estrus females copulate a large number of times. Among lions, however, because it appears that most estrous periods are nonovulatory and because lions are presumed to be induced ovulators, large numbers of copulations are required for fertilization; the behavioral *consequences* of these physiological requirements may have been to minimize male competition. Do Barbary macaque females mate at high frequencies with many different males because they are choosing to become pregnant? If there were differential fertility among males, and if females were unable to detect infertile males, or if females required multiple inseminations for fertilization, then females might benefit from mating at high frequencies with several males so that they would insure fertilization. Most estrous periods of primates are ovulatory (e.g. Jewett and Dukelow, 1972, estimated that only 10 percent of the cycles for *Macaca fascicularis* females were anovulatory), and primates are spontaneous ovulators—females do not require multiple intromissions for ovulation to occur. There is no evidence to suggest that males are differentially fertile, (e.g. laboratory research on reproduction has shown that naturally occurring male sterility is rare). Neither is there evidence to suggest that a fertile male that sexually monopolizes a female has a diminished capacity to fertilize her due to a possible depletion of sperm reserves from frequent matings. Assuming that males are capable of detecting gross changes in a female's sexual condition—a possibility that is enhanced among Barbary macaques due to the females' highly visible morphological changes relative to cycling—it would be advantageous for a male to monopolize sexual access to an estrous female thereby insuring that he copulated with her during ovulation at the same time he prevented rivals from doing so. It seems certain that Barbary macaque females do not copulate at high frequencies with multiple males simply to insure that they will become pregnant.

There are fundamental differences between the mating strategies of female lions and those of female Barbary macaques. For example, although the females of a lion pride may mate indiscriminately with the pride's males, it is not because the estrous female deliberately seeks them out for copulations. The first male that happens upon an unattached estrous female monopolizes her for a day or so, but because it requires several estrous periods to become pregnant, several of the pride's males may copulate with a single female before fertilization occurs. As shown, Barbary macaque females copulate at high rates with multiple males because they themselves continually and deliberately solicit these different males. The differences in female strategies of these two species notwithstanding, among Barbary macaque males, as with lions, there appears to be little overt, combative competition over estrous females. Why is there so litte intermale competition and is its absence a consequence of female choice? This is a critical question for which there are two likely answers. One of these is dependent on and a result of female choice and is a modification of the concept of "economic defensibility" (Brown, 1964); the other is based on the implications of kinship selection (Hamilton, 1964) as this theory has been applied to the mating system of lions (Bertram, 1976).

Brown (1964) has argued convincingly that defense of a territory will evolve when the resources within the territory are economically defendable; i.e. when the cost of the defense if less than the benefits accrued from defending the resource from competitors. A crucial feature of this concept is the *balance* between the advantages and disadvantages of territorial behavior as it relates to ecological competition over crucial resources to reproductive fitness. Implicit in the balance between costs and benefits is the requirement that the resources be "defendable." Thus, aerial or oceanic food resources cannot be economically defended, even if they are crucial and in short supply; hence avian species exploiting such resource bases have not evolved territorial behavior because these resources cannot be defended—the costs of such behavior relative to benefits gained are too great. The cost of active or combative competition—i.e. defending the resource from competitors—which may involve escalating degrees of physical combat as the resources become more limited, are necessarily high. For example, they require large expenditures of energy, and they always carry the risk of serious injury and perhaps death (LeBoeuf and Peterson, 1969; Hrdy, 1974). but they may not insure acquisition and/or exclusive access to the resource. Estrous Barbary macaque females are not an economically defensible resource by virtue of the particular form of their mate selection behavior. As demonstrated, regardless of the strategies used by a male, the behavior of the female—whereby she deliberately terminates, after a brief period, each consociation with all males to solicit new partners—dictates that one male cannot monopolize her (i.e. defend her as a resource) to the exclusion of his rivals. Consequently, the cost of fighting to defend the estrous female from rivals would be greater than the derived benefits—no amount of energy expended by the male could insure him exclusive rights to the resource, since the female does not allow herself to be sequestered from access by rival males. Under these prevailing conditions intensive intermale competition over estrous females would be selected against.

Similarly, the absence of intensive intermale competition may occur, as Bertram (1975b) suggests it does among lions, because of the reduced "value" of a copulation to males in terms of the potential genetic return via that copulation. If the estrous Barbary macaque female copulates with a majority of the group's males around the day or two of ovulation, and if females copulate every 12 to 30 minutes during that period, then even the most active and successful male may mate with the female only a few times, and not in significantly greater numbers than several of his rivals. Data from this study suggest that, in fact, these conditions occur in the Barbary macaques' mating system. Thus, as with lions, a single copulation may be of small potential value in producing impregnation, and the selective pressures on males to fight for an opportunity to copulate is concordantly small; consequently, reduced competition among males would be favored. Simply stated, the potential costs of fighting for a chance to copulate are greater than the probability for fertilization induced by that copulation, especially since the female will seek out most males for copulation. If the male could sequester the female (i.e. economically defend this resource) the "genetic value" of each copulation (the probability that one would produce fertilization) would increase greatly, and selective pressures might favor increased intermale competition. But again, it is the direct consequence of female choice that the estrous female is not a defendable resource.

Implicit in the highly competitive situation seen among males in most mating systems that involve female choice, e.g. in lek systems, is the fact that the female will choose one mate and reject the others. The cost of competition in these situations carries with it the large potential benefits of mate acquisition and rival exclusion, and the potential for between-male variance in R.S. is great. In contrast, the female choice of Barbary macaques is to purposefully choose all mates and reject none, *minimizing* the variance among males in differential R.S. Thus it appears that one consequence of this particular form of female choice has been to favor a reduction of intermale competition for reproductive resources.

Bertram (1976) has presented the provocative argument that among lions, communal suckling of cubs, male tolerance of cubs at kill-sites, and the absence of male competition for estrous females may have been favored through kinship selection processes. By making certain limiting assumptions, Bertram has calculated the average degree of relatedness among all members of a typical lion pride. Since every female born in a pride remains there for life, and males of the same cohort, although they emigrate from their natal group, remain together, members of male and female lineages within a pride are each closely related. Thus the social structure of lions is composed of ". . . a permanent group of closely interrelated females and a smaller group of separately interrelated coequal males, which are with the females for a short period" (Bertram, 1975a, pp. 57). Consequently, because of the resulting high average degree of relatedness between all males and between all females of the pride, cooperative and altruistic acts are favored due to the cumulative benefits of such behavior to each individual's "inclusive fitness." Bertram readily concedes that kin selection, which surely must be operating among lions, is only *one* of a great array of selection processes that may be operating. As

Bertram's arguments relate specifically to reduced competition among males, similar arguments may be advanced for Barbary macaques. As with lions and most nonhuman primates, the Barbary macaque female is born and raised in her natal group, where she remains for life; consequently, females of such groups tend to be closely related. Unlike lions and most nonhuman primates, however, Barbary macaque males do not regularly leave the natal group but rather appear similar to females because they too remain in the natal group. Consequently, all of the members of the Barbary macaque group may be more closely related than are members of primate groups where male migration is the rule. If it is true that adult males of Barbary macaque groups are as closely related to one another (and to other group members as well) as the nonmigrating females of most primate groups appear to be among themselves (Kurland, 1977), then one male's progeny will share a high proportion (the proportion depending on the average degree of relatedness between them) of his genes with all males of the group. This situation would be reinforced by female choice as it occurs among Barbary macaques because, as shown, the sexual performances of the several most successful males are coequal, and these males may, therefore, be producing equal numbers of progeny. Since all males are presumed to be closely related, they are also producing progeny "by proxy" via their successful male relatives, albeit at a reduced level. In terms of genes contributed, all males are making some contribution to the next generation and are thereby increasing their "inclusive fitnesses." The difference in the gene frequencies passed directly by the progenitors of offspring and those passed by proxy by virtue of being a closely related companion, will determine the strength of the selective pressures favoring a reduction of intermale competition. The smaller the difference (i.e. the more closely related are the males) the stronger the pressure favoring an absence of male competition. In this way, kinship selection could be a factor influencing the apparent low level of male competition for estrous females among Barbary macaques.

Thus far we have considered mainly the consequences of Barbary macaque female choice on the behavior of the males, especially as it relates to the apparent reduction of competition among them. It may now be asked why Barbary macaque females choose consort partners as they do? Wherein lies the advantage of such patterned behavior for the females, above and beyond those benefits that might be obtained due to an absence of intermale competition? As shown, the female choice of Barbary macaques is fundamentally different from that described for most species, where the female chooses only one mate from among many competing ones. Because Barbary macaque females do not select mating partners in this way, but rather choose many males indiscriminately, this type of female courtship behavior does not fit into any of the established models of female choice (Orians, 1969).

Wilson (1975, pp. 324) has pointed out that "... the ultimate basis of sexual selection is greater variance in mating success within one sex." Among most animals, with few exceptions, male reproductive success varies more than does female reproductive success (Trivers, 1972). Unlike female mammals, whose prepartum investment in gametogenesis and gestation is relatively great, males generally invest little during and after mating. Consequently, male reproductive potential is high, and competition among males is the rule; some males are excluded from mating as a consequence of intermale competition and polygyny is the most common mammalian mating system. Even with the addition of male parental care, polygyny among mammals tends to be favored because postnatal investment in progeny between the sexes is seldom equal (Trivers, 1972; Wilson, 1975). Trivers (1972) has shown that this situation reverses only when the males do devote more effort to the rearing of offspring than do females. There may be situations however, in which this does not hold true. Assume that, as with females, males are ultimately limited in the amount and quality of care they can contribute to offspring during any reproductive season. If it is then the case that it is the quality of male care that is critical for offspring survival, then even if the male contributes smaller absolute amounts of care than does the female, the male will become the limiting resource. If males do become the limiting resource, then selective forces should favor the evolution of female behaviors that enable them to secure that resource. This should be true for any resource the male represents, whether ecological or behavioral.

Individual nonhuman primate males are not resources that directly benefit a female's reproductive success, except for monogamous primates and the one-male "harem" type of social organization, in which males may hold or represent a resource analogous to those among monogamous birds. An exception to this is the Barbary macaque male. Unlike most mammals and all nonmonogamous primates, Barbary macaque males are involved extensively with infants throughout the infants' first year of life. Several aspects of infant care-taking behavior by male Barbary macaques critical to the current discussion are summarized here (see Taub, 1978, for a detailed discussion of male-infant interactions):

- All males regularly interact with infants in a wide variety of care-taking activities during the infants' first year of life.
- Although among themselves males are not equally involved with infants, most confine their attention to one or two specific infants; consequently, all infants receive attention from several different males, but in total amounts that are not equal among them.
- The only case of infant mortality was the one and only infant that never interacted with adult or subadult males.

Although Barbary macaque males are extensively involved with infants and there is some evidence that this care is essential for infant survival, can the care-taking activities of Barbary macaque males be considered true "parental investment?" Trivers (1972) has defined parental investment as any behavior toward offspring that increases the offspring's chance of survival at the cost of the parent's ability to invest in other offspring. The pratical application of this definition, and the measurement of costs in terms of "reproductive fitness" are difficult, indeed, perhaps impossible with the present techniques of field research. If it is necessary to measure male care and demonstrate its effects on an infant's fitness before one can assert that any behavior toward infants is "parental investment," it cannot be said that Barbary macaque male care-taking behavior is "male investment." Such measurements were not, and for the most part could not be, collected during this brief study. Yet, the cost of gametogenesis and gestation among female mammals, for example, has not been measured precisely, but no one questions that the female has made an investment. Indeed, it is the assumption that this female investment is large that is the keystone of the presumed differential in potential male and female reproductive success, upon which rest most current models of the evolution of mating systems and parental behavior (Trivers, 1972; Wilson, 1975; Brown, 1975). Barbary macaque males take time and expend energy in their inter-actions with infants, time and energy which could conceivably be directed toward other self-maintenance activities. Sometimes, these activities may involve substantial risk, as for example, when males retrieve infants from the threat of a predator. I will therefore, for the sake of the following arguments, assume that the Barbary macaque males' care-taking activities are "investments" in infants. If it is so that male care is true "investment" and is crucial to enhanced neonatal survival and/or ultimate fitness (however defined), then males of this species should be valuable, perhaps indispensible resources for both male and female reproductive success. If these assumptions regarding male care are valid, then the following speculations may be ad-vanced regarding the advantages of the apparently unique form of the Barbary macaque female's sexual strategy.

Let us consider the possible consequences to the Barbary macaque female that adopts an alternative reproductive strategy, say, of allowing one male to monopolize her. Although this male now has exclusive access to this "deviant" female, he would not restrict his own sexual activities to this female alone. Because all of the other females of the group would be using the common strategy of soliciting multiple males, it is likely that this male would also copulate with some other females. Hence, it is probable that this male would interact not only with the deviant female's offspring, but also with offspring of some of the other females with whom he copulated. Of more fundamental importance, however, is the fact that it would be highly unlikely that any other male of the group besides the exclusive consort would interact with the deviant female's infant, and this male's interests would be divided among several infants. Although it is unknown whether care from only one male would be insufficient for infant survival or for obtaining other benefits, the amount of care received from just the one male would clearly be considerably less than the total care received by the other infants, each of whom would receive attention from several different males. If it is assumed that the benefits to an infant of male care are proportional to the number of males contributing to it, then it follows that the reproductive success of the deviant female would be less than that of a female using the common strategy (where each offspring would receive care from several males).

What would be the consequence if all Barbary macaque females adopted the deviant female's strategy of allowing one or two males to sexually monopolize them? If each female adopted this strategy, then all the offspring would be sired by the same, few males. While this might appear at first to be advantageous to the female, because it minimizes paternity uncertainty and would likely promote more male care-taking behavior, it is actually disadvantageous to both males and females as the following example illustrates. Let us assume that each male can adequately invest in two offspring.[5] Let us assume further, for illustrative purposes, that three dominant males of a group of six share equally all the offspring of 12 females of one social group (i.e. each male produces four offspring in a year). After the progeny's birth, these reproductively active males would be faced with the following "choices": (a) if the males invest the maximum amount possible in the two infants they can adequately care for, then they must invest nothing in two of their other offspring; (b) if the males invest equally in all progeny they produce, then they must reduce the amount contributed to each infant from 50 percent (i.e. the amount divided equally between two infants) to 25 percent (i.e. the amount divided equally between four infants)—a reduction of 50 percent! The effect of a male adopting choice "a" is to reduce individual *female* reproductive success (and secondarily, male R.S. too) since some females' offspring would receive adequate male care while others would receive no male care, and under the constraints of the model's assumptions would not survive. Consequently, there should be strong counterselective pressure on females to obtain, by whatever means, exclusive rights to the male's care-taking services, or otherwise insure that he made an adequate investment in her offspring even if the male also invested in a rival female's offspring. If males adopted choice "b," then each offspring would receive only half the male care received if a male restricted his investments to just two infants. If the benefits derived from male care are proportional to the amount (or quality) received, these infants would obtain only half the benefits

[5] This value is chosen because males that are most heavily involved with infants restrict their interactions primarily to two infants, suggesting that male ability to invest in infants is finite and may be adequate for support of two infants (Taub, 1978).

that the infants would receive from choice "a," and their survival chances and/or fitness would be reduced accordingly. In either situation here, and the previous case of the single deviant female, average female R.S. (and to a lesser degree, male R.S. too) would decline. As a consequence of this lowered reproductive fitness, such female strategies should be selected against. It is suggested here that the strategy which Barbary macaque females use in reality increases their individual R.S. over that of the alternatives just presented, and in the following way.

Extending the three-male analogy used above, let us assume now (as is actually the case among Barbary macaques) that each of the 12 females of the hypothetical group copulates with all six males and the reproductive performances of the three dominant males, instead of being restricted equally, but exclusively, to three females each, is now distributed equally among all of them. This should have the effect of minimizing variance in male sexual performance (and also R.S.?), while at the same time insuring that all males have some real probability of siring offspring. Hence selective pressures should favor some investment in offspring by all males, presumably in relation to their respective probabilities of fertilizing the female, although the exact proximate mechanisms involved are not known. Regardless of how the three dominant males now distribute their care-taking among the progeny, the other three males, previously contributing nothing, will now also contribute some care to infants. Consequently it no longer matters, as it did in the previous hypothetical situation, whether a single male contributes less than 50 percent of his care-taking potential to a single infant. Since more than one male now contributes care to an infant, the *cumulative total amount* obtained by each—now received from several different males—will, on average, counterbalance what care might or might not be received from a single male. Under these conditions, *all* infants of the group would now not only receive some care, but could receive as much male care[6] as only the select few (i.e. two of four) would receive under the predictions of the previous hypothetical situation. Thus individual female R.S. (and male R.S. too, since twice as many offspring as before will likely survive) will be greater than that deriving from the alternative conditions; hence the female strategy promoting this situation will be selected over the other, less successful, alternatives. In a polygynous but socially cohesive species such as the Barbary macaque, for whom male care is assumed to be crucial to offspring survival, the critical choice for the female is not which male's care to choose over all potential males, but rather whether to receive some male care or no male care for her offspring. Choosing to allow all

---

[6] Ideally assuming equal distribution of care from all males to all infants—i.e. six males invest equally (one-twelfth of their potential) in each of 12 infants equals .5 units of care per infant, for all infants. One male investing equally (half of his potential) in two infants equals .5 units of care per infant.

males to copulate with her—indeed, by actively seeking out multiple males to insure that many of them, in fact, do so, and therefore offering them all a chance to sire offspring, may be the most effective means for females to induce some care from many males and thereby obtain a maximum cumulative investment in her offspring. Simply stated, by manipulating her sexual associations so that many males participate in mating, the females promote more total male care that is available for contribution to infants.

In the evolution of a social system, numerous selective pressures operate on the individuals of a species. Some of these diverse and multiple selection forces doubtless work in concert while others may work in opposition, producing counterselective forces. Sexual selection is only one of these processes. Because animals that live together in permanent social groups are often genetically related to one another, selective pressures that operate through the kinship network probably operate in nonhuman primate social systems (Kurland, 1977). W. D. Hamilton has developed (1964) and refined (1971, 1972) an important theory to explain the evolution of altruistic behavior, which Maynard-Smith (1964) has called "kin selection." Central to this theory are "degree of relatedness," which is the probability of sharing genes through common descent, and "inclusive fitness," which is the ". . . sum of an individual's own fitness plus the sum of all the effects it causes to the related parts of the fitnesses of all its relatives" (Wilson, 1975; pp. 118). Thus, an altruistic act will evolve if the cost to the actor's inclusive fitness is, on the average, less than that gained by the beneficiary of the altruistic act, devalued by how closely the two are related. Hamilton and others have shown through rigorous mathematical models that certain altruistic social behavior is adaptive because it maximizes the inclusive fitness of the altruist. Kurland (1977) has been the first to test directly the predicitions of this kin selection model in primates, and he argues convincingly that certain primate social behavior has been selected through kinship processes. Parental care is probably a subset of kinship altruism (Brown, 1970, 1975; Taub, 1975; Dawkins, 1976; Hrdy, 1976), and, to be sure, some males are interacting with infants that are their biological offspring; such a situation could be easily understood within a kinship selection framework (Trivers, 1972). It may be possible, however, that some Barbary macaque subadult males take care of certain infants because the infants are their full or half siblings (or even more distant relatives, e.g. "cousins") instead of being their potential offspring. In fact, in terms only of degree of relatedness (ignoring the difference in future reproductive potential, i.e. "reproductive value"), an actor's full sibling is as valuable as his offspring, since both are related to the actor by half. Similarly, it may also be possible that some of the adult males are interacting with the progeny of their own siblings (i.e. with "nephews" and "neices") rather than with their own progeny. Consequently, care-taking behavior towards neonates among at least some

Barbary macaque males could have been selected through kin selection processes. Kinship relations most probably explain why sexually immature Barbary macaque males of all ages show exactly the same kinds of care-taking behavior toward selected infants that the adult males do; it is difficult to reconcile the care-taking behavior of sexually immature males in a sexual selection framework.

Because emigration and immigration appear to be minimal among groups of Barbary macaques, inbreeding may be high. Hence, the average degree of relatedness between any group members may be substantially higher than that between members of primate groups where males typically migrate from their natal groups. Bertram (1976) has shown that for lions the average degree of relatedness among all sexually mature males of the pride is quite high—about equal, on average, to that of half siblings (range .22 to .15). Bertram argues that because of this close genetic relationship, cooperation among males and a reduction of competition among them has evolved, in part, through the operation of kinship selection. In a similar fashion, among closely related Barbary macaque males, an investment in an offspring that may not be one's own could have been favored through the aegis of kinship selection because it enhanced one's inclusive fitness.

The speculations offered in concluding this chapter on the Barbary macaques' mating system are not intended to be definitive tests of sociobiological models. Such tests are not possible since this study did not set out to test these hypotheses and the data are clearly insufficient to do so. But the arguments presented here are suggestive, and they do generally fit predictions from the theoretical models. Long-term studies which quantify and measure male investment as it relates to selective mating behavior of females and male sexual performance (and ultimately reproductive success) are needed. In that regard, this macaque is probably the best primate species for testing directly sociobiological models of mating strategies, reproductive success and male care-taking behavior.

ACKNOWLEDGMENTS

With deepest gratitude, I wish to acknowledge financial support from the following individuals and institutions: My wife, Pam Taub, generously provided support from her personal savings. Grants-in-aid were provided by The National Science Foundation (GB-37497), The Fauna Preservation Society of Great Britain (Oryx 100% Fund), The New York Zoological Society, The Rockefeller Foundation, Sigma Xi, and the Explorers Club of New York.

I wish to thank the Forestry Service of Morocco for authorizing my research, and in particular M. Zaki, Director, M. Hessim and M. Benjalloun.

Dr. William A. Mason was instrumental in the production of the funding proposal which permitted this research to be undertaken; Dr. John M. Deag shared unselfishly his experiences about Barbary macaques with me; Dr. Peter S. Rodman and Dr. Donald G. Lindburg both made many, detailed and invaluable suggestions about this material during its preparation for my doctoral dissertation; Dr. Thomas B. Clarkson kindly provided support throughout the preparation of this material; Dr. Jeffery A. Kurland introduced me to the literature of sociobiology, and many of the theoretical ideas expressed in this paper have grown from seeds planted and germinated from his fertile intellect. I thank them all for their generosity.

I also wish to warmly thank this book's editor, Dr. D. G. Lindburg, for the opportunity of presenting such a detailed report on my field study of Barbary macaques.

REFERENCES

Altmann, J. Observational study of behavior; sampling methods. *Behaviour*, **49**: 227–267 (1974).

Altmann, S. A. A field study of the sociobiology of rhesus monkeys, *Macaca mulatta. Annals of the New York Academy of Science*, **102**: 338–435 (1962).

Bateman, A. J. Intrasexual selection in *Drosophila. Heredity*, **2**: 349–368 (1948).

Bernstein, I. S. Primate status hierarchies. In L. A. Rosenblum, (ed.), *Primate Behavior: Developments in Field and Laboratory Research*, Vol 1. pp. 71–109, New York: Academic Press (1970).

Bertram, B. C. R. Lion population regulation. *East African Wildlife Journal*, **11**: 215–225 (1973).

_____. The social system of lions, *Scientific American*, **223**: 54–65 (1975a).

_____. Social factors influencing reproduction in wild lions. *Journal of Zoology, London*, **177**: 463–482 (1975b).

_____. Kin selection in lions and evolution. In P.P.G. Bateson and R.A. Hinde, (eds.) *Growing Points in Ethology*, pp. 281–302. Cambridge: Cambridge University Press (1976).

Blakley, G. A., Blaine, C. R. and Morton, W. R. Correlation of perineal detumescence and ovulation in the pigtail macaque (*Macaca nemestrina*). *Laboratory Animal Science*, **27** (3): 352–355 (1977).

Boelkins, R. C., and Wilson, A. P. Intergroup social dynamics of the Cayo Santiago rhesus (*Macaca mulatta*) with special references to changes in group membership by males. *Primates*, **13**: 125–140 (1972).

Brown, J. L. The evolution of diversity in avian territorial systems. *Wilson Bulletin*, **6**: 160–169 (1964).

_____. Cooperative breeding and altruistic behavior in the Mexican jay, *Alhelocoma ultramarina. Animal Behaviour*, **18**: 366–378 (1970).

_____. Communal feeding of nestlings in the Mexican jay (*Alhelocoma ultramarina*): interflock comparisons. *Animal Behaviour*, **20**: 395–403 (1972).

_____. *The Evolution of Behavior*. New York: Norton (1975).

Breuggeman, J. A. Parental care in a group of free-ranging rhesus monkeys (*Macaca mulatta*). *Folia Primatologica*, **20**: 178–210 (1973).

Bullock, D. W., Paris, C. A. and Goy, R. W. Sexual behaviour, swelling of the sex skin and plasma progesterone in the pigtail macaque. *Journal of Reproduction and Fertility*, **31**: 225–236 (1972).

Burton, F. D. The integration of biology and behavior in the socialization of *Macaca sylvana* of Gibraltar. In F. Poirier, (ed.), *Primate Socialization*, pp. 26–62. New York: Random House (1972).

Cockrum, E. L. *Introduction to Mammalogy*. New York: Ronald Press (1962).

Darwin, C. *The Descent of Man, and Selection in Relation to Sex*. 1871. 2 Vols. New York: International Publishing Service (1969).

Dawkins, R. *The Selfish Gene*. Oxford: Oxford University Press (1976).

Deag, J. M. A study of the social behaviour and ecology of the wild Barbary macaque, *Macaca sylvanus* L. Ph. D. Dissertation. University of Bristol, England (1974).

_____. Aggression and submission in monkey societies. *Animal Behaviour*, **25 (2)**: 465–474 (1977).

Deag, J. M., and Crook, J. H. Social behaviour and 'agonistic buffering' in the wild Barbary macaque, *Macaca sylvana* L. *Folia Primatologica*, **15**: 183–200 (1971).

DeVore, I. Male dominance and mating behaviour in baboons. In F. A Beach, (ed.), *Sex and Behavior*, pp. 266–298. New York: John Wiley (1965).

DeVore, I., and Hall, K. R. L. Baboon ecology. In I. DeVore, (ed), *Primate Behavior*, pp. 20–52. New York: Holt, Rinehart and Winston (1965).

Dixson, A. F., Everitt, B. J., Herbert, J., Rugman, S. M. and Scruton, D. M. Hormonal and other determinants of sexual attractiveness and receptivity in rhesus and talapoin monkeys. *Symposium of the Fourth International Congress of Primatology*, Vol 2. pp. 36–63. Basel: S. Karger (1973).

Dunbar, R.I. M., and Dunbar, E. P. Social dynamics of gelada baboons. *Contributions to Primatology*, Vol 1. Basel: S. Karger (1975).

Eaton, G. G. Male dominance and aggression in Japanese macaque reproduction. In W. Montagna, and W. A. Sadler, (eds.), *Reproductive Behavior*, pp. 287–298. New York: Plenum Press (1974).

Ghiselin, M. T. *The Economy of Nature and the Evolution of Sex*. Berkeley: University of California Press (1974).

Goss-Custard, J. D., Dunbar, R. I. M. and Aldrich-Blake, F.P. G. Survival, mating and rearing strategies in the evolution of primate social structure. *Folia Primatologica*, **17**: 1–19 (1972).

Graham, C. E., Keeling, M., Chapman, C., Cummins, C. L. B. and Haynie, J. Method of endoscopy in the chimpanzee: relations of ovarian anatomy, endometrial histology and sexual swelling. *American Journal of Physical Anthropology*, **38**: 211–216 (1973).

Hamilton, W. D. The genetical evolution of social behaviour. *Journal of Theoretical Biology*, **7**: 1–51 (1964).

_____. Selection of selfish and altruistic behavior in some extreme models. In J. P. Eisenberg, and W. S. Dillon, (eds.), *Man and Beast*, pp. 57–91. Washington, D. C.: Smithsonian Institution Press (1971).

_____. Altruism and related phenomena mainly in social insects. *Annual Review of Ecology and Systematics*, **3**: 193–232 (1972).

Hanby, J. P., and Brown, C. E. The development of sociosexual behaviors in Japanese macaques, *Macaca fuscata*. *Behaviour*, **49**: 152–196 (1974).

Hanby, J. P., Robertson, L.T., and Phoenix, C. The sexual behavior of a confined troop of Japanese macaques. *Folia Primatologica*, **16**: 123–143 (1971).

Hausfater, G. Dominance and reproduction in baboons (*Papio cynocephalus*). *Contributions to Primatology*, Vol 7. Basel: S. Karger (1975).

Hendrickx, A., and Kraemer, D. G. Reproduction. In A. Hendrickx, (ed.), *Embryology of the Baboon*. pp. 3–30. Chicago: University of Chicago Press (1971).

Hrdy, S. B. Male-male competition and infanticide among langurs (*Presbytis entellus*) of Abu, Rajasthan. *Folia Primatologica*, **22**: 19–58 (1974).

_____. Care and exploitation of nonhuman primate infants by conspecifics other than the mother. *Advances in the Study of Behavior.*, **6**: 101–158 (1976).

Jewett, D. A., and Dukelow, W. R. Cyclicity and gestation length of *M. fascicularis*. *Primates*, **13**: 327–330 (1972).

Kaufmann, J. H. A three year study of mating behavior in a free ranging band of rhesus monkeys. *Ecology*, **46**: 500–512 (1965).

Koyama, N. Changes in dominance and division of a wild Japanese monkey troop in Arashiyama. *Primates*, **8**: 189–216 (1970).

Kummer, H. *Social Organisation of Hamadryas Baboons*. Chicago: University of Chicago Press (1968).

_____. Dominance versus possession. *Symposium of the Fourth International Congress of Primatology*, Vol 1: *Precultural Primate Behavior*. pp. 226–231. Basel: S. Karger (1973).

Kummer, H., Gotz, W., and Angst, W. Triadic differentiation: an inhibitory process protecting pair bonds in baboons. *Behaviour*, **49**: 62–87 (1974).

Kurland, J. A. Kin selection in the Japanese monkey. *Contributions to Primatology*, Vol 12. Basel: S. Karger (1977).

Lahiri, R. K., and Southwick, C. H. Paternal care in *Macaca sylvana*. *Folia Primatologica*, **4**: 257–264 (1966).

LeBoeuf, B. J., and Peterson, R. S. Social status and mating activity in elephant seals. *Science*, **163**: 91–93 (1969).

Lindburg, D. G. The rhesus monkey in North India. In L.A. Rosenblum, (ed.), *Primate Behavior*, Vol. 2. pp. 1–106. New York: Academic Press (1971).

_____. Mate selection in the rhesus monkey, *Macaca mulatta*. Paper presented at the 44th annual meetings of the *American Association of Physical Anthropologists*, Denver, Colorado, (1975).

Loy, J. Peri-menstrual sexual behaviour among rhesus monkeys. *Folia Primatologica*, **13**: 286–297 (1970).

MacRoberts, M. H. The social organization of the Barbary Apes (*Macaca sylvana*) on Gibraltar. *American Journal of Physical Anthropology*, **33**: 83–100 (1970).

Maynard-Smith, J. Parental investment: a prospective analysis. *Animal Behaviour*, **25**: 1–9 (1977).

Mayr, E. Sexual selection and natural selection. In B. Campbell, (ed.), *Sexual Selection and the Descent of Man*. pp. 87–104. Chicago: Aldine (1972).

Modahl, K. B., and Eaton, G. G. Display behaviour in a confined troop of Japanese macaques (*Macaca fuscata*). *Animal Behaviour*, **25 (3)**: 525–535 (1977).

Orians, G. H. On the evolution of mating systems in birds and mammals. *American Naturalist*, **103**: 589–603 (1969).

Petter-Rousseaux, A. Reproductive physiology and behavior of the Lemuroidea. In J. Buettner-Janusch, (ed.), *Evolutionary and Genetic Biology of Primates*, Vol 2. pp. 91–132. New York: Academic Press (1966).

Ransom, T. W. Ecology and social behavior of baboons in Gombe Stream National Park. Ph. D. Dissertation. University of California, Berkeley (1971).

Rowell, T. E. Hierarchy in the organization of a captive baboon group. *Animal Behaviour*, **14**: 430–443 (1966).

_____. Baboon menstrual cycles affected by social environment. *Journal of Reproduction and Fertility*, **21**: 133–141 (1970).

Saayman, G. S. The menstrual cycle and sexual behaviour in a troop of free-ranging chacma baboons, *Papio ursinus*. *Folia Primatologica*, **12**: 81–110 (1970).

Sade, D. S. Determinants of dominance in a group of free-ranging rhesus monkeys. In S. A. Altmann, (ed.), *Social Communication Among Primates*, pp. 99–114. Chicago: University of Chicago Press (1967).

Schaller, G. *The Serengeti Lion: A Study of Predator-Prey Relations*. Chicago: University of Chicago Press (1972).

Stephenson, G. R. Social structure of mating activity in Japanese macaques. S. Kondo, M. Kawai, A. Ehara, and S. Kawamura, (eds.), *Symposium of the Fifth Congress of the International Primatological Society*. Tokyo: Japan Science Press (1975).

Struhsaker, T. T. Behavior of vervet monkeys and other cercopithecines. *Science*, **156**: 1197–1203 (1967).

Taub, D. M. Paternalism in free-ranging Barbary macaques, *Macaca sylvanus* L. Paper presented at the 44th annual meetings of the *American Association of Physical Anthropologists*. Denver, Colorado, (1975).

_____. Geographic distribution and habitat diversity of the Barbary macaque, *Macaca sylvanus* L. *Folia Primatologica,* **27**: 108–133 (1977).

_____. Aspects of the Biology of the Wild Barbary Macaque (Primates, Cercopithecinae, *Macaca sylvanus* L 1758): Biogeography, the Mating System and Male-infant Associations. Ph. D. Dissertation. University of California, Davis (1978).

Tokuda, K. A study on the sexual behavior in the Japanese monkey troop. *Primates,* 3: 1–40 (1961).

Trivers, R. L. Parental investment and sexual selection. In B. Campbell, (ed.), *Sexual Selection and the Descent of Man,* pp. 136–179. Chicago: Aldine Publishing Company (1972).

Van Lawick-Goodall, J. The behavior of free-living chimpanzees in the Gombe Stream Reserve. *Animal Behaviour Monographs* **1**: 161–311 (1968).

White, R. J., Blaine, C. R. and Blakley, G. A. Detecting ovulation in *Macaca nemestrina* by correlation of vaginal cytology, body temperature and perineal tumescence with laparoscopy. *American Journal of Physical Anthropology,* **38**: 189–194 (1973).

Wilson, E. O. *Sociobiology.* Cambridge: Belknap Press (1975).

Chapter 13
Ontogenetic and Psychobiological Aspects of the Mating Activities of Male Macaca radiata

Barbara Beckerman Glick

INTRODUCTION

Sexual relations among group-living bonnet monkeys (*Macaca radiata*) have been characterized as differing from those of other species of macaque in a number of fundamental ways. Field observers have emphasized the active role of bonnet males, which contrasts sharply with the passive nature of female participation. Only rarely does the bonnet female actively solicit copulation; she is thought to be remarkably undiscriminating, mating freely with males of all ages and ranks (Simonds, 1965; Rahaman and Parthasarathy, 1969; Sugiyama, 1971). Among rhesus (*M. mulatta*), pigtail (*M. nemestrina*), stumptail (*M. arctoides*), and Japanese (*M. fuscata*) macaques, females typically solicit the sexual advances of males, often showing preferential interest in some males over others. The females of these species also frequently experience sex skin color variations or swellings corresponding to their ovulatory cycles (Southwick *et al.*, 1965; Kuehn *et al.*, 1965; Tokuda *et al.*, 1968; Bertrand, 1969; Imanishi, 1963). Sexually mature bonnet females, however, show no overt changes reflecting their reproductive status, and receptivity is thought to be communicated to males primarily via olfactory signals (Rahaman and Parthasarathy, 1971). More often than not, high-ranking rhesus (Lindburg, 1971) and Japanese (Hanby *et al.*, 1971) macaque males have priority of access to receptive females, and subordinate males must sometimes escape to the periphery of the group to copulate without interference. In contrast, field observations of bonnet monkeys suggest that all males have comparable sexual access to females, and that young or low-ranking males are not inhibited by the presence of fully adult or high-ranking males.

Laboratory investigations of *paired* bonnet monkeys, however, have indicated that certain aspects of the relationship between a given male and his female partner can inhibit or facilitate successful sexual activity. Familiarity tends to reduce aggression, enhance grooming behavior, and therefore foster sexual relations (Nadler and Rosenblum, 1969). The age and dominance status of mates in the laboratory are also important factors. When placed with young or relatively small receptive females, two-year-old bonnet males perform all of the elements of the adult copulatory pattern. When placed with fully mature females, however, these same males may show no sexual behavior. Low status in a male vis-à-vis his female partner can also exert a potentially disruptive effect on mating (Rosenblum and Nadler, 1971). Since socially restrictive conditions during early development have been shown to impair sexual behavior in rhesus monkeys (Mason, 1960), it is important to note that the subjects used in these bonnet monkey studies were born and reared for one to one-and-a-half years with their mothers and other conspecific adults, as well as peers. Subsequent housing conditions provided social contact with a number of animals other than mothers. These findings suggest that there are optimal social parameters for successful reproduction in this species, and further, that the bonnet female may in fact play a much more significant role in sexual interactions than previously thought.

The extent to which these laboratory data can be generalized to explain differences in sexual behavior between individual males within social groups of bonnet monkeys is unknown. Rosenblum and Nadler (1971) postulate that the possible dominance-related advantages conferred by the presence of familiar group members may obscure the effect of the rank differential between mates to some extent. They note that ". . . in the group setting, some males of a given age may be seen engaging in interactions with females of considerable maturity, whereas under dyadic testing conditions in the laboratory, such behavior might be absent" (p. 398). This could be particularly true in this species, where high rank reportedly provides no basis for controlling the sexual activities of others. However, the data on sexual behavior of group-living bonnet monkeys are mainly descriptive in nature, and no systematic investigation of this problem has yet been reported.

This chapter examines the mating activities of captive group-living male bonnet monkeys aged 2.3 to 12.5 years. Data on overall levels of sexual activity, access to specific classes of females throughout the mating season, access to conceiving females, ability to form consort relationships, and reciprocity of social relations with dominant females, are described in relation to male age, dominance status, and testosterone level. The purposes here are threefold: (1) to test the null hypothesis that there are no differences between bonnet males in ability to gain sexual access to females; (2) to present a cross-sectional view of the ontogeny of certain sociological aspects of bonnet male sexual activity; and (3) to describe the contributions of specific psychobiological factors to mating success during three distinct stages in the life cycle of the male.

METHOD

Subjects and Observation Site

Fifteen male bonnet macaques (*M. radiata*), including six adults, five adolescents, and four juveniles were selected as the focal subjects of study. The subjects live together in a large, well-established social group of natural composition in an outdoor enclosure at the California Primate Research Center, University of California, Davis. The enclosure is a one-half-acre wire-mesh construction, measuring 60.96 x 30.48 x 2.44 meters. At the onset of the study, the group consisted of 63 individuals, including 27 adults (6 male, 21 female), 12 adolescents (5 male, 7 female), 14 juveniles (10 male, 4 female), and 10 infants (6 male, 4 female). Table 13-1 provides specific focal subject information including ages, body weights, and serum testosterone levels determined at the peak of the mating season, in October 1977. Rosenblum and Nadler (1971) place both the two- and three-year-old bonnet male in the prepubertal stage of development. In this study, however, three-year-olds were considered as having reached puberty. Among rhesus males, a cyclical enlargement and regression of the testes which commences at age three years has been shown to be reflective of spermatogenic function (van Wagenen and Simpson, 1954; Conaway and Sade, 1965). A marked seasonality of testicular size was clearly evident in all of the bonnet males in this study who were three or more years old during the 1977-78 breeding season. Testicular cyclicity was not observed in any of the bonnet males aged two to three years (Glick, in press). Additionally, the three age-classes differed significantly in both body weight and testosterone level, and thus represented three distinct maturational stages. All of the adult males except the oldest (#99) were captive born. Males #99, #29, and #32 were introduced into their current enclosure in October 1970; male #61 was born in the enclosure in April 1971; males #22 and #24 were introduced in July 1971. The adolescent and juvenile males were born and raised with their mothers in the enclosure, and developed in the context of a social group including diverse age-sex classes of individuals.

Table 13-1. Focal Subject Information: Ages, Body Weights, and Serum Testosterone Levels as of October 1977.

AGE-CLASS	MALE	AGE (YEARS-MONTHS)	BODY WEIGHT (KG)	SERUM TESTOSTERONE LEVEL (NG/ML)
Adult males	99	12–6	11.6	5.4
	22	8–4	9.9	9.0
	24	8–3	9.5	10.8
	29	7–8	11.5	7.8
	32	7–8	11.8	13.4
	61	6–6	11.1	12.2
	($\bar{x}$)	(8–6)	(10.9)	(9.8)
Adolescent males	15	5–0	6.3	4.1
	79	4–5	6.8	4.4
	91	3–5	4.7	3.9
	98	3–5	5.5	2.1
	95	3–4	4.0	2.2
	($\bar{x}$)	(3–11)	(5.5)	(3.3)
Juvenile males	52	2–6	3.3	2.0
	53	2–6	4.1	2.0
	55	2–5	3.5	1.2
	56	2–4	3.1	1.8
	($\bar{x}$)	(2–5)	(3.5)	(1.8)

Data Collection

Behavioral Observations. Behavioral data were collected from May 1977 through April 1978. This paper reports on the information collected over the six-month period during which mating activities were observed, from September 1977 through February 1978. Data on sexual behavior were collected using two separate routines. Copulations were recorded for all animals using a group scan method. Group scanning was done in balanced morning and afternoon blocks for a total of 40 hours per month. Only those intromissions followed by ejaculation were counted as copulations. Data obtained in this manner are used here for comparisons between age-classes in frequencies of copulations with females of various ranks and ages. Specific measures of sexual behavior and social interactions were obtained with a one-zero focal animal sampling technique (Altmann, 1974). Sampling periods consisted of 150 seconds divided into ten 15-second intervals. Active and passive roles in each category of behavior were noted and recorded on prepared checksheets. A pocket-sized intervalometer equipped with an earplug and timed to mark 15-second intervals was used. Proximity and orientation were measured using focal animal bull's-eye maps for instantaneous sampling. Animals in contact with, within specific distances of, and either facing or being faced by the focal subject were recorded. A predetermined random sampling order of subjects was followed. Proximity and orientation maps were completed for each subject immediately prior to the onset of the 150-second sampling period. Ten maps and 10 checksheets were completed for each subject per month. Thus, the analysis was based on 60 maps and 60 checksheets per subject. Since map data were collected instantaneously, the quantified unit of proximity and orientation was the individual map. The quantified unit of behavior, however, was the interval. Six hundred intervals of observation, then, comprised each subject's total behavioral sample. The following measures of activity and association were assessed:

Sexual activity:
  Presentation—raising of hindquarters with tail lifted, sometimes accompanied by lowering of forelimbs; typically occurs in response to male solicitation.
  Genital inspection—visual exploration, touching, or sniffing (sometimes accompanied by oral contact) of another's anogenital region.
  Mount—ventral—dorsal climbing upon another with foot-clasping.
  Pelvic thrust—self-explanatory.
  Ejaculation—self-explanatory.
  Ejaculate ingestion—consumption of ejaculate from one's own or one's partner's genitals.
  Masturbation—auto-stimulation of the genitals.
Proximity:
  Contact—touching of any kind with any body part, exclusive of other categories such as grooming.
  Within five feet—sitting or standing within five feet of another animal.
Social orientation:
  Facing—looking directly at another animal.
  Being faced—being looked at by another animal.
Social interaction:

Grooming—methodical parting of fur with hands with or without the removal of dirt and ectoparasites.

Aggression—including threatening by facial (staring, brow movements, exposure of teeth) or body (inflicting physical harm) gestures, and attack-chasing.

Submission—including withdrawal or flight from another, lipsmacking and grimacing.

Testosterone Determination. Two cc of blood were extracted from either the brachial or saphenous vein of each subject and serum testosterone levels were assayed according to the method of Barkely and Goldman (1977), modified for bonnet macaque male serum determination. The mean values of duplicate measures on each subject are reported. The precision of the radioimmunoassay technique used was evaluated according to the method employed by Bernard *et al.* (1975). Intra-assay variation was estimated at 7.6 percent, and interassay variation was estimated at 11.3 percent.

SEXUAL BEHAVIOR OF BONNET MALES

A stereotypic pattern of copulation was observed among the adult males. This pattern consisted of the male approaching a given female who either presented her hindquarters in response to receiving a genital inspection, or was abruptly positioned into a presenting posture by the male. During his approach, the male sometimes displayed the clonic jaw movement with retracted lips and movements of the tongue (Kaufman and Rosenblum, 1966). He then proceeded to mount the female by clasping her lower legs with his feet and grasping her hips with his hands. When the male was appropriately positioned and balanced, the female supported his weight. At the first intromission, she frequently reached back to touch the male while grimacing and lipsmacking at him. A series of between 5 and 21 thrusts preceded ejaculation, which was almost invariably followed by ingestion of some of the ejaculate. Ejaculation was almost always achieved in a single mount sequence. Double-mount copulations resulted from loss of balance during the first mount. During copulation, the male sometimes slapped the female. The alpha male, for example, was observed slapping two dominant females (one adult and one adolescent) during and after copulation. When a female refused the advances of an adult male, she was sometimes forced into copulation. Forced copulations were characterized by the aggressive pursuit and restraint of a screaming female, whose efforts to escape the male were unsuccessful. A pair frequently copulated several times in sequence, and both males and females were seen copulating with two or more partners within minutes. These sequential matings involved individuals of all ages and ranks.

Adolescent and juvenile males typically displayed most of the elements of the adult copulatory pattern. They were never, however, observed to exhibit the jaw and tongue movements sometimes seen among the fully adult males. The older and larger adolescents typically achieved sufficient balance upon first mountings to complete copulation, but the smaller adolescents and juveniles had some difficulty in obtaining the proper foot-clasp position and were therefore less successful in achieving copulation. Dominant or elder males did not hinder the sexual activities of young males. Females protested the advances of some young males more often than they did those of others; nevertheless, the majority of the young males were observed in multiple copulations with the same or several females.

A quantitative assessment was made of general differences between age-classes in levels of sexual activities. Table 13-2 presents the mean frequencies of intervals during which all major elements of sexual behavior were observed. Adolescents displayed adult levels of sexual activity and differed reliably from juveniles in all behaviors except masturbation. While adults exhibited slightly greater mean levels of nearly all behaviors than did adolescents, the Student's t test showed that the differences were not significant ($p > .10$). In comparisons with juveniles, adolescents received more presentations from females ($p < .05$), and displayed more genital inspecting ($p < .05$) mounting ($p < .005$), pelvic thrusting ($p < .05$), ejaculation ($p < .05$), and ejaculate ingesting ($p < .005$).

SEXUAL ACCESS TO FEMALES THROUGHOUT THE MATING SEASON

Matings with all 28 sexually mature females were examined in order to assess age-related differences between males in sexual access to specific classes of females throughout the mating season. Females were classified into two

Table 13-2.   Measures of Sexual Activity.*

SPECIFIC ELEMENTS	ADULT MALES	ADOLESCENT MALES	JUVENILE MALES
Genital inspections	18.67 ± 7.15	17.00 ± 13.15	2.75 ± 2.36
Sexual presents received	11.00 ± 5.76	6.40 ± 7.16	0.25 ± 0.50
Mounts	6.83 ± 3.19	6.00 ± 1.87	1.25 ± 1.50
Pelvic thrusts	7.00 ± 2.10	5.00 ± 3.08	1.50 ± 1.73
Ejaculations	5.00 ± 2.68	3.80 ± 3.56	0.75 ± 1.50
Ejaculate ingestion	11.67 ± 8.45	8.00 ± 4.12	0.25 ± 0.50
Masturbation	1.33 ± 1.51	3.40 ± 4.72	0.25 ± 0.50

* Values represent mean numbers of intervals per male ± standard deviations.

Table 13-3.   Ages, Weights, and Parity of Females as of October 1977.

AGE-CLASS	N	AGE (YRS)		BODY WEIGHT (KG)		PARITY
		RANGE	MEAN	RANGE	MEAN	
Adult females	21	5.6–12.6	8.5	4.3–9.6	6.4	0–5
Adolescent females	7	3.4–4.5	3.9	3.8–5.2	4.2	0

categories of age and rank. Their ages, weights, and parity are summarized in Table 13-3. Ranks were assigned on the basis of dyadic relationships between females, defined by approach-withdrawal frequencies. A few of the adolescent females clearly ranked above many of the adult females, and a single hierarchy was therefore established. Of the 28 females, the top 10 were selected for the analysis of matings with dominant females; all others were referred to as subdominant. The Student's t test and a parametric two-way analysis of variance with repeated measures on one factor were used to evaluate the effects of the age and rank of female partners on matings.

Table 13-4 presents the mean copulation frequencies for each age group of males with females classified in this way. On the average, adult and adolescent

**Table 13-4.   Total Copulations with Females
in Relation to Their Age and Rank.***

MEASURE	ADULT MALES	ADOLESCENT MALES	JUVENILE MALES
No. of female partners	12.00 ± 6.32	12.80 ± 5.12	7.25 ± 5.74
Total copulations	29.00 ± 19.62	28.60 ± 23.00	16.75 ± 15.88
Copulations with adult females:			
No. of partners	8.83 ± 4.58	7.40 ± 3.44	4.00 ± 2.94
Frequency	20.67 ± 14.11	17.20 ± 18.59	6.25 ± 5.56
Copulations with adolescent females:			
No. of partners	3.17 ± 2.04	5.40 ± 1.82	3.25 ± 2.99
Frequency	8.33 ± 8.62	11.40 ± 5.22	10.50 ± 11.09
Copulations with dominant females:			
No. of partners	6.00 ± 3.22	4.80 ± 1.64	2.00 ± 1.83
Frequency	17.83 ± 13.12	11.80 ± 13.72	4.00 ± 4.08
Copulations with subdominant females:			
No. of partners	6.00 ± 3.41	8.00 ± 4.36	5.25 ± 4.03
Frequency	11.17 ± 7.03	16.80 ± 9.68	12.75 ± 11.95

* Values represent mean frequencies per male ± standard deviations.

males copulated with comparable numbers of females at about the same frequency, while juvenile males had fewer partners and engaged in fewer copulations than either of the older classes of males. A linear relationship emerged between male age and the ability to gain sexual access to the older and higher-ranking females (Fig. 13-1). The same effect of the age of females was found for adult and adolescent males, who mated significantly more often with adult than with adolescent females ($p < .05$). The juvenile males differed markedly from adolescents in this respect. Adolescents were far more successful than juveniles in gaining access to adult females ($p < .05$), and juveniles were observed in more matings with younger than with older females. Examination of matings with females of differing ranks revealed a different relationship between age-classes of males. While adult males mated more frequently with dominant relative to subdominant females, the reverse was true for adolescent males ($p < .025$). Adolescents were more successful than juveniles in gaining access to dominant females. However, the effect of female rank was the same for all nonadult males, who were observed in significantly more matings with subdominant than with dominant females ($p < .025$). These differences were not without exception, however, and while the main effect of male age was significant, individual variation and overlap between

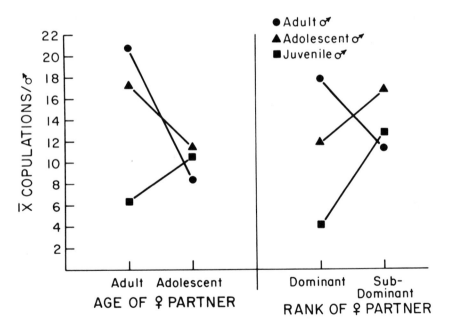

Fig. 13-1. Total copulations in relation to the age and rank of female partners.

classes did occur. Two of the adolescent males (# 15 and # 79), for example, were observed in more matings with adult and dominant females than were two of the adult males (# 22 and # 24).

SEXUAL ACCESS TO FEMALES AT CONCEPTION TIME

While the adult males might have copulated more often with dominant females than did the adolescent males, the latter were as sexually active and copulated with as many females as the adults throughout the mating season. This fact raised the issue of the younger males' potential reproductive success. Age-class comparisons were therefore made of sexual access to females during their respective conception months. Approximate conception dates were calculated by counting back 5.5 months from the known date of delivery for each female. All live- and stillbirths of all females were included, providing a total of 24 known conceptions. In order to account for individual variability in gestation periods, this analysis utilized a "conception month" for each female. The conception month encompassed the two-week periods preceding and following the approximate conception date.

The temporal distribution of matings of all males with females during their conception months were examined in relation to the distribution of known conceptions. Conceptions peaked in October, were moderately frequent in September and November, relatively infrequent in December and January, and absent in February (Fig. 13-2). Adult males were seen mating most often with conceiving females during October, the peak conception month. At this time, they mated nearly twice as often as adolescent males. During September

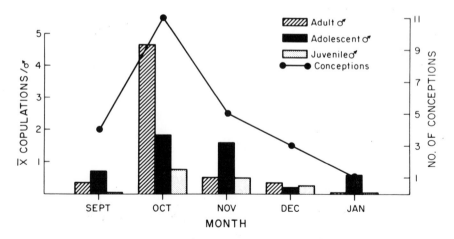

Fig. 13-2. Distribution of copulations with conceiving females in relation to known conceptions.

Table 13-5. Copulations with Females During Their Month of Conception in Relation to Their Age and Rank.*

MEASURE	ADULT MALES	ADOLESCENT MALES	JUVENILE MALES
No. of female partners	4.00 ± 3.03	3.00 ± 2.54	1.00 ± 1.41
Total copulations	6.00 ± 5.10	4.40 ± 3.85	1.25 ± 1.89
Copulations with adult females:			
No. of partners	2.83 ± 2.23	1.60 ± 1.34	0
Frequency	4.33 ± 4.23	2.40 ± 2.30	0
Copulations with adolescent females:			
No. of partners	1.17 ± 1.21	1.40 ± 1.52	1.00 ± 1.41
Frequency	1.67 ± 1.97	2.00 ± 2.35	1.25 ± 1.89
Copulations with dominant females:			
No. of partners	2.50 ± 2.17	1.00 ± 1.00	0.50 ± 0.58
Frequency	4.33 ± 4.27	1.80 ± 2.17	0.50 ± 0.58
Copulations with subdominant females:			
No. of partners	1.50 ± 1.22	2.00 ± 1.73	0.50 ± 1.00
Frequency	1.67 ± 1.37	2.60 ± 2.07	0.75 ± 1.50

* Values represent mean frequencies per male ± standard deviations.

and November, however, adult males mated relatively infrequently, and adolescent males were seen mating more than twice as often with conceiving females as were adult males. Over the five-month period during which conceptions occurred, adult males accounted for 49.2 percent, adolescent males accounted for 42.3 percent, and juvenile males accounted for 8.5 percent of all observed matings with conceiving females.

In terms of total copulations with all females during their respective months of conception, the Student's t test showed that adult males neither had significantly more mating partners ($p > .10$) nor engaged in significantly more copulations ($p > .10$) than did adolescent males. Table 13-5 presents these data along with the copulation frequencies for each age-class of males with specific classes of conceiving females. The differences found between males in matings with females of differing ages and ranks throughout the season prevailed at conception time as well. Fig. 13-3 illustrates that adult males had the greatest access to older and higher ranking females at their conception times. Adult males showed strong preferences for adult over adolescent and for dominant over subdominant conceiving females. Adolescent males had nearly equal access to conceiving adult and adolescent females, but dominant females were

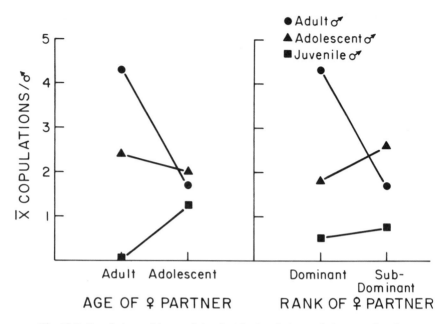

Fig. 13-3. Copulations with conceiving females in relation to their age and rank.

less accessible than were subdominant females to adolescent males. An analysis of variance indicated that the interaction of male age and rank of conceiving female partner was highly significant; adult males mated far more frequently with dominant relative to subdominant conceiving females, while the reverse was true for adolescent males ($p < .025$). There were two occasions when conceiving adult females demonstrated a sexual preference for adolescent over adult males. A midranking female was seen refusing the advances of male # 22 and immediately afterward copulating with male # 79. On a second occasion, a low-ranking female withdrew from the solicitation of the beta male (# 29) and immediately accepted that of male # 15. Juvenile males were never observed mating with adult females at conception time. They mated primarily with subdominant adolescent females, and did so only very infrequently.

SOCIAL RELATIONS WITH DOMINANT FEMALES

The relatively greater sexual access to dominant over subdominant females was the one factor that consistently distinguished the adult from nonadult males. As a rule, the bonnet females did not solicit sexual interactions; they merely responded to the initiative of males. It was difficult, therefore, to separate the influences of male initiative and female responsiveness in examining the differential success of males in gaining sexual access to specific females. It was not clear, for example, to what extent dominant females might be responding preferentially to older males. Thus, the reciprocal nature of certain social relations between males and dominant females was studied in order to provide a measure of female initiative. Comparisons of proximity scores were also made to determine whether any differences existed between age-classes in spatial access to dominant females. Spatial relations and reciprocity of orientation, grooming, and aggression were examined by analysis of variance and the Student's t test. The measures of facing and being faced comprised orientation; grooming and aggression were divided into initiations and receipts. Frequencies of contact and maintaining not more than five feet of distance between focal males and females provided the measures of proximity. The reciprocal nature of these measures could not be assessed because it was frequently not possible to identify the initiator or recipient of the association.

Mean frequencies of associations and interactions with dominant females for each age-class of males are presented in Table 13-6. Of particular interest here are the differences between adult and adolescent males. Adult males were

Table 13-6.   Social Relations with Dominant Females Throughout the Mating Season.

MEASURE	ADULT MALES	ADOLESCENT MALES	JUVENILE MALES
Proximity*			
In contact	11.33 ± 4.68	6.40 ± 2.97	3.25 ± 1.50
Within 5′	20.00 ± 5.10	17.00 ± 10.17	8.00 ± 2.31
Orientation*			
Facing	14.67 ± 4.37	9.80 ± 4.32	3.75 ± 2.06
Being faced	13.50 ± 6.09	5.00 ± 2.55	3.00 ± 3.37
Grooming**			
Initiated	15.33 ± 12.68	11.40 ± 9.18	3.50 ± 1.29
Received	26.67 ± 19.56	8.00 ± 10.65	4.00 ± 5.42
Aggression**			
Initiated	4.17 ± 3.19	3.00 ± 2.65	0.50 ± 1.00
Received	0.67 ± 1.21	3.40 ± 2.30	4.75 ± 4.11

* Values represent mean numbers of occurrences per male ± standard deviations.

** Values represent mean numbers of intervals per male ± standard deviations.

observed in more frequent bouts of contact with dominant females and were found within five feet of them more than were adolescent males, but these differences were not statistically significant. The same effect of reciprocity of orientation was found for all males. On the average, males faced females more than they were faced by them ($p < .01$). However, adult males were observed in overall orientation with females significantly more often than were adolescent males ($p < .025$), and females faced adult males far more than they faced adolescents ($p < .025$).

Adult males were generally involved in more grooming with dominant females than were adolescent males. Adults initiated slightly more and received on the average three times as much grooming from females as did adolescents. The differences between age-classes in grooming females did not approach statistical significance, but those in being groomed by females fell just short of it ($p = .089$). The reciprocal nature of grooming did differ significantly between age-classes. Adults received more grooming from females than they initiated, while adolescents initiated more than they received ($p < .05$). Although males of both age-classes engaged in similar overall amounts of aggression with females, the latter directed far more aggression toward adolescent than toward adult males ($p < .05$). A significant interaction of the age of males and reciprocity of aggression was also found. Adult males received less aggression from females than they initiated, while the reverse was true for the adolescent males ($p < .025$).

Juvenile males differed significantly from adult males on every measure of social relations with dominant females. Marked differences were also found between juveniles and adolescents, but one juvenile consistently fell within the range of the adolescents and caused the differences between these two age-classes to fall below significance. In sum, a linear relationship emerged between male age and amounts of proximity, orientation, and grooming with dominant females, and an inverse relationship was found between male age and aggression received from females throughout the mating season. On the average, dominant females showed considerably more interest in adult than in adolescent or juvenile males. These females faced and groomed adult males more and directed far less aggression toward adult than toward younger males.

CONSORT RELATIONSHIPS

Consort relationships were observed between females and adult males only. Further, they were seen most frequently among the dominant adult males. Females who were engaged in consortships generally mated only with their partners; males, however, often remained promiscuous regardless of the availability of their consort female. The consort relationship typically involved the male taking on the active role of maintaining close physical proximity and frequent contact with his female partner. He appeared to visually monitor her actions and whereabouts. It was not unusual, for example, for a male to follow his current mate throughout the day, allowing no more than 5 to 10 feet of distance between them. He generally approached her repeatedly for ano-genital inspection, and frequently but not invariably initiated copulation.

During the mating period, the alpha male (#99) consorted with at least four different adult females. These relationships were of varying durations, lasting anywhere from a few hours to 10 days. It is noteworthy that male #99 was never observed to form a consortship with any of the adolescent females, regardless of the fact that three of them ranked among the 10 most dominant females. He engaged in consortships with two high-ranking adult females during their respective months of conception. He monopolized their sexual activities; each female was observed to mate only once with one other male during her conception month. Male #99 was seen copulating with the alpha female repeatedly during her month of conception. Although they did not maintain the spatial proximity characteristic of consort pairs between copulations, no other male was observed to copulate with this female at this or any other time during the mating season.

The beta male (#29) engaged in a number of consortships with adult females as well as with one high-ranking adolescent female. He was overtly possessive of one high-ranking adult female partner, and actually slapped both an adult and an adolescent male who attempted to copulate with her during their consort association. The only female observed to play an active role in maintaining a consortship was an exceedingly high-ranking adolescent. During her lengthy liaison with male #29, this female appeared to be remarkably assertive in establishing close contact with him, and eagerly attempted to catch his attention on numerous occasions. She interfered in his interactions with other females and sometimes displaced them with facial threats. The impression she gave was of being a very jealous mate.

The oldest and highest ranking adolescent male (#15) was observed in a rudimentary kind of consortship. He was seen following a midranking female at a short distance for three days. Although they copulated on occasion, each of them engaged in frequent copulations with others, and the relationship was not of a true consort nature. Similarly, male #79 displayed a strong preference for one particular high-ranking female; however, their copulations were neither preceded nor followed by efforts at maintaining close proximity. Throughout a two-month period this pair copulated very frequently, more frequently in fact than any other single pair (22 times). Their sexual interactions were abrupt, aggressive, and typically followed by immediate separation. Both of them also copulated often with others during this period. Interestingly, however, he was the only male with whom she was seen copulating during her month of conception.

MALE SOCIAL RANK AND TESTOSTERONE LEVEL

In addition to the general differences in sexual activity between age-classes of males, there were also marked variations between individuals at each stage of development. A few males in each class were clearly the most sexually active and had the greatest access to females. These interindividual differences were examined in relation to dominance status and testosterone level. A linear dominance hierarchy was established for each age-class on the basis of dyadic relationships, determined by the combined profiles of initiations and elicitations of aggression and submission (see Table 13-7). Where frequencies of such interactions were equal between two males, interactions with a third male were examined to determine status relations. The following five measures of

Table 13-7. Dominance Relations Within Age-Classes, as Determined by Aggressive and Submissive Interactions.*

		AGGRESSION						SUBMISSION					
		ADULT MALES											
		99	29	61	32	22	24	99	29	61	32	22	24
INITIATOR	99	–	4	8	2	5	1	–	0	0	0	0	0
	29	0	–	2	2	2	6	5	–	1	5	0	0
	61	0	0	–	0	2	2	15	2	–	5	11	4
	32	1	0	0	–	4	1	2	4	13	–	7	2
	22	0	0	3	4	–	0	19	8	13	16	–	2
	24	0	0	2	2	0	–	7	16	17	20	4	–
		ADOLESCENT MALES											
		15	79	91	98	95		15	79	91	98	95	
INITIATOR	15	–	1	1	2	1		–	0	1	1	0	
	79	0	–	5	0	1		5	–	1	0	0	
	91	1	0	–	4	1		1	9	–	5	0	
	98	0	0	0	–	1		5	2	26	–	1	
	95	0	0	0	0	–		2	8	3	3	–	
		JUVENILE MALES											
		55	53	56	52			55	53	56	52		
INITIATOR	55	–	4	3	4			–	1	0	0		
	53	2	–	2	2			3	–	1	0		
	56	1	0	–	2			2	4	–	0		
	52	0	0	0	–			3	2	1	–		

* Values represent number of intervals.

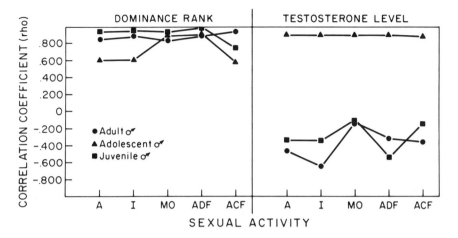

Fig. 13-4. Correlation of dominance rank and testosterone level with sexual activities (A = arousal, I = initiative, MO = mating options, ADF = access to dominant females, ACF = access to conceiving females).

behavior were selected for this analysis: (1) arousal, as indicated by overall amounts of all sexual activities (see Table 13-2 for description); (2) initiative, as indicated by frequencies of genital inspections and mounts initiated; (3) mating options, as provided by frequencies of presents received from females in response to solicitation; (4) sexual access to dominant females; and (5) sexual access to conceiving females. Spearman rho analysis was used to evaluate the relationship of dominance and testosterone to these measures of sexual activity within each age-class. The results are illustrated in Fig. 13-4.

Adult Males

Dominance status was found to be the major determinant of sexual behavior in the adult males. High-ranking males showed significantly greater sexual arousal ($p < .05$) and initiative ($p < .05$) than did low-ranking males. The two highest-ranking males engaged in twice the overall amount of sexual activity and initiated twice the amount of genital inspections and mounts as did the two lowest ranking males. Dominant males were also afforded significantly more mating options ($p = .05$) and greater access to dominant females ($p < .05$), and were observed to mate with females much more frequently during their month of conception than were subordinate males ($p = .01$). Only the alpha male was observed to mate with the alpha female, and the two lowest ranking adult males were never observed to mate with any of the top five ranking females. Status relations among the adult males were also correlated with their affiliative relations with dominant females. High-ranking adults were observed in close proximity ($p < .05$) and orientation ($p < .05$), and engaged in grooming ($p < .05$) with dominant females far more often than low-ranking adults.

Testosterone level was not related to any measure of adult sexual behavior. The male displaying the lowest testosterone (#99) was actually observed in twice the amount of sexual activity as the one displaying the highest testosterone (#32).

Adolescent Males

A complex interaction of dominance rank, testosterone level, and sexual behavior was found among the adolescent males. Testosterone was highly correlated with every aspect of adolescent male sexual behavior. The most sexually aroused and initiatory adolescent males displayed the highest testosterone levels ($p = .05$). The adolescent displaying the highest testosterone (#79) was observed in more than four times the amount of overall sexual activity as the adolescent who displayed the lowest testosterone (#98). Male #79 initiated 32 times the amount of genital inspections and received 18 times the amount of sexual presents as male #98 did. It is significant that the more sexually active male was one year older than the less active male. However, in comparisons between male #98 and another male of exactly the same age (#91), a similar relationship between sexual activity and testosterone level was found. Despite their common age, male #91 displayed higher testosterone and significantly more sexual activity than male #98. Male #91 initiated 26 times the amount of genital inspections and received six times the amount of sexual presentations as male #98 did. Those who mated most frequently with conceiving females also displayed the highest testosterone levels, but this correlation fell just slightly short of significance. Interestingly, testosterone and dominance were not significantly correlated with each other, but high-ranking males with high testosterone had the most mating options ($p = .05$) and the greatest access to dominant females ($p = .05$). Dominance status bore no statistically reliable relationship to arousal, initiative, or access to conceiving females among the adolescent males. Social relations with dominant females varied greatly, but inconsistently, between adolescent males.

Juvenile Males

Similar to the adult males, dominance was found to be the most important correlate of sexual activity for the juvenile males. Dominant juveniles showed much higher levels of arousal, initiative, and mating options than subordinate juveniles. The small number of subjects in this sample required a perfect correlation for statistical significance, and these correlations fell slightly short of 1.0, but high-ranking juveniles did have significantly greater access to dominant females than did low-ranking juveniles ($p < .05$). In contrast to the adolescent males, testosterone did not correlate positively with any measure of sexual activity among the juvenile males. The most sexually active juvenile male (#55) displayed the lowest testosterone, and the least active male (#52) displayed the highest testosterone. Male #55 initiated six times as many genital inspections and engaged in seven times as many copulations as male #52. Finally, social relations between dominant females and all juvenile males were exceedingly infrequent.

DISCUSSION

The sociology of mating activities observed in this group of bonnet monkeys is both consistent and inconsistent with that previously reported for the species. The findings that males of all ages actively participate in sexual interactions within the group context and that young or low-ranking males engaging in mating are not harassed by more mature or high-ranking males agree with the field observations of Simonds (1965) and Sugiyama (1971). With regard to the formation of consort relationships, reports vary from a total absence of consortships (Rahaman and Parthasarathy, 1969), to the observation of a single consort relation between the alpha male and female (Simonds, 1965), to an age-graded pattern in which dominant or elder males tend to engage in consortships more often than do younger males (Sugiyama, 1971). In this study, consort pairs were clearly observed, and followed the age-graded pattern.

The data reported here run contrary to the suggestions from the field literature that bonnet females are equally accessible to all males and that males do not vary in their access to females at different stages of estrus (Simonds, 1965; Sugiyama, 1971). The relative absence of female solicitation and aggressive competition between males for estrous females may have fostered these conclusions prematurely. Moreover, the mating activities of these free-ranging males were not examined in relation to the probable conception dates of females. In the cross-sectional sampling of sexual behavior presented here, very young bonnet males mated primarily with young or subdominant females, and with increasing maturity came greater access to older and then to higher ranking females, both throughout the mating season, as well as at conception time. Further, the sexual activities of individual adult males were directly related to their social status. High-ranking adult males were far more successful than low-rankers in gaining access to conceiving females and to dominant females, the latter probably facilitating this by their stronger affiliative relations with dominant than with subordinate males. These findings are consistent with those of Silk *et al.* (1978), who studied the 1977–78 mating season interactions of the same six adult bonnet males with six of the adult females who themselves differed widely in rank.

Interestingly, adult and adolescent males were equally active with conceiving females, thus potentiating a relatively high level of reproductive success for the young males of this species. Adult males did, however, have greater access to dominant females during their months of conception. Dominant females did not actively solicit copulations from any males, but they may have encouraged the sexual advances of adult males by the nature of their nonsexual interactions. If high rank confers increased reproductive success among bonnet females as it appears to do so among males, the adult males would then maintain some advantage over the adolescent males.

The relative contributions of dominance status and testosterone level to sexual behavior were found to differ at specific stages of development. Testosterone level proved to be the more reliable index of differences between individual adolescent males in sexual activity. For the juvenile males, however, sexual activities bore no relation to testosterone, and were generally but not quite statistically significantly correlated with status relations. These data suggest that prior to puberty, assertiveness may facilitate the expression of sexual activity in a social context, and that during puberty, the inductive forces of increased testosterone exert a strong influence on sexual behavior which can supersede status. Thus, the more sexually motivated pubertal males (i.e., those with higher testosterone) may be more active. At neither of these stages of development do the social responses of females reflect differences between individual males in rank or testosterone, whereas in adulthood, well-established dominance relations between males do promote fairly predictable patterns of sexual and social interactions with females.

Recent experimental studies on rhesus monkeys provide increasing evidence that social and hormonal factors interact to shape psychosexual development (Goy and Goldfootn 1974). Yet, there are few data on the naturally occurring behavioral and hormonal changes attendant on sexual maturation, and there are no data on the relative contributions of specific psychobiological factors to the sexual behavior of intact animals at various stages of development. While conclusions drawn from the present findings are admittedly preliminary, a consistent profile of differential testosterone sensitivity during puberty and adult life does emerge when these findings are considered in the context of the results of studies which have manipulated endocrine status.

Among adult rhesus males, testosterone has at least a facilitative effect on sexual behavior. While castration typically reduces a male's motivation for mounting, intromission, and ejaculation (Wilson et al., 1972), some males do continue to copulate long after castration (Wilson and Vessey, 1968; Phoenix, 1978). The lack of correlation between absolute testosterone level and specific measures of the sexual activity of the adult bonnet males in this study agrees with previous findings on adult rhesus males. Testosterone level has been found to be unrelated to the number of ejaculations, intromissions, or mounts displayed by rhesus castrates either before or after testosterone replacement therapy (Resko and Phoenix, 1972; Phoenix, 1973), and the sexual activities of intact males may actually decline significantly with advanced age without a concomitant change in testosterone level (Robinson *et al.*, 1975), perhaps as a result of a decrease in central neural or peripheral structure sensitivity (Phoenix, 1977).

Furthermore, castration has been shown to produce more profound sexual deficits in maturing than in fully adult rhesus males (Michael *et al.*, 1973; Loy, 1976). Michael *et al.* (1975) note that,

> Following castration, there was a marked decline in the number of ejaculations and intromissions per test for both groups . . .Whereas ejaculations ceased entirely in the sub-adult group about 12 weeks after castration, they persisted in some adult males, as did intromissions, throughout the 32 week period to which these observations refer. In contrast, a marked decline in mounting activity was only observed during this time in sub-adults, and mounting continued more or less unchanged in the adult group. (p. 281)

It is likely that the social reinforcement consequent to previous sexual experience is an important factor in the continued sexual prowess of some adult castrates. This, in addition to what Michael *et al.* (1975) call an "androgen-dependent maturational change," may well account for the more delicate balance between testosterone and sexual behavior during puberty than adulthood. One might speculate, then, that puberty may be the period of greatest postnatal sensitivity to changes in endocrine status and, further, that the "organizational" (Young *et al.*, 1964) influences of androgen on the primate central nervous system may not be entirely confined to the period of prenatal development.

A number of fundamental similarities between bonnet and other macaque males can be concluded from the data presented here on the sociological aspects of sexual activity. Perhaps more interesting, however, are the suggested intrageneric differences in the manifestation of sexual maturity, and the possible relationship of these differences to known variations in social adaptation. The chronological consistency between the attainment and manifestation of sexual maturity in the bonnet male contrasts sharply with the discrepant pattern of sexual maturation in at least two other species of macaques. In rhesus groups, only males aged five years and over are consistently actively involved in sexual interactions, while three- and four-year-olds remain relatively inactive and peripheral (Conaway and Koford, 1965; Boelkins and Wilson, 1972). Among Japanese macaques, males do not display the full complement of adult sexual behavior with ejaculation until at least 4.5 years of age (Hanby and Brown, 1974; Wolfe, 1978). However, there is increasing evidence that the onset of reproductive fertility in the male macaque may occur at least one year earlier than these behavioral observations indicate. By the age of 3–3.5 years, rhesus males display the seasonal variations in testosterone, testicular size, and spermatogenic function characteristic of fully mature males (van Wagenen and Simpson, 1954), and recent reports of laboratory stumptail and rhesus monkeys have shown that males can sire offspring as early as 3.3 years of age (Trollope and Blurton Jones, 1972; Maple *et al.*, 1973). Bonnet males begin to engage in sexual interactions at age two years. The fact that the two-year-old bonnet male can mount with intromission and ejaculation and does so routinely in the presence of older and high-ranking males seems to be unique to the species. At three years plus, bonnet males are equally as sexually active as adult males with as many females throughout the mating season as well as at conception time. The differences between pubertal and adult males lie not in the amount of mating, but in the ability to gain access to dominant females; this ability seems to depend largely on having attained full social maturity and high status. Three-year-old bonnet males are also reproductively fertile, since they begin to exhibit the cyclical enlargement and regression of the testes presumably reflective of seasonal spermatogenesis (Glick, in press), and sperm are detectable in their ejaculate (Rosenblum and Nadler, 1971).

While the effects of contextual variables on the physiological status of the developing primate have not been investigated, several studies have illustrated the effects of social stimuli on the endocrine balance of the adult male. For example, stress has been shown to reduce testosterone levels in adult human males (Kreuz *et al.*, 1972), and dominance interactions can influence the testosterone level of the adult male rhesus monkey (Rose *et al.*, 1972). A recent report on adolescent testosterone and behavior does note that the number of adult males in a group can influence the pattern of testosterone increase during adolescence. A comparison of three social groups revealed that the presence of fewer adult males facilitated earlier and more marked testosterone increases in pubertal males (Rose *et al.*, 1978). Thus, it is reasonable to hypothesize that any of a number of intragroup social variables might influence patterns of sexual development, thereby implicating species-characteristic social attributes as one major determinant of these patterns.

Among rhesus and Japanese macaques, young males develop under restrictive social conditions, and are typically found to be both spatially and behaviorally peripheral (Nishida, 1966; Lindburg, 1969). Such a constrained social environment may foster a somewhat retarded manifestation of sexual maturity. In contrast, bonnet society is characterized by a highly integrated male component, and bonnet males in general tend to be extensively involved in group social dynamics. The young bonnet male's position in society is more flexible and relaxed, and adolescents do not appear to experience the social exclusion routinely experienced by adolescent rhesus and Japanese macaque males (Rahaman and Parthasarathy, 1969; Simonds, 1973). For the bonnet male, fewer social restrictions may impose fewer biological restraints; maturity is not retarded and the manifestation of the full complement of adult sexual behavior occurs relatively early on. The close taxonomic affinity between the Japanese, rhesus, and bonnet macaques, and the modest ecological differences between the latter two, make the apparent differences in their maturational profiles particularly interesting and serve to reiterate the complex nature of the interaction between phylogenetic and environmental factors in the determination of behavior.

ACKNOWLEDGMENTS

I am greatly indebted to Dr. William Mason for his invaluable help and encouragement throughout this study, and to Dr. Donald Lindburg for his support and critical evaluation of the manuscript. I also thank Dr. Ethel Sassenrath for making the endocrine laboratory of the Behavioral Biology Unit available to me, and Mr. Tom Madley for running the hormone assays. This research was supported in part by National Institutes of Health Grant #RR00169 to the California Primate Research Center, and by a University of California, Los Angeles, Anthropology Research Grant and a Sigma Xi National Research Society Grant to the author.

REFERENCES

Altmann, J. Observational study of behavior: Sampling methods. *Behaviour*, **49**: 227–267 (1974).

Barkely, M. S. and Goldman, B. D. A quantitative study of serum testosterone, sex accessory organ growth and the development of intermale aggression in the mouse. *Horm. Behav.*, **8**: 208–218 (1977).

Bernard, G.J. R., Hennam, J. F. and Collins, W. P. Further studies on radioimmunoassay systems for plasma oestradiol. *J. Steroid Biochem.*, **6**: 107–116 (1975).

Bertrand, M. The behavioral repertoire of the stumptail macaque. *Bibliotheca Primatologica*, No. 11. Basel: S. Karger (1969).

Boelkins, R. C. and Wilson, A. P. Intergroup social dynamics of the Cayo Santiago rhesus (*Macaca mulatta*) with special reference to changes in group membership by males. *Primates*, **13**: 125–140 (1972).

Conaway, C. H., and Koford, C. B. Estrous cycles and mating behavior in a free-ranging band of rhesus monkeys. *J. Mammal.*, **45**: 577–588 (1965).

Conaway, C. H. and Sade, D. S. The seasonal spermatogenic cycle in free ranging rhesus monkeys. *Folia Primatol.*, **3**: 1–12 (1965).

Glick. B. B. Testicular size, testosterone level, and body weight in male *Macaca radiata*: Maturational and seasonal effects. *Folia primatol.* (In press).

Goy, R. W. and Goldfoot, D. A. Experimental and hormonal factors influencing development of sexual behavior in male rhesus monkeys. In F. O. Schmitt and F. G. Worden, *The Neurosciences, Third Study Program*, Chapter 50, pp. 571–581. Cambridge, Mass. MIT Press (1974).

Hanby, J. P. and Brown, C. E. The development of sociosexual behaviours in Japanese macaques *Macaca fuscata. Behaviour*, **49**: 152–196 (1974).

Hanby, J. P. Robertson, L. T., and Phoenix, C. H. The sexual behavior of a confined troop of Japanese macaques. *Folia Primatol.*, **16**: 123–143 (1971).

Imanishi, K. Social behavior in Japanese monkeys, *Macaca fuscata. Psychologia*, **1**: 47–54 (1957).

Kaufman, I. C. and Rosenblum, L. A. A behavioral taxonomy for *Macaca nemestrina* and *Macaca radiata*: Based on longitudinal observation of family groups in the laboratory. *Primates*, **7**: 206–258 (1966).

Kruez, L. E., Rose, R. M. and Jennings, J. Suppression of plasma testosterone levels and psychological stress: A longitudinal study of young men in officer candidate school. *Arch. Gen. Psychiatry*, **26**: 479–482 (1972).

Kuehn, R. E., Jensen, G. D. and Morrill, R. K. Breeding *Macaca nemestria*: A program of birth engineering. *Folia Primatol.*, **3**: 251–262 (1965).

Lindburg, D. G. Rhesus monkeys: Mating season mobility of adult males. *Science*, **166**: 1176–1178 (1969).

_____. The rhesus monkey of north India: An ecological and behavioral study. In L.A. Rosenblum, (ed.), *Primate Behavior: Developments in Field and Laboratory Research*, pp. 1–106. New York: Academic Press (1971).

Loy, J. Behavior of gonadectomized rhesus monkeys. Paper presented at the 75th annual meetings of the American Anthropological Association, Washington, D. C., November 1976.

Maple, T., Erwin, J. and Mitchell, G. Age of sexual maturity in laboratory-born pairs of rhesus monkeys (*Macaca mulatta*). *Primates*, **14**: 427–428 (1973).

Mason, W. A. The effects of social restriction on the behavior of rhesus monkeys. I. Free social behavior. *J. Comp. Physiol. Psychol.*, **53**: 582–589 (1960).

Michael, R. P., Wilson, M., and Plant, T. M. Sexual behaviour of male primates and the role of testosterone. In R. P. Michael and J. H. Crook, (eds.), *Comparative Ecology and Behaviour of Primates*, pp. 235–313. New York: Academic Press (1975).

Nadler, R. D., and Rosenblum, L. A. Sexual behavior of male bonnet monkeys in the laboratory. *Brain Behav. Evol.*, **2**: 482–497 (1969).

Nishida, T. A sociological study of solitary male monkeys. *Primates*, **7**: 141–204 (1966).

Phoenix, C. H. 1973. The role of testosterone in the sexual behavior of the laboratory male rhesus. In *Symp. IVth Int. Congr. Primat.*, Vol. 2, pp. 99–122. Basel: S. Karger.

_____. Factors influencing sexual performance in male rhesus monkeys. *J. Comp. Physiol. Psychol.*, **91**: 697–710 (1977).

_____. Steroids and sexual behavior in castrated male rhesus monkeys. *Horm. Behav.*, **10**: 1–9 (1978).

Rahaman, H. and Parthasarathy, M. D. Studies on the social behaviour of bonnet monkeys. *Primates*, **10**: 149–162 (1969).

_____. The role of the olfactory signals in the mating behavior of bonnet monkeys (*Macaca radiata*). *Commun. Behav. Biol.*, **6**: 97–104 (1971).

Resko, J. A. and Phoenix, C. H. Sexual behavior and testosterone concentrations before and after castration. *Endocrinology*, **91**: 499–503 (1972).

Robinson, J. A., Scheffler, G., Eisele, S. G. and Goy, R. W. Effects of age and season on sexual behavior and plasma testosterone and dihydrotestosterone concentrations of laboratory-housed male rhesus monkeys (*Macaca mulatta*). *Biol. Reprod.*, **13**: 203–210 (1975).

Rose, R. M., Bernstein, I. S., Gordon, T. P. and Lindsley, J. G. Changes in testosterone and behavior during adolescence in the male rhesus monkey. *Psychosom. Med.*, **40**: 60–70 (1978).

Rose, R. M., Gordon, T. P. and Bernstein, I. S. Plasma testosterone levels in the male rhesus: Influences of sexual and social stimuli. *Science*, **178**: 643–645 (1972).

Rosenblum, L. A., and Nadler, R. D. The ontogeny of sexual behavior in male bonnet macaques. In *Influence of Hormones on the Nervous System. Proceedings of the International Society for Psychoneuroendocrinology*, Brooklyn, 1970. pp. 388–400, Basel: S. Karger (1971).

Silk, J. B., Rodman, P. S. and Samuels, A. Social and sexual relationships between male and female bonnet macaques during their breeding season. Paper presented at the 2nd annual meeting of the American Society of Primatologists, Atlanta, Georgia (1978).

Simonds, P. The bonnet macaque in south India. In I. DeVore, (ed.), *Primate Behavior*, pp. 175–196. New York: Holt, Rinehart and Winston (1965).

_____. Outcast males and social structure among bonnet macaques *(Macaca radiata). Am. J. Phys. Anthropol.,* **38**: 599–604 (1973).

Southwick, C. H., Beg, M. A. and Siddiqi, M. R. Rhesus monkeys in north India. In I. DeVore, (ed.), *Primate Behavior*, pp. 111–159. New York: Holt, Rinehart and Winston (1965).

Sugiyama, Y. Characteristics of the social life of bonnet macaques (*Macaca radiata*). *Primates*, **12**: 247–266 (1971).

Tokuda, K., Simons, R. C. and Jensen, G. D. Sexual behavior in a captive group of pigtailed monkeys (*Macaca nemestrina*). *Primates*, **9**: 283–294 (1968).

Trollope, J. and Blurton Jones, N. G. Age of sexual maturity in the stumptailed macaque (*Macaca arctoides*): A birth from laboratory born parents. *Primates*, **13**: 229–230 (1972).

van Wagenen, G. and Simpson, M. E. Testicular development in the rhesus monkey. *Anat. Rec.*, **118**: 231–251 (1954).

Wilson, A. P. and Vessey, S. H. Behavior of free-ranging, castrated rhesus monkeys. *Folia Primatol.*, **9**: 1–14 (1968).

Wilson, M., Plant, T. M. and Michael, R. P. Androgens and the sexual behavior of male rhesus monkeys. *J. Endocrinol.*, **52**: ii (1972).

Wolfe, L. Age and sexual behavior of Japanese macaques (*Macaca fuscata*). *Arch. Sex. Behav.*, **7**: 55–68 (1978).

Young, W. C., Goy, R. W. and Phoenix, C. H. Hormones and sexual behavior. *Science*, **143**: 212–217 (1964).

AUTHOR INDEX

SUBJECT INDEX

activity patterns, 159, 221, 240, 254-258
adaptability, 151
affiliative behavior, 269, 272
age classification, 265, 266
age-graying, 97
age-related sexual performance, 351-356, 361-365
agonistic behavior, 259, 260, 268, 269, 270, 272, 278, 279, 280, 281,
albinism, 97
Allen's rule, 6
altitudinal zonation, 162, 164, 165, 166, 179
Alouatta caraya, 273, 275, 276
Alouatta palliata, 278, 281
Alouatta seniculus, 273, 275, 281
amino acid sequencing, 37, 39
arboreality, 69, 164, 165, 170, 177, 192, 194-196, 199, 200, 237, 242
arctoides group, 1-3, 5-7, 22, 26, 52, 53, 65
Asian fossil macaques, 19-21, 52-83
Ateles, 190
Ateles belzebuth, 241

behavioral isolating mechanisms, 126, 127, 132, 136, 137, 138, 145, 146

Bergmann's rule, 6
birth interval, 174, 175, 238

castration, 364, 365
Cercocebus, 16, 32
Cercocebus albigena, 34, 242
Cercocebus torquatus, 34
Cercopithecus, 34, 264
Cercopithecus aethiops, 33, 43, 274, 278, 281
Cercopithecus asnoti, 20
Cercopithecus mitis, 190, 241
character displacement, 97
chromosome banding, 32, 33, 39, 67
chromosomes, 32, 33, 58, 67
cladistics, 16, 37, 44, 45, 46, 47, 48, 49, 53
classification, 1, 2, 3, 6, 7, 8, 10, 20, 22, 23, 24, 31, 32, 33, 36, 42, 43, 48, 52, 53, 65, 68, 69, 84, 85, 86, 91, 105, 107, 111, 112, 113, 114-116, 125, 155, 182, 189, 264, 367
cold adaptation, 6, 65, 70
Colobus, 34
conservation, 119-122
consort behavior, 298-311, 358, 359
copulation, 2, 21, 22, 53, 313-324
cranial morphology, 11, 14, 15, 20, 21, 22, 23, 48, 58, 65, 105, 112, 116